An Introduction to Botany

An Introduction to Botany

Edited by
Diago Bell

Larsen & Keller
www.larsen-keller.com

An Introduction to Botany
Edited by Diago Bell
ISBN: 978-1-63549-052-7 (Hardback)

Larsen & Keller
Published by Larsen and Keller Education,
5 Penn Plaza,
19th Floor,
New York, NY 10001, USA

Cataloging-in-Publication Data

An introduction to botany / edited by Diago Bell.
p. cm.
Includes bibliographical references and index.
ISBN 978-1-63549-052-7
1. Botany. 2. Plants. I. Bell, Diago.
QK51 .I58 2017
581--dc23

Printed and bound in China.

Table of Contents

Preface

This book provides comprehensive insights into the field of botany. It talks in detail about the fundamental uses of botany and the various techniques that fall under this field. Botany, as a branch of biological sciences, refers to the study of the structure, classification, economic importance, genetics and physiology of plants. Through this text, we aim to provide students with the relevant and important information about the vast field of botany. This book will serve as a reference to a broad spectrum of readers. Most of the topics introduced in the textbook cover new techniques and the applications of plant biology. It is appropriate for students and for those seeking detailed information in this area.

To facilitate a deeper understanding of the contents of this book a short introduction of every chapter is written below:

Chapter 1- Botany is the study of biology which studies plant life. A botanist is a person who specializes in this field. It is integral in the cultivation and propagation of plant life. Modern botany is vast and assists various allied fields and is crucial for upholding biodiversity on Earth. The chapter on botany offers an insightful focus, keeping in mind the subject matter.

Chapter 2- Botany is an interdisciplinary subject. Some of the branches of botany are bryology, plant physiology, plant morphology, plant genetics, paleobotany and economic botany. Bryology is the scientific study of bryophytes whereas plant physiology concerns itself with the functioning of plants. The following content will provide a glimpse of related branches of botany.

Chapter 3- The key concepts of botany discussed within the text are flowers, seed, bud, cell wall, tree and forest. Each section is an integral whole as well as part of a composite evolutionary process. The section strategically encompasses and incorporates the major components and key concepts of botany, providing a complete understanding.

Chapter 4- The technology of using plants for food and fuel is agronomy. Agronomy combines within it subjects like biology, chemistry, economics and Earth science. Agronomists deal with a number of issues such as providing food, managing environmental issues and the extraction of energy from plants. The following unit focuses on the principles and the processes of botany.

Chapter 5- Plants belong to the kingdom of Plantae. The science that identifies plants and names them is known as plant taxonomy. Plant taxonomy can also be used as a synonym for plant systematics. Bryophyte, embryophyte, spermatophyte, flowering plant and moss are numerous aspects of a plant that have been elaborately explained the chapter. The aspects elucidated in this chapter are of vital importance, and provide a better understanding of botany.

Chapter 6- Plant anatomy is the study of the structure of a plant. Plant anatomy investigates plant cell, pollen and meristem. Plant anatomy is an emerging field of study. Advances in technology have allowed great progress to be made in this field. The text helps the reader in developing a profound understanding on the topic.

I would like to share the credit of this book with my editorial team who worked tirelessly on this book. I owe the completion of this book to the never-ending support of my family, who supported me throughout the project.

Editor

1

Introduction to Botany

Botany is the study of biology which studies plant life. A botanist is a person who specializes in this field. It is integral in the cultivation and propagation of plant life. Modern botany is vast and assists various allied fields and is crucial for upholding biodiversity on Earth. The chapter on botany offers an insightful focus, keeping in mind the subject matter.

Botany, also called plant science(s), plant biology or phytology, is the science of plant life and a branch of biology. A botanist or plant scientist is a scientist who specializes in this field. The term "botany" comes from the Ancient Greek word βοτάνη (*botanē*) meaning "pasture", "grass", or "fodder"; βοτάνη is in turn derived from βόσκειν (*boskein*), "to feed" or "to graze". Traditionally, botany has also included the study of fungi and algae by mycologists and phycologists respectively, with the study of these three groups of organisms remaining within the sphere of interest of the International Botanical Congress. Nowadays, botanists study approximately 400,000 species of living organisms of which some 260,000 species are vascular plants and about 248,000 are flowering plants.

Botany originated in prehistory as herbalism with the efforts of early humans to identify – and later cultivate – edible, medicinal and poisonous plants, making it one of the oldest branches of science. Medieval physic gardens, often attached to monasteries, contained plants of medical importance. They were forerunners of the first botanical gardens attached to universities, founded from the 1540s onwards. One of the earliest was the Padua botanical garden. These gardens facilitated the academic study of plants. Efforts to catalogue and describe their collections were the beginnings of plant taxonomy, and led in 1753 to the binomial system of Carl Linnaeus that remains in use to this day.

In the 19th and 20th centuries, new techniques were developed for the study of plants, including methods of optical microscopy and live cell imaging, electron microscopy, analysis of chromosome number, plant chemistry and the structure and function of enzymes and other proteins. In the last two decades of the 20th century, botanists exploited the techniques of molecular genetic analysis, including genomics and proteomics and DNA sequences to classify plants more accurately.

Modern botany is a broad, multidisciplinary subject with inputs from most other areas of science and technology. Research topics include the study of plant structure, growth and differentiation, reproduction, biochemistry and primary metabolism, chemical products, development, diseases, evolutionary relationships, systematics, and plant taxonomy. Dominant themes in 21st century plant science are molecular genetics and epigenetics, which are the mechanisms and control of gene expression during differentiation of plant cells and tissues. Botanical research has diverse applications in providing staple foods, materials such as timber, oil, rubber, fibre and drugs, in modern horticulture, agriculture and forestry, plant propagation, breeding and genetic modification, in the synthesis of chemicals and raw materials for construction and energy production, in environmental management, and the maintenance of biodiversity.

History

Early Botany

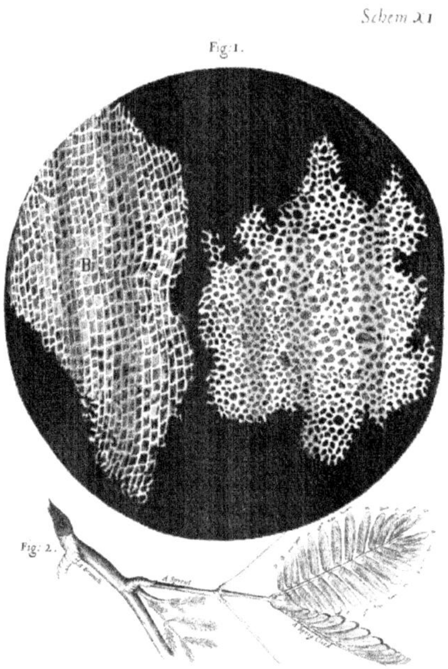

An engraving of the cells of cork, from Robert Hooke's *Micrographia*, 1665

Botany originated as herbalism, the study and use of plants for their medicinal properties. Many records of the Holocene period date early botanical knowledge as far back as 10,000 years ago. This early unrecorded knowledge of plants was discovered in ancient sites of human occupation within Tennessee, which make up much of the Cherokee land today. The early recorded history of botany includes many ancient writings and plant classifications. Examples of early botanical works have been found in ancient texts from India dating back to before 1100 BC, in archaic Avestan writings, and in works from China before it was unified in 221 BC.

Modern botany traces its roots back to Ancient Greece, specifically to Theophrastus (c. 371–287 BC), a student of Aristotle who invented and described many of its principles and is widely regarded in the scientific community as the "Father of Botany". His major works, *Enquiry into Plants* and *On the Causes of Plants*, constitute the most important contributions to botanical science until the Middle Ages, almost seventeen centuries later.

Another work from Ancient Greece that made an early impact on botany is *De Materia Medica*, a five-volume encyclopedia about herbal medicine written in the middle of the first century by Greek physician and pharmacologist Pedanius Dioscorides. *De Materia Medica* was widely read for more than 1,500 years. Important contributions from the medieval Muslim world include Ibn Wahshiyya's *Nabatean Agriculture*, Abū Ḥanīfa Dīnawarī's (828–896) the *Book of Plants*, and Ibn Bassal's *The Classification of Soils*. In the early 13th century, Abu al-Abbas al-Nabati, and Ibn al-Baitar (d. 1248) wrote on botany in a systematic and scientific manner.

In the mid-16th century, "botanical gardens" were founded in a number of Italian universities – the Padua botanical garden in 1545 is usually considered to be the first which is still in its original location. These gardens continued the practical value of earlier "physic gardens", often associated with monasteries, in which plants were cultivated for medical use. They supported the growth of botany as an academic subject. Lectures were given about the plants grown in the gardens and their medical uses demonstrated. Botanical gardens came much later to northern Europe; the first in England was the University of Oxford Botanic Garden in 1621. Throughout this period, botany remained firmly subordinate to medicine.

German physician Leonhart Fuchs (1501–1566) was one of "the three German fathers of botany", along with theologian Otto Brunfels (1489–1534) and physician Hieronymus Bock (1498–1554) (also called Hieronymus Tragus). Fuchs and Brunfels broke away from the tradition of copying earlier works to make original observations of their own. Bock created his own system of plant classification.

Physician Valerius Cordus (1515–1544) authored a botanically and pharmacologically important herbal *Historia Plantarum* in 1544 and a pharmacopoeia of lasting importance, the *Dispensatorium* in 1546. Naturalist Conrad von Gesner (1516–1565) and herbalist John Gerard (1545–c. 1611) published herbals covering the medicinal uses of plants. Naturalist Ulisse Aldrovandi (1522–1605) was considered the *father of natural history*, which included the study of plants. In 1665, using an early microscope, Polymath Robert Hooke discovered cells, a term he coined, in cork, and a short time later in living plant tissue.

Early Modern Botany

The Linnaean Garden of Linnaeus' residence in Uppsala, Sweden, was planted according to his *Systema sexuale*.

During the 18th century, systems of plant identification were developed comparable to dichotomous keys, where unidentified plants are placed into taxonomic groups (e.g. family, genus and species) by making a series of choices between pairs of characters. The choice and sequence of the characters may be artificial in keys designed purely for identification (diagnostic keys) or more closely related to the natural or phyletic order of the taxa in synoptic keys. By the 18th century, new plants for study were arriving in Europe in increasing numbers from newly discovered countries and the European colonies worldwide. In 1753 Carl von Linné (Carl Linnaeus) published his Species Plantarum, a hierarchical classification of plant species that remains the reference point for modern botanical nomenclature. This established a standardised binomial or two-part naming scheme where the first name represented the genus and the second identified the species within the genus. For the purposes of identification, Linnaeus's *Systema Sexuale* classified plants into 24 groups according to the number of their male sexual organs. The 24th group, *Cryptogamia*, included all plants with concealed reproductive parts, mosses, liverworts, ferns, algae and fungi.

Increasing knowledge of plant anatomy, morphology and life cycles led to the realisation that there were more natural affinities between plants than the artificial sexual system of Linnaeus had indicated. Adanson (1763), de Jussieu (1789), and Candolle (1819) all proposed various alternative natural systems of classification that grouped plants using a wider range of shared characters and were widely followed. The Candollean system reflected his ideas of the progression of morphological complexity and the later classification by Bentham and Hooker, which was influential until the mid-19th century, was influenced by Candolle's approach. Darwin's publication of the *Origin of Species* in 1859 and his concept of common descent required modifications to the Candollean system to reflect evolutionary relationships as distinct from mere morphological similarity.

Botany was greatly stimulated by the appearance of the first "modern" textbook, Matthias Schleiden's *Grundzüge der Wissenschaftlichen Botanik*, published in English in 1849 as *Principles of Scientific Botany*. Schleiden was a microscopist and an early plant anatomist who co-founded the cell theory with Theodor Schwann and Rudolf Virchow and was among the first to grasp the significance of the cell nucleus that had been described by Robert Brown in 1831. In 1855, Adolf Fick formulated Fick's laws that enabled the calculation of the rates of molecular diffusion in biological systems.

Echeveria glauca in a Connecticut greenhouse. Botany uses Latin names for identification, here, glauca means blue.

Modern Botany

Building upon the gene-chromosome theory of heredity that originated with Gregor Mendel (1822–1884), August Weismann (1834–1914) proved that inheritance only takes place through gametes. No other cells can pass on inherited characters. The work of Katherine Esau (1898–1997) on plant anatomy is still a major foundation of modern botany. Her books *Plant Anatomy* and *Anatomy of Seed Plants* have been key plant structural biology texts for more than half a century.

Micropropagation of transgenic plants

Biologist and statistician Ronald Fisher

The discipline of plant ecology was pioneered in the late 19th century by botanists such as Eugenius Warming, who produced the hypothesis that plants form communities, and his mentor and successor Christen C. Raunkiær whose system for describing plant life forms is still in use today. The concept that the composition of plant communities such as temperate broadleaf forest changes by a process of ecological succession was developed by Henry Chandler Cowles, Arthur Tansley and Frederic Clements. Clements is credited with the idea of climax vegetation as the most complex vegetation that an environment can support and Tansley introduced the concept of ecosystems to biology. Building on the extensive earlier work of Alphonse de Candolle, Nikolai Vavilov (1887–1943) produced accounts of the biogeography, centres of origin, and evolutionary history of economic plants.

Particularly since the mid-1960s there have been advances in understanding of the physics of plant physiological processes such as transpiration (the transport of water within plant tissues), the temperature dependence of rates of water evaporation from the leaf surface and the molecular diffusion of water vapour and carbon dioxide through stomatal apertures. These developments, coupled with new methods for measuring the size of stomatal apertures, and the rate of photosynthesis have enabled precise description of the rates of gas exchange between plants and the atmosphere. Innovations in statistical analysis by Ronald Fisher, Frank Yates and others at Rothamsted Experimental Station facilitated rational experimental design and data analysis in botanical research. The discovery and identification of the auxin plant hormones by Kenneth V. Thimann in 1948 enabled regulation of plant growth by externally applied chemicals. Frederick Campion Steward pioneered techniques of micropropagation and plant tissue culture controlled by plant hormones. The synthetic auxin 2,4-Dichlorophenoxyacetic acid or 2,4-D was one of the first commercial synthetic herbicides.

20th century developments in plant biochemistry have been driven by modern techniques of organic chemical analysis, such as spectroscopy, chromatography and electrophoresis. With the rise of the related molecular-scale biological approaches of molecular biology, genomics, proteomics and metabolomics, the relationship between the plant genome and most aspects of the biochemistry, physiology, morphology and behaviour of plants can be subjected to detailed experimental analysis. The concept originally stated by Gottlieb Haberlandt in 1902 that all plant cells are totipotent and can be grown *in vitro* ultimately enabled the use of genetic engineering experimentally to knock out a gene or genes responsible for a specific trait, or to add genes such as GFP that report when a gene of interest is being expressed. These technologies enable the biotechnological use of whole plants or plant cell cultures grown in bioreactors to synthesise pesticides, antibiotics or other pharmaceuticals, as well as the practical application of genetically modified crops designed for traits such as improved yield.

Modern morphology recognizes a continuum between the major morphological categories of root, stem (caulome), leaf (phyllome) and trichome. Furthermore, it emphasizes structural dynamics. Modern systematics aims to reflect and discover phylogenetic relationships between plants. Modern Molecular phylogenetics largely ignores morphological characters, relying on DNA sequences as data. Molecular analysis of DNA sequences from most families of flowering plants enabled the Angiosperm Phylogeny Group to publish in 1998 a phylogeny of flowering plants, answering many of the questions about relationships among angiosperm families and species. The theoretical possibility of a practical method for identification of plant species and commercial varieties by DNA barcoding is the subject of active current research.

Scope and Importance

The study of plants is vital because they underpin almost all animal life on Earth by generating a large proportion of the oxygen and food that provide humans and other organisms with aerobic respiration with the chemical energy they need to exist. Plants, algae and cyanobacteria are the major groups of organisms that carry out photosynthesis, a process that uses the energy of sunlight to convert water and carbon dioxide into sugars that can be used both as a source of chemical energy and of organic molecules that are used in the structural components of cells. As a by-product of photosynthesis, plants release oxygen into the atmosphere, a gas that is required by nearly all living things to carry out cellular respiration. In addition, they are influential in the global carbon and water cycles and plant roots bind and stabilise soils, preventing soil erosion. Plants are crucial to the future of human society as they provide food, oxygen, medicine, and products for people, as well as creating and preserving soil.

Botany involves the recording and description of plants, such as this herbarium specimen of the lady fern *Athyrium filix-femina.*

Historically, all living things were classified as either animals or plants and botany covered the study of all organisms not considered animals. Botanists examine both the internal functions and processes within plant organelles, cells, tissues, whole plants, plant populations and plant communities. At each of these levels, a botanist may be concerned with the classification (taxonomy), phylogeny and evolution, structure (anatomy and morphology), or function (physiology) of plant life.

The strictest definition of "plant" includes only the "land plants" or embryophytes, which include seed plants (gymnosperms, including the pines, and flowering plants) and the free-sporing cryptogams including ferns, clubmosses, liverworts, hornworts and mosses. Embryophytes are multicellular eukaryotes descended from an ancestor that obtained its energy from sunlight by photosynthesis. They have life cycles with alternating haploid and diploid phases. The sexual haploid phase of embryophytes, known as the gametophyte, nurtures the developing diploid embryo sporophyte within its tissues for at least part of its life, even in the seed plants, where the gametophyte itself is nurtured by its parent sporophyte. Other groups of organisms that were previously studied by botanists include bacteria (now studied in bacteriology), fungi (mycology) – including lichen-forming fungi (lichenology), non-chlorophyte algae (phycology), and viruses (virology). However, attention is still given to these groups by botanists, and fungi (including lichens) and photosynthetic protists are usually covered in introductory botany courses.

Paleobotanists study ancient plants in the fossil record to provide information about the evolutionary history of plants. Cyanobacteria, the first oxygen-releasing photosynthetic organisms on Earth, are thought to have given rise to the ancestor of plants by entering into an endosymbiotic relationship with an early eukaryote, ultimately becoming the chloroplasts in plant cells. The new photosynthetic plants (along with their algal relatives) accelerated the rise in atmospheric oxygen started by the cyanobacteria, changing the ancient oxygen-free, reducing, atmosphere to one in which free oxygen has been abundant for more than 2 billion years.

Among the important botanical questions of the 21st century are the role of plants as primary producers in the global cycling of life's basic ingredients: energy, carbon, oxygen, nitrogen and water, and ways that our plant stewardship can help address the global environmental issues of resource management, conservation, human food security, biologically invasive organisms, carbon sequestration, climate change, and sustainability.

Human Nutrition

The food we eat comes directly or indirectly from plants such as rice.

Virtually all staple foods come either directly from primary production by plants, or indirectly from animals that eat them. Plants and other photosynthetic organisms are at the base of most food chains because they use the energy from the sun and nutrients from the soil and atmosphere, converting them into a form that can be used by animals. This is what ecologists call the first trophic level. The modern forms of the major staple foods, such as maize, rice, wheat and other cereal grasses, pulses, bananas and plantains, as well as flax and cotton grown for their fibres, are the outcome of prehistoric selection over thousands of years from among wild ancestral plants with the most desirable characteristics. Botanists study how plants produce food and how to increase yields, for example through plant breeding, making their work important to mankind's ability to feed the world and provide food security for future generations. Botanists also study weeds, which are a considerable problem in agriculture, and the biology and control of plant pathogens in agriculture and natural ecosystems. Ethnobotany is the study of the relationships between plants and people. When applied to the investigation of historical plant–people relationships ethnobotany may be referred to as archaeobotany or palaeoethnobotany. Some of the earliest plant-people relationships arose between the indigenous people of Canada in identifying edible plants from inedible plants. This relationship the indigenous people had with plants was recorded by ethnobotanists.

Plant Biochemistry

Plant biochemistry is the study of the chemical processes used by plants. Some of these processes are used in their primary metabolism like the photosynthetic Calvin cycle and crassulacean acid metabolism. Others make specialized materials like the cellulose and lignin used to build their bodies, and secondary products like resins and aroma compounds.

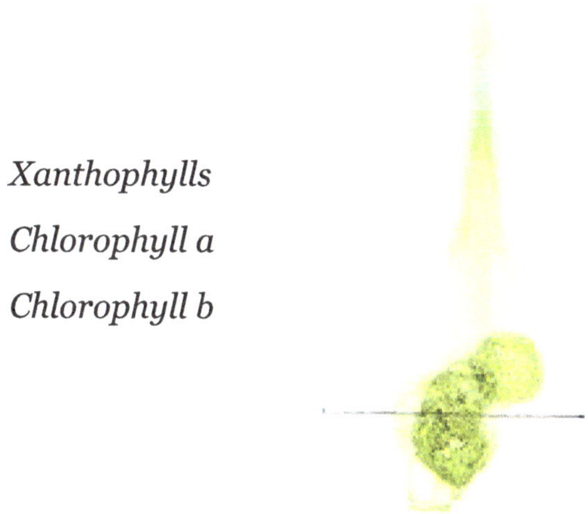

Plants make various photosynthetic pigments, some of which can be seen here through paper chromatography.

Plants and various other groups of photosynthetic eukaryotes collectively known as "algae" have unique organelles known as chloroplasts. Chloroplasts are thought to be descended from cyanobacteria that formed endosymbiotic relationships with ancient plant and algal ancestors. Chloroplasts and cyanobacteria contain the blue-green pigment chlorophyll *a*. Chlorophyll *a* (as well as its plant and green algal-specific cousin chlorophyll *b*)[a] absorbs light in the blue-violet and orange/red parts of the spectrum while reflecting and transmitting the green light that we see as the characteristic colour of these organisms. The energy in the red and blue light that these pigments absorb is used by chloroplasts to make energy-rich carbon compounds from carbon dioxide and water by oxygenic photosynthesis, a process that generates molecular oxygen (O_2) as a by-product.

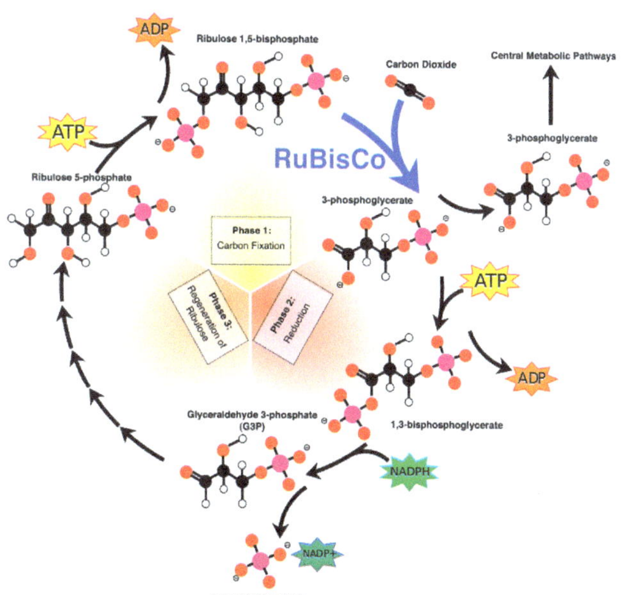

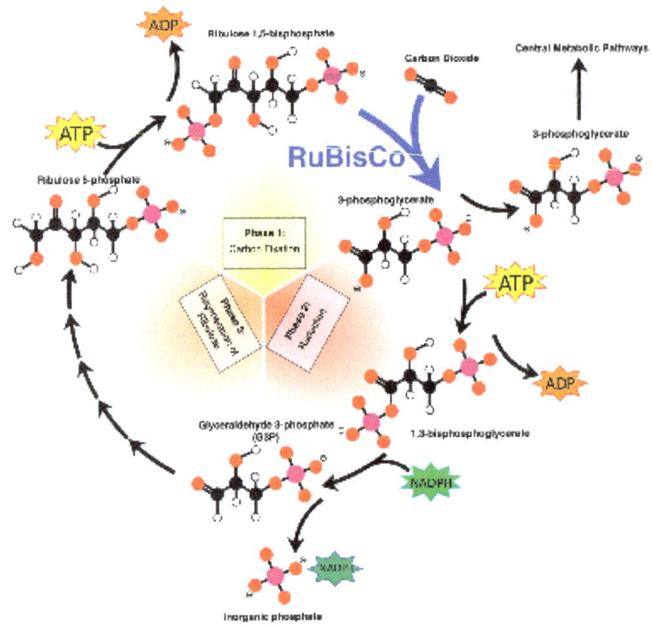

The Calvin cycle *(Interactive diagram)* The Calvin cycle incorporates carbon dioxide into sugar molecules.

The light energy captured by chlorophyll *a* is initially in the form of electrons (and later a proton gradient) that's used to make molecules of ATP and NADPH which temporarily store and transport energy. Their energy is used in the light-independent reactions of the Calvin cycle by the enzyme rubisco to produce molecules of the 3-carbon sugar glyceraldehyde 3-phosphate (G3P). Glyceraldehyde 3-phosphate is the first product of photosynthesis and the raw material from which glucose and almost all other organic molecules of biological origin are synthesized. Some of the glucose is converted to starch which is stored in the chloroplast. Starch is the characteristic energy store of most land plants and algae, while inulin, a polymer of fructose is used for the same purpose in the sunflower family Asteraceae. Some of the glucose is converted to sucrose (common table sugar) for export to the rest of the plant.

Unlike in animals (which lack chloroplasts), plants and their eukaryote relatives have delegated many biochemical roles to their chloroplasts, including synthesizing all their fatty acids, and most amino acids. The fatty acids that chloroplasts make are used for many things, such as providing material to build cell membranes out of and making the polymer cutin which is found in the plant cuticle that protects land plants from drying out.

Plants synthesize a number of unique polymers like the polysaccharide molecules cellulose, pectin and xyloglucan from which the land plant cell wall is constructed. Vascular land plants make lignin, a polymer used to strengthen the secondary cell walls of xylem tracheids and vessels to keep them from collapsing when a plant sucks water through them under water stress. Lignin is also used in other cell types like sclerenchyma fibers that provide structural support for a plant and is a major constituent of wood. Sporopollenin is a chemically resistant polymer found in the outer cell walls of spores and pollen of land plants responsible for the survival of early land plant spores and the pollen of seed plants in the fossil record. It is widely regarded as a marker for the start of land plant evolution during the Ordovician period. The concentration of carbon dioxide in the atmosphere today is much lower than it was when plants emerged onto land during the Ordovician and

Silurian periods. Many monocots like maize and the pineapple and some dicots like the Asteraceae have since independently evolved pathways like Crassulacean acid metabolism and the C_4 carbon fixation pathway for photosynthesis which avoid the losses resulting from photorespiration in the more common C_3 carbon fixation pathway. These biochemical strategies are unique to land plants.

Medicine and Materials

Tapping a rubber tree in Thailand

Phytochemistry is a branch of plant biochemistry primarily concerned with the chemical substances produced by plants during secondary metabolism. Some of these compounds are toxins such as the alkaloid coniine from hemlock. Others, such as the essential oils peppermint oil and lemon oil are useful for their aroma, as flavourings and spices (e.g., capsaicin), and in medicine as pharmaceuticals as in opium from opium poppies. Many medicinal and recreational drugs, such as tetrahydrocannabinol (active ingredient in cannabis), caffeine, morphine and nicotine come directly from plants. Others are simple derivatives of botanical natural products. For example, the pain killer aspirin is the acetyl ester of salicylic acid, originally isolated from the bark of willow trees, and a wide range of opiate painkillers like heroin are obtained by chemical modification of morphine obtained from the opium poppy. Popular stimulants come from plants, such as caffeine from coffee, tea and chocolate, and nicotine from tobacco. Most alcoholic beverages come from fermentation of carbohydrate-rich plant products such as barley (beer), rice (sake) and grapes (wine). Native Americans have used various plants as ways of treating illness or disease for thousands of years. This knowledge Native Americans have on plants has been recorded by enthnobotanists and then in turn has been used by pharmaceutical companies as a way of drug discovery.

Plants can synthesise useful coloured dyes and pigments such as the anthocyanins responsible for the red colour of red wine, yellow weld and blue woad used together to produce Lincoln green, indoxyl, source of the blue dye indigo traditionally used to dye denim and the artist's pigments gamboge and rose madder.

Sugar, starch, cotton, linen, hemp, some types of rope, wood and particle boards, papyrus and paper, vegetable oils, wax, and natural rubber are examples of commercially important materials made from plant tissues or their secondary products. Charcoal, a pure form of carbon made by pyrolysis of wood, has a long history as a metal-smelting fuel, as a filter material and adsorbent and as an artist's material and is one of the three ingredients of gunpowder. Cellulose, the world's most abundant organic polymer, can be converted into energy, fuels, materials and chemical feedstock. Products made from cellulose include rayon and cellophane, wallpaper paste, biobutanol and gun cotton. Sugarcane, rapeseed and soy are some of the plants with a highly fermentable sugar or oil content that are used as sources of biofuels, important alternatives to fossil fuels, such as biodiesel. Sweetgrass was used by NativeAmericanse to ward of bugs like mosquitoes. These bug repelling properties of sweetgrass were later found by the American Chemical Society in the molecules phytol and coumarin.

Plant Ecology

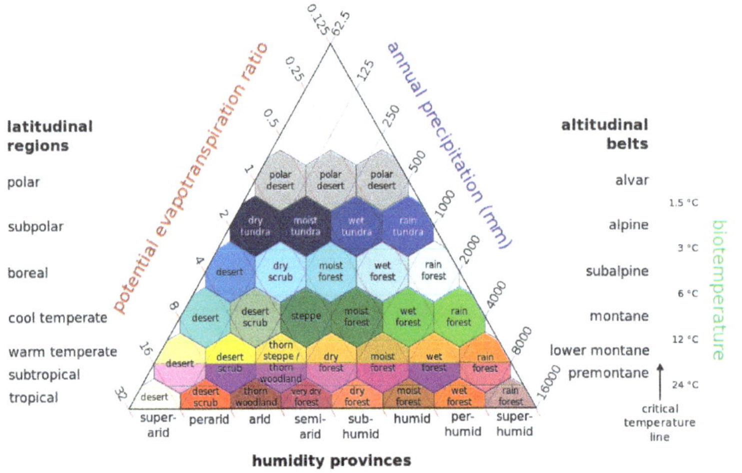

Holdridge life zones model relationships between vegetation type, moisture availability and temperature.

Plant ecology is the science of the functional relationships between plants and their habitats—the environments where they complete their life cycles. Plant ecologists study the composition of local and regional floras, their biodiversity, genetic diversity and fitness, the adaptation of plants to their environment, and their competitive or mutualistic interactions with other species. Some ecologists even rely on empirical data from indigenous people that is gathered by ethnobotanists. This information can relay a great deal of information on how the land once was thousands of years ago and how it has changed over that time. The goals of plant ecology are to understand the causes of their distribution patterns, productivity, environmental impact, evolution, and responses to environmental change.

Plants depend on certain edaphic (soil) and climatic factors in their environment but can modify these factors too. For example, they can change their environment's albedo, increase runoff interception, stabilize mineral soils and develop their organic content, and affect local temperature. Plants compete with other organisms in their ecosystem for resources. They interact with their neighbours at a variety of spatial scales in groups, populations and communities that collectively constitute vegetation. Regions with characteristic vegetation types and dominant plants as well as similar abiotic and biotic factors, climate, and geography make up biomes like tundra or tropical rainforest.

The nodules of *Medicago italica* contain the nitrogen fixing bacterium *Sinorhizobium meliloti*. The plant provides the bacteria with nutrients and an anaerobic environment, and the bacteria fix nitrogen for the plant.

Herbivores eat plants, but plants can defend themselves and some species are parasitic or even carnivorous. Other organisms form mutually beneficial relationships with plants. For example, mycorrhizal fungi and rhizobia provide plants with nutrients in exchange for food, ants are recruited by ant plants to provide protection, honey bees, bats and other animals pollinate flowers and humans and other animals act as dispersal vectors to spread spores and seeds.

Plants, Climate and Environmental Change

Plant responses to climate and other environmental changes can inform our understanding of how these changes affect ecosystem function and productivity. For example, plant phenology can be a useful proxy for temperature in historical climatology, and the biological impact of climate change and global warming. Palynology, the analysis of fossil pollen deposits in sediments from thousands or millions of years ago allows the reconstruction of past climates. Estimates of atmospheric CO_2 concentrations since the Palaeozoic have been obtained from stomatal densities and the leaf shapes and sizes of ancient land plants. Ozone depletion can expose plants to higher levels of ultraviolet radiation-B (UV-B), resulting in lower growth rates. Moreover, information from studies of community ecology, plant systematics, and taxonomy is essential to understanding vegetation change, habitat destruction and species extinction.

Genetics

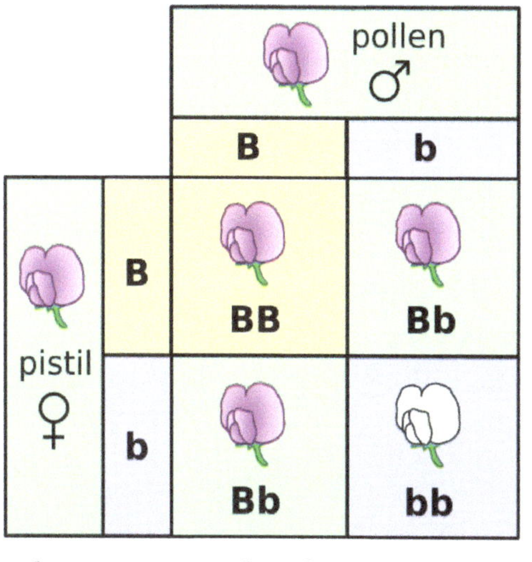

A Punnett square depicting a cross between two pea plants heterozygous for purple (B) and white (b) blossoms

Inheritance in plants follows the same fundamental principles of genetics as in other multicellular organisms. Gregor Mendel discovered the genetic laws of inheritance by studying inherited traits such as shape in *Pisum sativum* (peas). What Mendel learned from studying plants has had far reaching benefits outside of botany. Similarly, "jumping genes" were discovered by Barbara McClintock while she was studying maize. Nevertheless, there are some distinctive genetic differences between plants and other organisms.

Species boundaries in plants may be weaker than in animals, and cross species hybrids are often possible. A familiar example is peppermint, *Mentha* × *piperita*, a sterile hybrid between *Mentha aquatica* and spearmint, *Mentha spicata*. The many cultivated varieties of wheat are the result of multiple inter- and intra-specific crosses between wild species and their hybrids. Angiosperms with monoecious flowers often have self-incompatibility mechanisms that operate between the pollen and stigma so that the pollen either fails to reach the stigma or fails to germinate and produce male gametes. This is one of several methods used by plants to promote outcrossing. In many land plants the male and female gametes are produced by separate individuals. These species are said to be dioecious when referring to vascular plant sporophytes and dioicous when referring to bryophyte gametophytes.

Unlike in higher animals, where parthenogenesis is rare, asexual reproduction may occur in plants by several different mechanisms. The formation of stem tubers in potato is one example. Particularly in arctic or alpine habitats, where opportunities for fertilisation of flowers by animals are rare, plantlets or bulbs, may develop instead of flowers, replacing sexual reproduction with asexual reproduction and giving rise to clonal populations genetically identical to the parent. This is one of several types of apomixis that occur in plants. Apomixis can also happen in a seed, producing a seed that contains an embryo genetically identical to the parent.

Most sexually reproducing organisms are diploid, with paired chromosomes, but doubling of their chromosome number may occur due to errors in cytokinesis. This can occur early in development to produce an autopolyploid or partly autopolyploid organism, or during normal processes of cellular differentiation to produce some cell types that are polyploid (endopolyploidy), or during gamete formation. An allopolyploid plant may result from a hybridisation event between two different species. Both autopolyploid and allopolyploid plants can often reproduce normally, but may be unable to cross-breed successfully with the parent population because there is a mismatch in chromosome numbers. These plants that are reproductively isolated from the parent species but live within the same geographical area, may be sufficiently successful to form a new species. Some otherwise sterile plant polyploids can still reproduce vegetatively or by seed apomixis, forming clonal populations of identical individuals. Durum wheat is a fertile tetraploid allopolyploid, while bread wheat is a fertile hexaploid. The commercial banana is an example of a sterile, seedless triploid hybrid. Common dandelion is a triploid that produces viable seeds by apomictic seed.

As in other eukaryotes, the inheritance of endosymbiotic organelles like mitochondria and chloroplasts in plants is non-Mendelian. Chloroplasts are inherited through the male parent in gymnosperms but often through the female parent in flowering plants.

Molecular Genetics

A considerable amount of new knowledge about plant function comes from studies of the molecular genetics of model plants such as the Thale cress, *Arabidopsis thaliana*, a weedy species in the mustard family (Brassicaceae). The genome or hereditary information contained in the genes of this species is encoded by about 135 million base pairs of DNA, forming one of the smallest genomes among flowering plants. *Arabidopsis* was the first plant to have its genome sequenced, in 2000. The sequencing of some other relatively small genomes, of rice (*Oryza sativa*) and *Brachypodium distachyon*, has made them important model species for understanding the genetics, cellular and molecular biology of cereals, grasses and monocots generally.

Thale cress, *Arabidopsis thaliana*, the first plant to have its genome sequenced, remains the most important model organism.

Model plants such as *Arabidopsis thaliana* are used for studying the molecular biology of plant cells and the chloroplast. Ideally, these organisms have small genomes that are well known or completely sequenced, small stature and short generation times. Corn has been used to study mechanisms of photosynthesis and phloem loading of sugar in C_4 plants. The single celled green alga *Chlamydomonas reinhardtii*, while not an embryophyte itself, contains a green-pigmented chloroplast related to that of land plants, making it useful for study. A red alga *Cyanidioschyzon merolae* has also been used to study some basic chloroplast functions. Spinach, peas, soybeans and a moss *Physcomitrella patens* are commonly used to study plant cell biology.

Agrobacterium tumefaciens, a soil rhizosphere bacterium, can attach to plant cells and infect them with a callus-inducing Ti plasmid by horizontal gene transfer, causing a callus infection called crown gall disease. Schell and Van Montagu (1977) hypothesised that the Ti plasmid could be a natural vector for introducing the Nif gene responsible for nitrogen fixation in the root nodules of legumes and other plant species. Today, genetic modification of the Ti plasmid is one of the main techniques for introduction of transgenes to plants and the creation of genetically modified crops.

Epigenetics

Epigenetics is the study of heritable changes in gene function that cannot be explained by changes in the underlying DNA sequence but cause the organism's genes to behave (or "express themselves") differently. One example of epigenetic change is the marking of the genes by DNA methylation which determines whether they will be expressed or not. Gene expression can also be controlled by repressor proteins that attach to silencer regions of the DNA and prevent that region of the DNA code from being expressed. Epigenetic marks may be added or removed from the DNA during programmed stages of development of the plant, and are responsible, for example, for the differences between anthers, petals and normal leaves, despite the fact that they all have the same underlying genetic code. Epigenetic changes may be temporary or may remain through successive cell divisions for the remainder of the cell's life. Some epigenetic changes have been shown to be heritable, while others are reset in the germ cells.

Epigenetic changes in eukaryotic biology serve to regulate the process of cellular differentiation. During morphogenesis, totipotent stem cells become the various pluripotent cell lines of the embryo, which in turn become fully differentiated cells. A single fertilized egg cell, the zygote, gives rise to the many different plant cell types including parenchyma, xylem vessel elements, phloem sieve tubes, guard cells of the epidermis, etc. as it continues to divide. The process results from the epigenetic activation of some genes and inhibition of others.

Unlike animals, many plant cells, particularly those of the parenchyma, do not terminally differentiate, remaining totipotent with the ability to give rise to a new individual plant. Exceptions include highly lignified cells, the sclerenchyma and xylem which are dead at maturity, and the phloem sieve tubes which lack nuclei. While plants use many of the same epigenetic mechanisms as animals, such as chromatin remodeling, an alternative hypothesis is that plants set their gene expression patterns using positional information from the environment and surrounding cells to determine their developmental fate.

Plant Evolution

The chloroplasts of plants have a number of biochemical, structural and genetic similarities to cyanobacteria, (commonly but incorrectly known as "blue-green algae") and are thought to be derived from an ancient endosymbiotic relationship between an ancestral eukaryotic cell and a cyanobacterial resident.

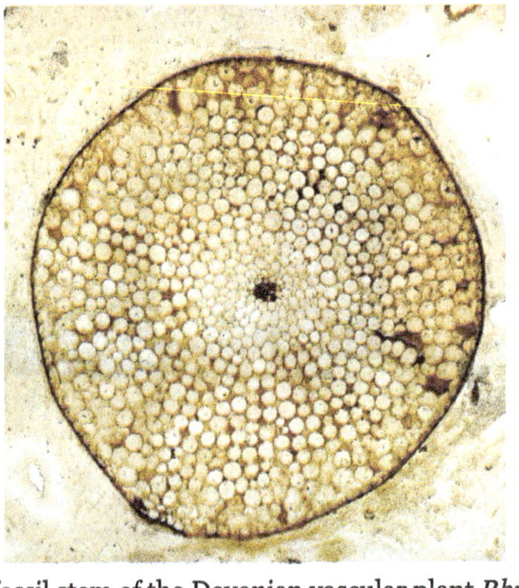

Transverse section of a fossil stem of the Devonian vascular plant *Rhynia gwynne-vaughani*

The algae are a polyphyletic group and are placed in various divisions, some more closely related to plants than others. There are many differences between them in features such as cell wall composition, biochemistry, pigmentation, chloroplast structure and nutrient reserves. The algal division

Charophyta, sister to the green algal division Chlorophyta, is considered to contain the ancestor of true plants. The Charophyte class Charophyceae and the land plant sub-kingdom Embryophyta together form the monophyletic group or clade Streptophytina.

Nonvascular land plants are embryophytes that lack the vascular tissues xylem and phloem. They include mosses, liverworts and hornworts. Pteridophytic vascular plants with true xylem and phloem that reproduced by spores germinating into free-living gametophytes evolved during the Silurian period and diversified into several lineages during the late Silurian and early Devonian. Representatives of the lycopods have survived to the present day. By the end of the Devonian period, several groups, including the lycopods, sphenophylls and progymnosperms, had independently evolved "megaspory" – their spores were of two distinct sizes, larger megaspores and smaller microspores. Their reduced gametophytes developed from megaspores retained within the spore-producing organs (megasporangia) of the sporophyte, a condition known as endospory. Seeds consist of an endosporic megasporangium surrounded by one or two sheathing layers (integuments). The young sporophyte develops within the seed, which on germination splits to release it. The earliest known seed plants date from the latest Devonian Famennian stage. Following the evolution of the seed habit, seed plants diversified, giving rise to a number of now-extinct groups, including seed ferns, as well as the modern gymnosperms and angiosperms. Gymnosperms produce "naked seeds" not fully enclosed in an ovary; modern representatives include conifers, cycads, *Ginkgo*, and Gnetales. Angiosperms produce seeds enclosed in a structure such as a carpel or an ovary. Ongoing research on the molecular phylogenetics of living plants appears to show that the angiosperms are a sister clade to the gymnosperms.

Plant Physiology

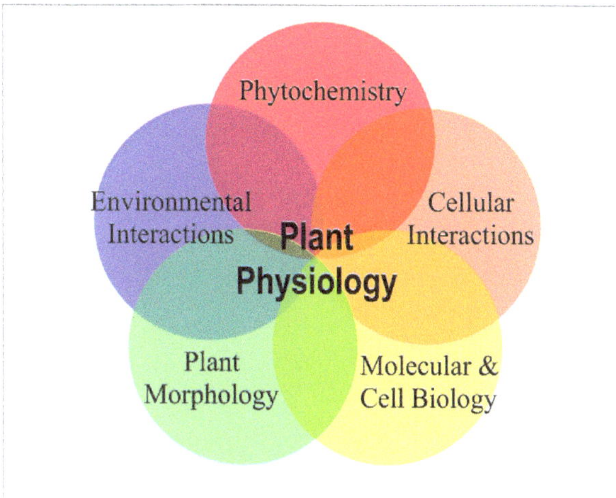

Five of the key areas of study within plant physiology

Plant physiology encompasses all the internal chemical and physical activities of plants associated with life. Chemicals obtained from the air, soil and water form the basis of all plant metabolism. The energy of sunlight, captured by oxygenic photosynthesis and released by cellular respiration, is the basis of almost all life. Photoautotrophs, including all green plants, algae and cyanobacteria gather energy directly from sunlight by photosynthesis. Heterotrophs including all animals, all fungi, all completely parasitic plants, and non-photosynthetic bacteria take in organic molecules produced by photoautotrophs and respire them or use them in the construction of cells and tissues.

Respiration is the oxidation of carbon compounds by breaking them down into simpler structures to release the energy they contain, essentially the opposite of photosynthesis.

Molecules are moved within plants by transport processes that operate at a variety of spatial scales. Subcellular transport of ions, electrons and molecules such as water and enzymes occurs across cell membranes. Minerals and water are transported from roots to other parts of the plant in the transpiration stream. Diffusion, osmosis, and active transport and mass flow are all different ways transport can occur. Examples of elements that plants need to transport are nitrogen, phosphorus, potassium, calcium, magnesium, and sulphur. In vascular plants, these elements are extracted from the soil as soluble ions by the roots and transported throughout the plant in the xylem. Most of the elements required for plant nutrition come from the chemical breakdown of soil minerals. Sucrose produced by photosynthesis is transported from the leaves to other parts of the plant in the phloem and plant hormones are transported by a variety of processes.

Plant Hormones

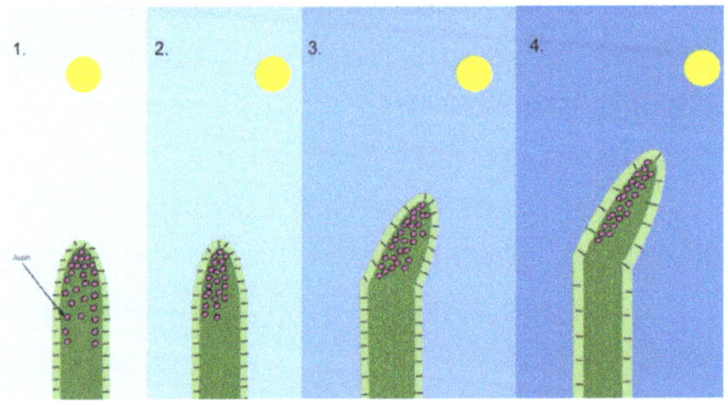

1 An oat coleoptile with the sun overhead. Auxin (pink) is evenly distributed in its tip.
2 With the sun at an angle and only shining on one side of the shoot, auxin moves to the opposite side and stimulates cell elongation there.
3 and 4 Extra growth on that side causes the shoot to bend towards the sun.

Plants are not passive, but respond to external signals such as light, touch, and injury by moving or growing towards or away from the stimulus, as appropriate. Tangible evidence of touch sensitivity is the almost instantaneous collapse of leaflets of *Mimosa pudica*, the insect traps of Venus flytrap and bladderworts, and the pollinia of orchids.

The hypothesis that plant growth and development is coordinated by plant hormones or plant growth regulators first emerged in the late 19th century. Darwin experimented on the movements of plant shoots and roots towards light and gravity, and concluded "It is hardly an exaggeration to say that the tip of the radicle . . acts like the brain of one of the lower animals . . directing the several movements". About the same time, the role of auxins (from the Greek auxein, to grow) in control of plant growth was first outlined by the Dutch scientist Frits Went. The first known auxin, indole-3-acetic acid (IAA), which promotes cell growth, was only isolated from plants about 50 years later. This compound mediates the tropic responses of shoots and roots towards light and gravity. The finding in 1939 that plant callus could be maintained in culture containing IAA, followed by the observation in 1947 that it could be induced to form roots and shoots by controlling the concentration of growth hormones were key steps in the development of plant biotechnology and genetic modification.

Cytokinins are a class of plant hormones named for their control of cell division or cytokinesis. The natural cytokinin zeatin was discovered in corn, *Zea mays*, and is a derivative of the purine adenine. Zeatin is produced in roots and transported to shoots in the xylem where it promotes cell division, bud development, and the greening of chloroplasts. The gibberelins, such as Gibberelic acid are diterpenes synthesised from acetyl CoA via the mevalonate pathway. They are involved in the promotion of germination and dormancy-breaking in seeds, in regulation of plant height by controlling stem elongation and the control of flowering. Abscisic acid (ABA) occurs in all land plants except liverworts, and is synthesised from carotenoids in the chloroplasts and other plastids. It inhibits cell division, promotes seed maturation, and dormancy, and promotes stomatal closure. It was so named because it was originally thought to control abscission. Ethylene is a gaseous hormone that is produced in all higher plant tissues from methionine. It is now known to be the hormone that stimulates or regulates fruit ripening and abscission, and it, or the synthetic growth regulator ethephon which is rapidly metabolised to produce ethylene, are used on industrial scale to promote ripening of cotton, pineapples and other climacteric crops.

Another class of phytohormones is the jasmonates, first isolated from the oil of *Jasminum grandiflorum* which regulates wound responses in plants by unblocking the expression of genes required in the systemic acquired resistance response to pathogen attack.

In addition to being the primary energy source for plants, light functions as a signalling device, providing information to the plant, such as how much sunlight the plant receives each day. This can result in adaptive changes in a process known as photomorphogenesis. Phytochromes are the photoreceptors in a plant that are sensitive to light.

Plant Anatomy and Morphology

A nineteenth-century illustration showing the morphology of the roots, stems, leaves and flowers of the rice plant *Oryza sativa*

Plant anatomy is the study of the structure of plant cells and tissues, whereas plant morphology is the study of their external form. All plants are multicellular eukaryotes, their DNA stored in nuclei. The characteristic features of plant cells that distinguish them from those of animals and fungi include a primary cell wall composed of the polysaccharides cellulose, hemicellulose and pectin, larger vacuoles than in animal cells and the presence of plastids with unique photosynthetic and biosynthetic functions as in the chloroplasts. Other plastids contain storage products such as starch (amyloplasts) or lipids (elaioplasts). Uniquely, streptophyte cells and those of the green algal order Trentepohliales divide by construction of a phragmoplast as a template for building a cell plate late in cell division.

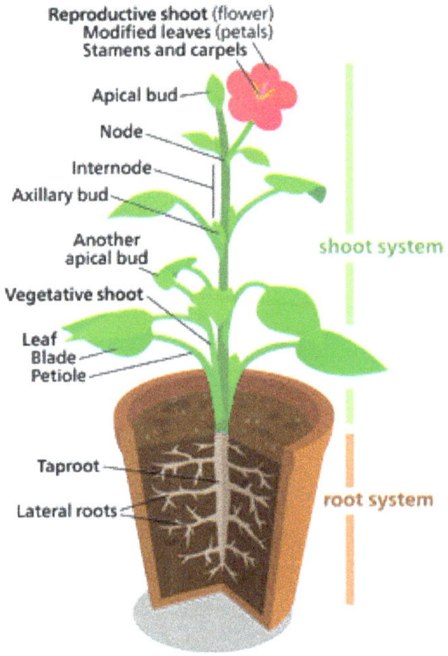

A diagram of a "typical" eudicot, the most common type of plant (three-fifths of all plant species). No plant actually looks exactly like this though.

The bodies of vascular plants including clubmosses, ferns and seed plants (gymnosperms and angiosperms) generally have aerial and subterranean subsystems. The shoots consist of stems bearing green photosynthesising leaves and reproductive structures. The underground vascularised roots bear root hairs at their tips and generally lack chlorophyll. Non-vascular plants, the liverworts, hornworts and mosses do not produce ground-penetrating vascular roots and most of the plant participates in photosynthesis. The sporophyte generation is nonphotosynthetic in liverworts but may be able to contribute part of its energy needs by photosynthesis in mosses and hornworts.

The root system and the shoot system are interdependent – the usually nonphotosynthetic root system depends on the shoot system for food, and the usually photosynthetic shoot system depends on water and minerals from the root system. Cells in each system are capable of creating cells of the other and producing adventitious shoots or roots. Stolons and tubers are examples of shoots that can grow roots. Roots that spread out close to the surface, such as those of willows, can produce shoots and ultimately new plants. In the event that one of the systems is lost, the other can often regrow it. In fact it is possible to grow an entire plant from a single leaf, as is the case with *Saintpaulia*, or even a single cell – which can dedifferentiate into a callus (a mass of unspecialised cells) that can grow into a new plant. In vascular plants, the xylem and phloem are the conductive tissues that transport resources between shoots and roots. Roots are often adapted to store food such as sugars or starch, as in sugar beets and carrots.

Stems mainly provide support to the leaves and reproductive structures, but can store water in succulent plants such as cacti, food as in potato tubers, or reproduce vegetatively as in the stolons of strawberry plants or in the process of layering. Leaves gather sunlight and carry out photosynthesis. Large, flat, flexible, green leaves are called foliage leaves. Gymnosperms, such as conifers, cycads, *Ginkgo*, and gnetophytes are seed-producing plants with open seeds. Angiosperms are seed-producing plants that produce flowers and have enclosed seeds. Woody plants, such as azaleas and oaks, undergo a secondary growth phase resulting in two additional types of tissues: wood (secondary xylem) and bark (secondary phloem and cork). All gymnosperms and many angiosperms are woody plants. Some plants reproduce sexually, some asexually, and some via both means.

Although reference to major morphological categories such as root, stem, leaf, and trichome are useful, one has to keep in mind that these categories are linked through intermediate forms so that a continuum between the categories results. Furthermore, structures can be seen as processes, that is, process combinations.

Systematic Botany

A botanist preparing a plant specimen for mounting in the herbarium

Systematic botany is part of systematic biology, which is concerned with the range and diversity of organisms and their relationships, particularly as determined by their evolutionary history. It involves, or is related to, biological classification, scientific taxonomy and phylogenetics. Biological classification is the method by which botanists group organisms into categories such as genera or species. Biological classification is a form of scientific taxonomy. Modern taxonomy is rooted in the work of Carl Linnaeus, who grouped species according to shared physical characteristics. These groupings have since been revised to align better with the Darwinian principle of common descent – grouping organisms by ancestry rather than superficial characteristics. While scientists do not always agree on how to classify organisms, molecular phylogenetics, which uses DNA sequences as data, has driven many recent revisions along evolutionary lines and is likely to continue to do so. The dominant classification system is called Linnaean taxonomy. It includes ranks and binomial nomenclature. The nomenclature of botanical organisms is codified in the International Code of Nomenclature for algae, fungi, and plants (ICN) and administered by the International Botanical Congress.

Kingdom Plantae belongs to Domain Eukarya and is broken down recursively until each species is separately classified. The order is: Kingdom; Phylum (or Division); Class; Order; Family; Genus (plural *genera*); Species. The scientific name of a plant represents its genus and its species within the genus, resulting in a single world-wide name for each organism. For example, the tiger lily is *Lilium columbianum*. *Lilium* is the genus, and *columbianum* the specific epithet. The combination is the name of the species. When writing the scientific name of an organism, it is proper to capitalise the first letter in the genus and put all of the specific epithet in lowercase. Additionally, the entire term is ordinarily italicised (or underlined when italics are not available).

The evolutionary relationships and heredity of a group of organisms is called its phylogeny. Phylogenetic studies attempt to discover phylogenies. The basic approach is to use similarities based on shared inheritance to determine relationships. As an example, species of *Pereskia* are trees or bushes with prominent leaves. They do not obviously resemble a typical leafless cactus such as an *Echinocactus*. However, both *Pereskia* and *Echinocactus* have spines produced from areoles (highly specialised pad-like structures) suggesting that the two genera are indeed related.

Two Cacti of Very Different Appearance

Pereskia aculeata

Echinocactus grusonii

Although *Pereskia* is a tree with leaves, it has spines and areoles like a more typical cactus, such as *Echinocactus*.

Judging relationships based on shared characters requires care, since plants may resemble one another through convergent evolution in which characters have arisen independently. Some euphorbias have leafless, rounded bodies adapted to water conservation similar to those of globular cacti, but characters such as the structure of their flowers make it clear that the two groups are not closely related. The cladistic method takes a systematic approach to characters, distinguishing between those that carry no information about shared evolutionary history – such as those evolved separately in different groups (homoplasies) or those left over from ancestors (plesiomorphies) – and derived characters, which have been passed down from innovations in a shared ancestor (apomorphies). Only derived characters, such as the spine-producing areoles of cacti, provide evidence for descent from a common ancestor. The results of cladistic analyses are expressed as cladograms: tree-like diagrams showing the pattern of evolutionary branching and descent.

From the 1990s onwards, the predominant approach to constructing phylogenies for living plants has been molecular phylogenetics, which uses molecular characters, particularly DNA sequences, rather than morphological characters like the presence or absence of spines and areoles. The difference is that the genetic code itself is used to decide evolutionary relationships, instead of being used indirectly via the characters it gives rise to. Clive Stace describes this as having "direct access to the genetic basis of evolution." As a simple example, prior to the use of genetic evidence, fungi were thought either to be plants or to be more closely related to plants than animals. Genetic evidence suggests that the true evolutionary relationship of multicelled organisms is as shown in the cladogram below – fungi are more closely related to animals than to plants.

In 1998 the Angiosperm Phylogeny Group published a phylogeny for flowering plants based on an analysis of DNA sequences from most families of flowering plants. As a result of this work, many questions, such as which families represent the earliest branches of angiosperms, have now been answered. Investigating how plant species are related to each other allows botanists to better understand the process of evolution in plants. Despite the study of model plants and increasing use of DNA evidence, there is ongoing work and discussion among taxonomists about how best to classify plants into various taxa. Technological developments such as computers and electron microscopes have greatly increased the level of detail studied and speed at which data can be analysed.

2

Diverse Branches of Botany

Botany is an interdisciplinary subject. Some of the branches of botany are bryology, plant physiology, plant morphology, plant genetics, paleobotany and economic botany. Bryology is the scientific study of bryophytes whereas plant physiology concerns itself with the functioning of plants. The following content will provide a glimpse of related branches of botany.

Bryology

Bryology (from Greek *bryon*, a moss, a liverwort) is the branch of botany concerned with the scientific study of bryophytes (mosses, liverworts, and hornworts). Bryologists are people who have an active interest in observing, recording, classifying or researching bryophytes. The field is often studied along with lichenology due to the similar appearance and ecological niche of the two organisms, even though bryophytes and lichens are not classified in the same kingdom.

Common bryophytes found in central Japan

History

Bryophytes were first studied in detail in the 18th century. The German botanist Johann Jacob Dillenius (1687–1747) was a professor at Oxford and in 1717 produced the work "Reproduction of the ferns and mosses." The beginning of bryology really belongs to the work of Johannes Hedwig, who clarified the reproductive system of mosses (1792, *Fundamentum historiae naturalist muscorum*) and arranged a taxonomy.

Areas of research include bryophyte taxonomy, bryophytes as bioindicators, DNA sequencing, and the interdependency of bryophytes and other plant and animal species. Among other things, scientists have discovered parasitic bryophytes such as Cryptothallus and potentially carnivorous liverworts such as Colura zoophaga and Pleurozia.

Centers of research in bryology include University of Bonn, Germany, the University of Helsinki, Finland and the New York Botanical Garden.

Plant Physiology

Plant physiology is a subdiscipline of botany concerned with the functioning, or physiology, of plants. Closely related fields include plant morphology (structure of plants), plant ecology (interactions with the environment), phytochemistry (biochemistry of plants), cell biology, genetics, biophysics and molecular biology.

Fundamental processes such as photosynthesis, respiration, plant nutrition, plant hormone functions, tropisms, nastic movements, photoperiodism, photomorphogenesis, circadian rhythms, environmental stress physiology, seed germination, dormancy and stomata function and transpiration, both parts of plant water relations, are studied by plant physiologists.

Aims

The field of plant physiology includes the study of all the internal activities of plants—those chemical and physical processes associated with life as they occur in plants. This includes study at many levels of scale of size and time. At the smallest scale are molecular interactions of photosynthesis and internal diffusion of water, minerals, and nutrients. At the largest scale are the processes of plant development, seasonality, dormancy, and reproductive control. Major subdisciplines of plant physiology include phytochemistry (the study of the biochemistry of plants) and phytopathology (the study of disease in plants). The scope of plant physiology as a discipline may be divided into several major areas of research.

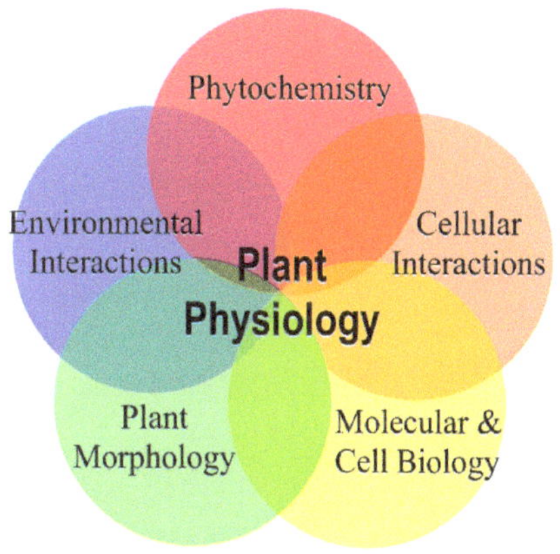

Five key areas of study within plant physiology.

First, the study of phytochemistry (plant chemistry) is included within the domain of plant physiology. To function and survive, plants produce a wide array of chemical compounds not found in other organisms. Photosynthesis requires a large array of pigments, enzymes, and other compounds to function. Because they cannot move, plants must also defend themselves chemically from herbivores, pathogens and competition from other plants. They do this by producing toxins and foul-tasting or smelling chemicals. Other compounds defend plants against disease, permit survival during drought, and prepare plants for dormancy, while other compounds are used to attract pollinators or herbivores to spread ripe seeds.

Secondly, plant physiology includes the study of biological and chemical processes of individual plant cells. Plant cells have a number of features that distinguish them from cells of animals, and which lead to major differences in the way that plant life behaves and responds differently from animal life. For example, plant cells have a cell wall which restricts the shape of plant cells and thereby limits the flexibility and mobility of plants. Plant cells also contain chlorophyll, a chemical compound that interacts with light in a way that enables plants to manufacture their own nutrients rather than consuming other living things as animals do.

Thirdly, plant physiology deals with interactions between cells, tissues, and organs within a plant. Different cells and tissues are physically and chemically specialized to perform different functions. Roots and rhizoids function to anchor the plant and acquire minerals in the soil. Leaves catch light in order to manufacture nutrients. For both of these organs to remain living, minerals that the roots acquire must be transported to the leaves, and the nutrients manufactured in the leaves must be transported to the roots. Plants have developed a number of ways to achieve this transport, such as vascular tissue, and the functioning of the various modes of transport is studied by plant physiologists.

Fourthly, plant physiologists study the ways that plants control or regulate internal functions. Like animals, plants produce chemicals called hormones which are produced in one part of the plant to signal cells in another part of the plant to respond. Many flowering plants bloom at the appropriate time because of light-sensitive compounds that respond to the length of the night, a phenomenon known as photoperiodism. The ripening of fruit and loss of leaves in the winter are controlled in part by the production of the gas ethylene by the plant.

Finally, plant physiology includes the study of plant response to environmental conditions and their variation, a field known as environmental physiology. Stress from water loss, changes in air chemistry, or crowding by other plants can lead to changes in the way a plant functions. These changes may be affected by genetic, chemical, and physical factors.

Biochemistry of Plants

Latex being collected from a tapped rubber tree.

The chemical elements of which plants are constructed—principally carbon, oxygen, hydrogen, nitrogen, phosphorus, sulfur, etc.—are the same as for all other life forms animals, fungi, bacteria and even viruses. Only the details of the molecules into which they are assembled differs.

Despite this underlying similarity, plants produce a vast array of chemical compounds with unique properties which they use to cope with their environment. Pigments are used by plants to absorb or detect light, and are extracted by humans for use in dyes. Other plant products may be used for the manufacture of commercially important rubber or biofuel. Perhaps the most celebrated compounds from plants are those with pharmacological activity, such as salicylic acid from which aspirin is made, morphine, and digoxin. Drug companies spend billions of dollars each year researching plant compounds for potential medicinal benefits.

Constituent Elements

Plants require some nutrients, such as carbon and nitrogen, in large quantities to survive. Such nutrients are termed macronutrients, where the prefix *macro-* (large) refers to the quantity needed, not the size of the nutrient particles themselves. Other nutrients, called micronutrients, are required only in trace amounts for plants to remain healthy. Such micronutrients are usually absorbed as ions dissolved in water taken from the soil, though carnivorous plants acquire some of their micronutrients from captured prey.

The following tables list element nutrients essential to plants. Uses within plants are generalized.

Macronutrients – necessary in large quantities		
Element	Form of uptake	Notes
Nitrogen	NO_3^-, NH_4^+	Nucleic acids, proteins, hormones, etc.
Oxygen	O_2 H_2O	Cellulose, starch, other organic compounds
Carbon	CO_2	Cellulose, starch, other organic compounds
Hydrogen	H_2O	Cellulose, starch, other organic compounds
Potassium	K^+	Cofactor in protein synthesis, water balance, etc.
Calcium	Ca^{2+}	Membrane synthesis and stabilization
Magnesium	Mg^{2+}	Element essential for chlorophyll
Phosphorus	$H_2PO_4^-$	Nucleic acids, phospholipids, ATP
Sulfur	SO_4^{2-}	Constituent of proteins

Micronutrients – necessary in small quantities		
Element	Form of uptake	Notes
Chlorine	Cl^-	Photosystem II and stomata function
Iron	Fe^{2+}, Fe^{3+}	Chlorophyll formation
Boron	HBO_3	Crosslinking pectin
Manganese	Mn^{2+}	Activity of some enzymes
Zinc	Zn^{2+}	Involved in the synthesis of enzymes and chlorophyll

Copper	Cu^{+}	Enzymes for lignin synthesis
Molybdenum	MoO_4^{2-}	Nitrogen fixation, reduction of nitrates
Nickel	Ni^{2+}	Enzymatic cofactor in the metabolism of nitrogen compounds

Pigments

Among the most important molecules for plant function are the pigments. Plant pigments include a variety of different kinds of molecules, including porphyrins, carotenoids, and anthocyanins. All biological pigments selectively absorb certain wavelengths of light while reflecting others. The light that is absorbed may be used by the plant to power chemical reactions, while the reflected wavelengths of light determine the color the pigment appears to the eye.

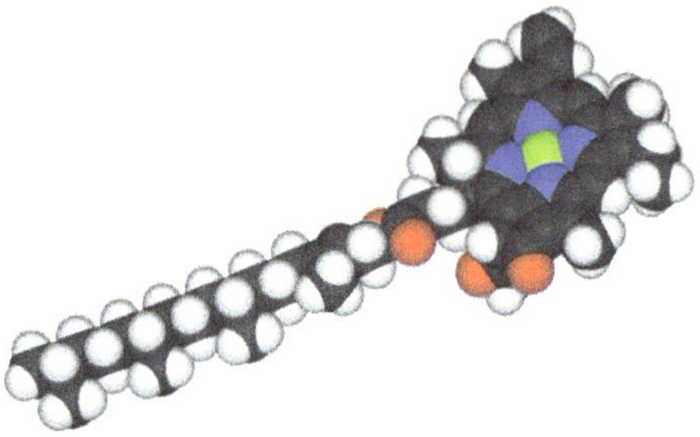

Space-filling model of the chlorophyll molecule.

Anthocyanin gives these pansies their dark purple pigmentation.

Chlorophyll is the primary pigment in plants; it is a porphyrin that absorbs red and blue wavelengths of light while reflecting green. It is the presence and relative abundance of chlorophyll that gives plants their green color. All land plants and green algae possess two forms of this pigment: chlorophyll *a* and chlorophyll *b*. Kelps, diatoms, and other photosynthetic heterokonts contain chlorophyll *c* instead of *b*, red algae possess chlorophyll *a* and " d". All chlorophylls serve as the primary means plants use to intercept light to fuel photosynthesis.

Carotenoids are red, orange, or yellow tetraterpenoids. They function as accessory pigments in plants, helping to fuel photosynthesis by gathering wavelengths of light not readily absorbed by chlorophyll. The most familiar carotenoids are carotene (an orange pigment found in carrots), lutein (a yellow pigment found in fruits and vegetables), and lycopene (the red pigment responsible for the color of tomatoes). Carotenoids have been shown to act as antioxidants and to promote healthy eyesight in humans.

Anthocyanins (literally "flower blue") are water-soluble flavonoid pigments that appear red to blue, according to pH. They occur in all tissues of higher plants, providing color in leaves, stems, roots, flowers, and fruits, though not always in sufficient quantities to be noticeable. Anthocyanins are most visible in the petals of flowers, where they may make up as much as 30% of the dry weight of the tissue. They are also responsible for the purple color seen on the underside of tropical shade plants such as *Tradescantia zebrina*. In these plants, the anthocyanin catches light that has passed through the leaf and reflects it back towards regions bearing chlorophyll, in order to maximize the use of available light

Betalains are red or yellow pigments. Like anthocyanins they are water-soluble, but unlike anthocyanins they are indole-derived compounds synthesized from tyrosine. This class of pigments is found only in the Caryophyllales (including cactus and amaranth), and never co-occur in plants with anthocyanins. Betalains are responsible for the deep red color of beets, and are used commercially as food-coloring agents. Plant physiologists are uncertain of the function that betalains have in plants which possess them, but there is some preliminary evidence that they may have fungicidal properties.

Signals and Regulators

A mutation that stops *Arabidopsis thaliana* responding to auxin causes abnormal growth (right)

Plants produce hormones and other growth regulators which act to signal a physiological response in their tissues. They also produce compounds such as phytochrome that are sensitive to light and which serve to trigger growth or development in response to environmental signals.

Plant Hormones

Plant hormones, known as plant growth regulators (PGRs) or phytohormones, are chemicals that regulate a plant's growth. According to a standard animal definition, hormones are signal molecules produced at specific locations, that occur in very low concentrations, and cause altered processes in target cells at other locations. Unlike animals, plants lack specific hormone-producing tissues or organs. Plant hormones are often not transported to other parts of the plant and production is not limited to specific locations.

Plant hormones are chemicals that in small amounts promote and influence the growth, development and differentiation of cells and tissues. Hormones are vital to plant growth; affecting processes in plants from flowering to seed development, dormancy, and germination. They regulate which tissues grow upwards and which grow downwards, leaf formation and stem growth, fruit development and ripening, as well as leaf abscission and even plant death.

The most important plant hormones are abscissic acid (ABA), auxins, ethylene, gibberellins, and cytokinins, though there are many other substances that serve to regulate plant physiology.

Photomorphogenesis

While most people know that light is important for photosynthesis in plants, few realize that plant sensitivity to light plays a role in the control of plant structural development (morphogenesis). The use of light to control structural development is called photomorphogenesis, and is dependent upon the presence of specialized photoreceptors, which are chemical pigments capable of absorbing specific wavelengths of light.

Plants use four kinds of photoreceptors: phytochrome, cryptochrome, a UV-B photoreceptor, and protochlorophyllide *a*. The first two of these, phytochrome and cryptochrome, are photoreceptor proteins, complex molecular structures formed by joining a protein with a light-sensitive pigment. Cryptochrome is also known as the UV-A photoreceptor, because it absorbs ultraviolet light in the long wave "A" region. The UV-B receptor is one or more compounds not yet identified with certainty, though some evidence suggests carotene or riboflavin as candidates. Protochlorophyllide *a*, as its name suggests, is a chemical precursor of chlorophyll.

The most studied of the photoreceptors in plants is phytochrome. It is sensitive to light in the red and far-red region of the visible spectrum. Many flowering plants use it to regulate the time of flowering based on the length of day and night (photoperiodism) and to set circadian rhythms. It also regulates other responses including the germination of seeds, elongation of seedlings, the size, shape and number of leaves, the synthesis of chlorophyll, and the straightening of the epicotyl or hypocotyl hook of dicot seedlings.

Photoperiodism

Many flowering plants use the pigment phytochrome to sense seasonal changes in day length, which they take as signals to flower. This sensitivity to day length is termed photoperiodism. Broadly speaking, flowering plants can be classified as long day plants, short day plants, or day neutral plants, depending on their particular response to changes in day length. Long day plants require a certain minimum length of daylight to starts flowering, so these plants flower in the spring or summer. Conversely, short day plants flower when the length of daylight falls below a certain critical level. Day neutral plants do not initiate flowering based on photoperiodism, though some may use temperature sensitivity (vernalization) instead.

The poinsettia is a short-day plant, requiring two months of long nights prior to blooming.

Although a short day plant cannot flower during the long days of summer, it is not actually the period of light exposure that limits flowering. Rather, a short day plant requires a minimal length of uninterrupted darkness in each 24-hour period (a short daylength) before floral development can begin. It has been determined experimentally that a short day plant (long night) does not flower if a flash of phytochrome activating light is used on the plant during the night.

Plants make use of the phytochrome system to sense day length or photoperiod. This fact is utilized by florists and greenhouse gardeners to control and even induce flowering out of season, such as the *Poinsettia*.

Environmental Physiology

Paradoxically, the subdiscipline of environmental physiology is on the one hand a recent field of study in plant ecology and on the other hand one of the oldest. Environmental physiology is the preferred name of the subdiscipline among plant physiologists, but it goes by a number of other names in the applied sciences. It is roughly synonymous with ecophysiology, crop ecology, horticulture and agronomy. The particular name applied to the subdiscipline is specific to the viewpoint and goals of research. Whatever name is applied, it deals with the ways in which plants respond to their environment and so overlaps with the field of ecology.

Phototropism in *Arabidopsis thaliana* is regulated by blue to UV light.

Environmental physiologists examine plant response to physical factors such as radiation (including light and ultraviolet radiation), temperature, fire, and wind. Of particular importance are water relations (which can be measured with the Pressure bomb) and the stress of drought or inundation, exchange of gases with the atmosphere, as well as the cycling of nutrients such as nitrogen and carbon.

Environmental physiologists also examine plant response to biological factors. This includes not only negative interactions, such as competition, herbivory, disease and parasitism, but also positive interactions, such as mutualism and pollination.

Tropisms and Nastic Movements

Plants may respond both to directional and non-directional stimuli. A response to a directional stimulus, such as gravity or sunlight, is called a tropism. A response to a nondirectional stimulus, such as temperature or humidity, is a nastic movement.

Tropisms in plants are the result of differential cell growth, in which the cells on one side of the plant elongates more than those on the other side, causing the part to bend toward the side with less growth. Among the common tropisms seen in plants is phototropism, the bending of the plant toward a source of light. Phototropism allows the plant to maximize light exposure in plants which require additional light for photosynthesis, or to minimize it in plants subjected to intense light and heat. Geotropism allows the roots of a plant to determine the direction of gravity and grow downwards. Tropisms generally result from an interaction between the environment and production of one or more plant hormones.

Nastic movements results from differential cell growth (e.g. epinasty and hiponasty), or from changes in turgor pressure within plant tissues (e.g., nyctinasty), which may occur rapidly. A familiar example is thigmonasty (response to touch) in the Venus fly trap, a carnivorous plant. The traps consist of modified leaf blades which bear sensitive trigger hairs. When the hairs are touched by an insect or other animal, the leaf folds shut. This mechanism allows the plant to trap and digest small insects for additional nutrients. Although the trap is rapidly shut by changes in internal cell pressures, the leaf must grow slowly to reset for a second opportunity to trap insects.

Plant Disease

Powdery mildew on crop leaves

Economically, one of the most important areas of research in environmental physiology is that of phytopathology, the study of diseases in plants and the manner in which plants resist or cope with infection. Plant are susceptible to the same kinds of disease organisms as animals, including viruses, bacteria, and fungi, as well as physical invasion by insects and roundworms.

Because the biology of plants differs with animals, their symptoms and responses are quite different. In some cases, a plant can simply shed infected leaves or flowers to prevent the spread of disease, in a process called abscission. Most animals do not have this option as a means of controlling disease. Plant diseases organisms themselves also differ from those causing disease in animals because plants cannot usually spread infection through casual physical contact. Plant pathogens tend to spread via spores or are carried by animal vectors.

One of the most important advances in the control of plant disease was the discovery of Bordeaux mixture in the nineteenth century. The mixture is the first known fungicide and is a combination of copper sulfate and lime. Application of the mixture served to inhibit the growth of downy mildew that threatened to seriously damage the French wine industry.

History

Early History

Jan Baptist van Helmont.

Sir Francis Bacon published one of the first plant physiology experiments in 1627 in the book, *Sylva Sylvarum*. Bacon grew several terrestrial plants, including a rose, in water and concluded that soil was only needed to keep the plant upright. Jan Baptist van Helmont published what is considered the first quantitative experiment in plant physiology in 1648. He grew a willow tree for five years in a pot containing 200 pounds of oven-dry soil. The soil lost just two ounces of dry weight and van Helmont concluded that plants get all their weight from water, not soil. In 1699, John Woodward published experiments on growth of spearmint in different sources of water. He found that plants grew much better in water with soil added than in distilled water.

Stephen Hales is considered the Father of Plant Physiology for the many experiments in the 1727 book; though Julius von Sachs unified the pieces of plant physiology and put them together as a discipline. His *Lehrbuch der Botanik* was the plant physiology bible of its time.

Researchers discovered in the 1800s that plants absorb essential mineral nutrients as inorganic ions in water. In natural conditions, soil acts as a mineral nutrient reservoir but the soil itself is not essential to plant growth. When the mineral nutrients in the soil are dissolved in water, plant roots absorb nutrients readily, soil is no longer required for the plant to thrive. This observation is the basis for hydroponics, the growing of plants in a water solution rather than soil, which has become a standard technique in biological research, teaching lab exercises, crop production and as a hobby.

Current Research

One of the leading journals in the field is *Plant Physiology*, started in 1926. All its back issues are available online for free. Many other journals often carry plant physiology articles, including *Physiologia Plantarum*, *Journal of Experimental Botany*, *American Journal of Botany*, *Annals of Botany*, *Journal of Plant Nutrition* and *Proceedings of the National Academy of Sciences*.

Economic Applications

Food Production

In horticulture and agriculture along with food science, plant physiology is an important topic relating to fruits, vegetables, and other consumable parts of plants. Topics studied include: *climatic* requirements, fruit drop, nutrition, ripening, fruit set. The production of food crops also hinges on the study of plant physiology covering such topics as optimal planting and harvesting times and post harvest storage of plant products for human consumption and the production of secondary products like drugs and cosmetics.

Plant Morphology

Inflorescences emerging from protective coverings

Plant morphology or phytomorphology is the study of the physical form and external structure of plants. This is usually considered distinct from plant anatomy, which is the study of the internal structure of plants, especially at the microscopic level. Plant morphology is useful in the visual identification of plants.

Scope

Asclepias syriaca showing complex morphology of the flowers.

Looking up into the branch structure of a *Pinus sylvestris* tree

Plant morphology "represents a study of the development, form, and structure of plants, and, by implication, an attempt to interpret these on the basis of similarity of plan and origin." There are four major areas of investigation in plant morphology, and each overlaps with another field of the biological sciences.

First of all, morphology is comparative, meaning that the morphologist examines structures in many different plants of the same or different species, then draws comparisons and formulates ideas about similarities. When structures in different species are believed to exist and develop as a result of common, inherited genetic pathways, those structures are termed homologous. For example, the leaves of pine, oak, and cabbage all look very different, but share certain basic structures and arrangement of parts. The homology of leaves is an easy conclusion to make. The plant morphologist goes further, and discovers that the spines of cactus also share the same basic structure and development as leaves in other plants, and therefore cactus spines are homologous to leaves as well. This aspect of plant morphology overlaps with the study of plant evolution and paleobotany.

Secondly, plant morphology observes both the vegetative (somatic) structures of plants, as well as the reproductive structures. The vegetative structures of vascular plants includes the study of the shoot system, composed of stems and leaves, as well as the root system. The reproductive structures are more varied, and are usually specific to a particular group of plants, such as flowers and seeds, fern sori, and moss capsules. The detailed study of reproductive structures in plants led to the discovery of the alternation of generations found in all plants and most algae. This area of plant morphology overlaps with the study of biodiversity and plant systematics.

Thirdly, plant morphology studies plant structure at a range of scales. At the smallest scales are ultrastructure, the general structural features of cells visible only with the aid of an electron microscope, and cytology, the study of cells using optical microscopy. At this scale, plant morphology overlaps with plant anatomy as a field of study. At the largest scale is the study of plant growth habit, the overall architecture of a plant. The pattern of branching in a tree will vary from species to species, as will the appearance of a plant as a tree, herb, or grass.

Fourthly, plant morphology examines the pattern of development, the process by which structures originate and mature as a plant grows. While animals produce all the body parts they will ever have from early in their life, plants constantly produce new tissues and structures throughout their life. A living plant always has embryonic tissues. The way in which new structures mature as they are produced may be affected by the point in the plant's life when they begin to develop, as well as by the environment to which the structures are exposed. A morphologist studies this process, the causes, and its result. This area of plant morphology overlaps with plant physiology and ecology.

A Comparative Science

A plant morphologist makes comparisons between structures in many different plants of the same or different species. Making such comparisons between similar structures in different plants tackles the question of *why* the structures are similar. It is quite likely that similar underlying causes of genetics, physiology, or response to the environment have led to this similarity in appearance. The result of scientific investigation into these causes can lead to one of two insights into the underlying biology:

1. Homology - the structure is similar between the two species because of shared ancestry and common genetics.
2. Convergence - the structure is similar between the two species because of independent adaptation to common environmental pressures.

Understanding which characteristics and structures belong to each type is an important part of understanding plant evolution. The evolutionary biologist relies on the plant morphologist to interpret structures, and in turn provides phylogenies of plant relationships that may lead to new morphological insights.

Homology

When structures in different species are believed to exist and develop as a result of common, inherited genetic pathways, those structures are termed *homologous*. For example, the leaves of pine, oak, and cabbage all look very different, but share certain basic structures and arrangement of parts. The homology of leaves is an easy conclusion to make. The plant morphologist goes further, and discovers that the spines of cactus also share the same basic structure and development as leaves in other plants, and therefore cactus spines are homologous to leaves as well.

Convergence

When structures in different species are believed to exist and develop as a result of common adaptive responses to environmental pressure, those structures are termed *convergent*. For example, the fronds of *Bryopsis plumosa* and stems of *Asparagus setaceus* both have the same feathery branching appearance, even though one is an alga and one is a flowering plant. The similarity in overall structure occurs independently as a result of convergence. The growth form of many cacti and species of *Euphorbia* is very similar, even though they belong to widely distant families. The similarity results from common solutions to the problem of surviving in a hot, dry environment.

Euphorbia obesa, a spurge

Astrophytum asterias, a cactus.

Vegetative and Wacteristics

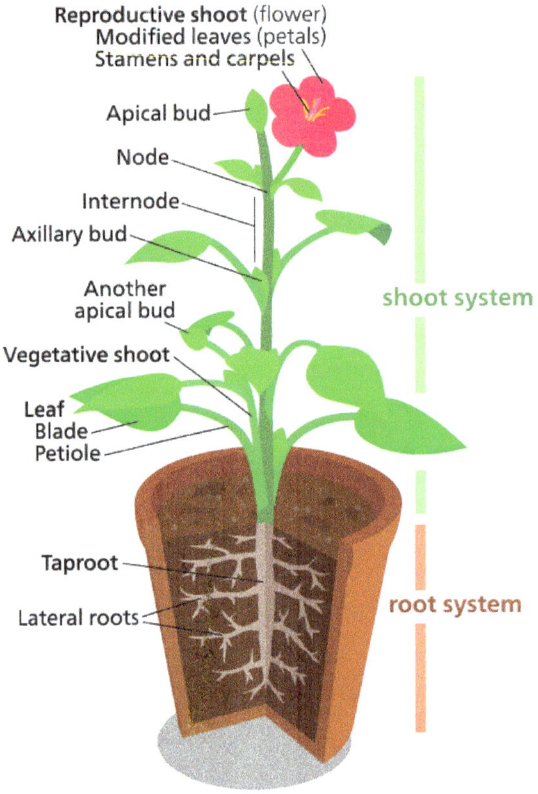

A diagram representing a "typical" eudicot.

Plant morphology treats both the vegetative structures of plants, as well as the reproductive structures.

The vegetative (somatic) structures of vascular plants include two major organ systems: (1) a shoot system, composed of stems and leaves, and (2) a root system. These two systems are common to nearly all vascular plants, and provide a unifying theme for the study of plant morphology.

By contrast, the reproductive structures are varied, and are usually specific to a particular group of plants. Structures such as flowers and fruits are only found in the angiosperms; sori are only found in ferns; and seed cones are only found in conifers and other gymnosperms. Reproductive characters are therefore regarded as more useful for the classification of plants than vegetative characters.

Use in Identification

Plant biologists use morphological characters of plants which can be compared, measured, counted and described to assess the differences or similarities in plant taxa and use these characters for plant identification, classification and descriptions.

When characters are used in descriptions or for identification they are called diagnostic or key characters which can be either qualitative and quantitative.

1. Quantitative characters are morphological features that can be counted or measured for example a plant species has flower petals 10–12 mm wide.

2. Qualitative characters are morphological features such as leaf shape, flower color or pubescence.

Both kinds of characters can be very useful for the identification of plants.

Alternation of Generations

The detailed study of reproductive structures in plants led to the discovery of the alternation of generations, found in all plants and most algae, by the German botanist Wilhelm Hofmeister. This discovery is one of the most important made in all of plant morphology, since it provides a common basis for understanding the life cycle of all plants.

Pigmentation in Plants

The primary function of pigments in plants is photosynthesis, which uses the green pigment chlorophyll along with several red and yellow pigments that help to capture as much light energy as possible. Pigments are also an important factor in attracting insects to flowers to encourage pollination.

Plant pigments include a variety of different kinds of molecule, including porphyrins, carotenoids, anthocyanins and betalains. All biological pigments selectively absorb certain wavelengths of light while reflecting others. The light that is absorbed may be used by the plant to power chemical reactions, while the reflected wavelengths of light determine the color the pigment will appear to the eye.

Development

Plant development is the process by which structures originate and mature as a plant grows. It is a subject studies in plant anatomy and plant physiology as well as plant morphology.

The process of development in plants is fundamentally different from that seen in vertebrate animals. When an animal embryo begins to develop, it will very early produce all of the body parts that it will ever have in its life. When the animal is born (or hatches from its egg), it has all its body parts and from that point will only grow larger and more mature. By contrast, plants constantly produce new tissues and structures throughout their life from meristems located at the tips of organs, or between mature tissues. Thus, a living plant always has embryonic tissues.

The properties of organization seen in a plant are emergent properties which are more than the sum of the individual parts. "The assembly of these tissues and functions into an integrated multicellular organism yields not only the characteristics of the separate parts and processes but also quite a new set of characteristics which would not have been predictable on the basis of examination of the separate parts." In other words, knowing everything about the molecules in a plant are not enough to predict characteristics of the cells; and knowing all the properties of the cells will not predict all the properties of a plant's structure.

Growth

A vascular plant begins from a single celled zygote, formed by fertilisation of an egg cell by a sperm cell. From that point, it begins to divide to form a plant embryo through the process of embryogenesis. As this happens, the resulting cells will organize so that one end becomes the first root, while the other end forms the tip of the shoot. In seed plants, the embryo will develop one or more "seed leaves" (cotyledons). By the end of embryogenesis, the young plant will have all the parts necessary to begin in its life.

Once the embryo germinates from its seed or parent plant, it begins to produce additional organs (leaves, stems, and roots) through the process of organogenesis. New roots grow from root meristems located at the tip of the root, and new stems and leaves grow from shoot meristems located at the tip of the shoot. Branching occurs when small clumps of cells left behind by the meristem, and which have not yet undergone cellular differentiation to form a specialized tissue, begin to grow as the tip of a new root or shoot. Growth from any such meristem at the tip of a root or shoot is termed primary growth and results in the lengthening of that root or shoot. Secondary growth results in widening of a root or shoot from divisions of cells in a cambium.

In addition to growth by cell division, a plant may grow through cell elongation. This occurs when individual cells or groups of cells grow longer. Not all plant cells will grow to the same length. When cells on one side of a stem grow longer and faster than cells on the other side, the stem will bend to the side of the slower growing cells as a result. This directional growth can occur via a plant's response to a particular stimulus, such as light (phototropism), gravity (gravitropism), water, (hydrotropism), and physical contact (thigmotropism).

Plant growth and development are mediated by specific plant hormones and plant growth regulators (PGRs) (Ross et al. 1983). Endogenous hormone levels are influenced by plant age, cold hardiness, dormancy, and other metabolic conditions; photoperiod, drought, temperature, and other external environmental conditions; and exogenous sources of PGRs, e.g., externally applied and of rhizospheric origin.

Morphological Variation

Plants exhibit natural variation in their form and structure. While all organisms vary from individual to individual, plants exhibit an additional type of variation. Within a single individual, parts are repeated which may differ in form and structure from other similar parts. This variation is most easily seen in the leaves of a plant, though other organs such as stems and flowers may show similar variation. There are three primary causes of this variation: positional effects, environmental effects, and juvenility.

Evolution of Plant Morphology

Transcription factors and transcriptional regulatory networks play key roles in plant morphogenesis and their evolution. During plant landing, many novel transcription factor families emerged and are preferentially wired into the networks of multicellular development, reproduction, and organ development, contributing to more complex morphogenesis of land plants.

Positional Effects

Variation in leaves from the giant ragweed illustrating positional effects. The lobed leaves come from the base of the plant, while the unlobed leaves come from the top of the plant.

Although plants produce numerous copies of the same organ during their lives, not all copies of a particular organ will be identical. There is variation among the parts of a mature plant resulting from the relative position where the organ is produced. For example, along a new branch the leaves may vary in a consistent pattern along the branch. The form of leaves produced near the base of the branch will differ from leaves produced at the tip of the plant, and this difference is consistent from branch to branch on a given plant and in a given species. This difference persists after the leaves at both ends of the branch have matured, and is not the result of some leaves being younger than others.

Environmental Effects

The way in which new structures mature as they are produced may be affected by the point in the plants life when they begin to develop, as well as by the environment to which the structures are exposed. This can be seen in aquatic plants and emergent plants.

Temperature

Temperature has a multiplicity of effects on plants depending on a variety of factors, including the size and condition of the plant and the temperature and duration of exposure. The smaller and more succulent the plant, the greater the susceptibility to damage or death from temperatures that are too high or too low. Temperature affects the rate of biochemical and physiological processes, rates generally (within limits) increasing with temperature. However, the Van't Hoff relationship for monomolecular reactions (which states that the velocity of a reaction is doubled or trebled by a temperature increase of 10 °C) does not strictly hold for biological processes, especially at low and high temperatures.

When water freezes in plants, the consequences for the plant depend very much on whether the freezing occurs intracellularly (within cells) or outside cells in intercellular (extracellular) spaces. Intracellular freezing usually kills the cell regardless of the hardiness of the plant and its tissues. Intracellular freezing seldom occurs in nature, but moderate rates of decrease in temperature, e.g., 1 °C to 6 °C/hour, cause intercellular ice to form, and this "extraorgan ice" may or may not be lethal, depending on the hardiness of the tissue.

At freezing temperatures, water in the intercellular spaces of plant tissues freezes first, though the water may remain unfrozen until temperatures fall below 7 °C. After the initial formation of ice intercellularly, the cells shrink as water is lost to the segregated ice. The cells undergo freeze-drying, the dehydration being the basic cause of freezing injury.

The rate of cooling has been shown to influence the frost resistance of tissues, but the actual rate of freezing will depend not only on the cooling rate, but also on the degree of supercooling and the properties of the tissue. Sakai (1979a) demonstrated ice segregation in shoot primordia of Alaskan white and black spruces when cooled slowly to 30 °C to -40 °C. These freeze-dehydrated buds survived immersion in liquid nitrogen when slowly rewarmed. Floral primordia responded similarly. Extraorgan freezing in the primordia accounts for the ability of the hardiest of the boreal conifers to survive winters in regions when air temperatures often fall to -50 °C or lower. The hardiness of the winter buds of such conifers is enhanced by the smallness of the buds, by the evolution of faster translocation of water, and an ability to tolerate intensive freeze dehydration. In boreal species of *Picea* and *Pinus*, the frost resistance of 1-year-old seedlings is on a par with mature plants, given similar states of dormancy.

Juvenility

Juvenility in a seedling of European beech. Notice the difference in shape between the first dark green "seed leaves" and the lighter second pair of leaves.

The organs and tissues produced by a young plant, such as a seedling, are often different from those that are produced by the same plant when it is older. This phenomenon is known as juvenility or heteroblasty. For example, young trees will produce longer, leaner branches that grow upwards more than the branches they will produce as a fully grown tree. In addition, leaves produced during early growth tend to be larger, thinner, and more irregular than leaves on the adult plant. Specimens of juvenile plants may look so completely different from adult plants of the same species that egg-laying insects do not recognize the plant as food for their young. Differences are seen in rootability and flowering and can be seen in the same mature tree. Juvenile cuttings taken from the base of a tree will form roots much more readily than cuttings originating from the mid to upper crown. Flowering close to the base of a tree is absent or less profuse than flowering in the higher branches especially when a young tree first reaches flowering age.

The transition from early to late growth forms is referred to as 'vegetative phase change', but there is some disagreement about terminology.

Some Recent Developments

Rolf Sattler has revised fundamental concepts of comparative morphology such as the concept of homology. He emphasized that homology should also include partial homology and quantitative homology. This leads to a continuum morphology that demonstrates a continuum between the morphological categories of root, shoot, stem (caulome), leaf (phyllome), and hair (trichome). How intermediates between the categories are best described has been discussed by Bruce K. Kirchoff et al.

Honoring Agnes Arber, author of the partial-shoot theory of the leaf, Rutishauser and Isler called the continuum approach Fuzzy Arberian Morphology (FAM). "Fuzzy" refers to fuzzy logic, "Arberian" to Agnes Arber. Rutishauser and Isler emphasized that this approach is not only supported by many morphological data but also by evidence from molecular genetics. More recent evidence from molecular genetics provides further support for continuum morphology. James (2009) concluded that "it is now widely accepted that... radiality [characteristic of most shoots] and dorsiventrality [characteristic of leaves] are but extremes of a continuous spectrum. In fact, it is simply the timing of the KNOX gene expression!." Eckardt and Baum (2010) concluded that "it is now generally accepted that compound leaves express both leaf and shoot properties."

Process morphology (dynamic morphology) describes and analyzes the dynamic continuum of plant form. According to this approach, structures do not *have* process(es), they *are* process(es). Thus, the structure/process dichotomy is overcome by "an enlargement of our concept of 'structure' so as to include and recognize that in the living organism it is not merely a question of spatial structure with an 'activity' as something over or against it, but that the concrete organism is a spatio-*temporal* structure and that this spatio-temporal structure is the activity itself."

For Jeune, Barabé and Lacroix, classical morphology (that is, mainstream morphology, based on a qualitative homology concept implying mutually exclusive categories) and continuum morphology are sub-classes of the more encompassing process morphology (dynamic morphology).

Plant Pathology

Plant pathology (also phytopathology) is the scientific study of diseases in plants caused by pathogens (infectious organisms) and environmental conditions (physiological factors). Organisms that cause infectious disease include fungi, oomycetes, bacteria, viruses, viroids, virus-like organisms, phytoplasmas, protozoa, nematodes and parasitic plants. Not included are ectoparasites like insects, mites, vertebrate, or other pests that affect plant health by consumption of plant tissues. Plant pathology also involves the study of pathogen identification, disease etiology, disease cycles, economic impact, plant disease epidemiology, plant disease resistance, how plant diseases affect humans and animals, pathosystem genetics, and management of plant diseases.

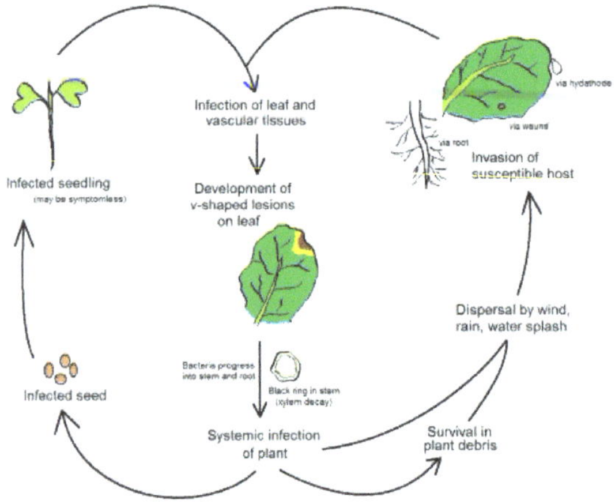

Life cycle of the black rot pathogen, *Xanthomonas campestris* pathovar *campes*

Overview

Control of plant diseases is crucial to the reliable production of food, and it provides significant reductions in agricultural use of land, water, fuel and other inputs. Plants in both natural and cultivated populations carry inherent disease resistance, but there are numerous examples of devastating plant disease impacts (see Irish potato famine, chestnut blight), as well as recurrent severe plant diseases (see rice blast, soybean cyst nematode, citrus canker). However, disease control is reasonably successful for most crops. Disease control is achieved by use of plants that have been bred for good resistance to many diseases, and by plant cultivation approaches such as crop rotation, use of pathogen-free seed, appropriate planting date and plant density, control of field moisture, and pesticide use. Across large regions and many crop species, it is estimated that diseases typically reduce plant yields by 10% every year in more developed settings, but yield loss to diseases often exceeds 20% in less developed settings. Continuing advances in the science of plant pathology are needed to improve disease control, and to keep up with changes in disease pressure caused by the ongoing evolution and movement of plant pathogens and by changes in agricultural practices. Plant diseases cause major economic losses for farmers worldwide. The Food and Agriculture Organization estimates indeed that pests and diseases are responsible for about 25% of crop loss. To solve this issue, new methods are needed to detect diseases and pests early, such as novel sensors that detect plant odours and spectroscopy and biophotonics that are able to diagnostic plant health and metabolism.

Plant Pathogens

Powdery mildew, a biotrophic fungus

Fungi

Most phytopathogenic fungi belong to the Ascomycetes and the Basidiomycetes.

The fungi reproduce both sexually and asexually via the production of spores and other structures. Spores may be spread long distances by air or water, or they may be soilborne. Many soil inhabiting fungi are capable of living saprotrophically, carrying out the part of their life cycle in the soil. These are known as facultative saprotrophs.

Fungal diseases may be controlled through the use of fungicides and other agriculture practices. However, new races of fungi often evolve that are resistant to various fungicides.

Biotrophic fungal pathogens colonize living plant tissue and obtain nutrients from living host cells. Necrotrophic fungal pathogens infect and kill host tissue and extract nutrients from the dead host cells. See the powdery mildew and rice blast images, below.

Rice blast, caused by a necrotrophic fungus

Significant fungal plant pathogens include:

Ascomycetes

- *Fusarium* spp. (causal agents of Fusarium wilt disease)
- *Thielaviopsis* spp. (causal agents of: canker rot, black root rot, *Thielaviopsis* root rot)
- *Verticillium* spp.
- *Magnaporthe grisea* (causal agent of rice blast)
- *Sclerotinia sclerotiorum* (causal agent of cottony rot)

Basidiomycetes

- *Ustilago* spp. (causal agents of smut)
- *Rhizoctonia* spp.

- *Phakospora pachyrhizi* (causal agent of soybean rust)
- *Puccinia* spp. (causal agents of severe rusts of virtually all cereal grains and cultivated grasses)
- *Armillaria* spp. (the so-called honey fungus species, which are virulent pathogens of trees and produce edible mushrooms)

Fungus-like Organisms

Oomycetes

The oomycetes are not true fungi but are fungus-like organisms. They include some of the most destructive plant pathogens including the genus *Phytophthora*, which includes the causal agents of potato late blight and sudden oak death. Particular species of oomycetes are responsible for root rot.

Despite not being closely related to the fungi, the oomycetes have developed very similar infection strategies. Oomycetes are capable of using effector proteins to turn off a plant's defenses in its infection process. Plant pathologists commonly group them with fungal pathogens.

Significant oomycete plant pathogens

- *Pythium* spp.
- *Phytophthora* spp., including the causal agent of the Great Irish Famine (1845–1849)

Phytomyxea

Some slime molds in Phytomyxea cause important diseases, including club root in cabbage and its relatives and powdery scab in potatoes. These are caused by species of *Plasmodiophora* and *Spongospora*, respectively.

Bacteria

Crown gall disease caused by Agrobacterium

Most bacteria that are associated with plants are actually saprotrophic and do no harm to the plant itself. However, a small number, around 100 known species, are able to cause disease. Bacterial diseases are much more prevalent in subtropical and tropical regions of the world.

Most plant pathogenic bacteria are rod-shaped (bacilli). In order to be able to colonize the plant they have specific pathogenicity factors. Five main types of bacterial pathogenicity factors are known: uses of cell wall–degrading enzymes, toxins, effector proteins, phytohormones and exopolysaccharides.

Pathogens such as *Erwinia* species use cell wall–degrading enzymes to cause soft rot. *Agrobacterium* species change the level of auxins to cause tumours with phytohormones. Exopolysaccharides are produced by bacteria and block xylem vessels, often leading to the death of the plant.

Bacteria control the production of pathogenicity factors via quorum sensing.

Vitis vinifera with "Ca. Phytoplasma vitis" infection

Significant bacterial plant pathogens:

- Burkholderia
- Proteobacteria
  - o *Xanthomonas* spp.
  - o *Pseudomonas* spp.
- Pseudomonas syringae pv. tomato causes tomato plants to produce less fruit, and it "continues to adapt to the tomato by minimizing its recognition by the tomato immune system."

Phytoplasmas ('Mycoplasma-like Organisms') and Spiroplasmas

Tobacco mosaic virus

Phytoplasma and *Spiroplasma* are a genre of bacteria that lack cell walls and are related to the mycoplasmas, which are human pathogens. Together they are referred to as the mollicutes. They also tend to have smaller genomes than most other bacteria. They are normally transmitted by sap-sucking insects, being transferred into the plants phloem where it reproduces.

Viruses, Viroids and Virus-like Organisms

There are many types of plant virus, and some are even asymptomatic. Under normal circumstances, plant viruses cause only a loss of crop yield. Therefore, it is not economically viable to try to control them, the exception being when they infect perennial species, such as fruit trees.

Most plant viruses have small, single-stranded RNA genomes. However some plant viruses also have double stranded RNA or single or double stranded DNA genomes. These genomes may encode only three or four proteins: a replicase, a coat protein, a movement protein, in order to allow cell to cell movement through plasmodesmata, and sometimes a protein that allows transmission by a vector. Plant viruses can have several more proteins and employ many different molecular translation methods.

Plant viruses are generally transmitted from plant to plant by a vector, but mechanical and seed transmission also occur. Vector transmission is often by an insect (for example, aphids), but some fungi, nematodes, and protozoa have been shown to be viral vectors. In many cases, the insect and virus are specific for virus transmission such as the beet leafhopper that transmits the curly top virus causing disease in several crop plants.

Nematodes

Root-knot nematode galls

Nematodes are small, multicellular wormlike animals. Many live freely in the soil, but there are some species that parasitize plant roots. They are a problem in tropical and subtropical regions of the world, where they may infect crops. Potato cyst nematodes (*Globodera pallida* and *G. rostochiensis*) are widely distributed in Europe and North and South America and cause $300 million worth of damage in Europe every year. Root knot nematodes have quite a large host range, whereas cyst nematodes tend to be able to infect only a few species. Nematodes are able to cause radical changes in root cells in order to facilitate their lifestyle.

Protozoa and Algae

There are a few examples of plant diseases caused by protozoa (e.g., *Phytomonas*, a kinetoplastid). They are transmitted as zoospores that are very durable, and may be able to survive in a resting state in the soil for many years. They have also been shown to transmit plant viruses.

When the motile zoospores come into contact with a root hair they produce a plasmodium and invade the roots.

Some colourless parasitic algae (e.g., *Cephaleuros*) also cause plant diseases.

Parasitic Plants

Parasitic plants such as mistletoe and dodder are included in the study of phytopathology. Dodder, for example, is used as a conduit either for the transmission of viruses or virus-like agents from a host plant to a plant that is not typically a host or for an agent that is not graft-transmissible.

Common pathogenic infection methods

- Cell wall-degrading enzymes: These are used to break down the plant cell wall in order to release the nutrients inside.
- Toxins: These can be non-host-specific, which damage all plants, or host-specific, which cause damage only on a host plant.
- Effector proteins: These can be secreted into the extracellular environment or directly into the host cell, often via the Type three secretion system. Some effectors are known to suppress host defense processes. This can include: reducing the plants internal signaling mechanisms or reduction of phytochemicals production. Bacteria, fungus and oomycetes are known for this function.

Physiological Plant Disorders

Significant abiotic disorders can be caused by:

Natural

Drought

Frost damage and breakage by snow and hail

Flooding and poor drainage

Nutrient deficiency

Salt deposition and other soluble mineral excesses (e.g., gypsum)

Wind (windburn and breakage by hurricanes and tornadoes)

Lightning and wildfire (also often man-made)

Man-made (arguably not abiotic, but usually regarded as such)

Soil compaction

Pollution of air, soil, or both

Salt from winter road salt application or irrigation

Herbicide over-application

Poor education and training of people working with plants (e.g. lawnmower damage to trees)

Vandalism

Orchid leaves with viral infections

Epidemiology

the study of phytopathology epidemics

Management

Quarantine

A diseased patch of vegetation or individual plants can be isolated from other, healthy growth. Specimens may be destroyed or relocated into a greenhouse for treatment or study. Another option is to avoid the introduction of harmful nonnative organisms by controlling all human traffic and activity (e.g., AQIS), although legislation and enforcement are crucial in order to ensure lasting effectiveness.

Cultural

Farming in some societies is kept on a small scale, tended by peoples whose culture includes farming traditions going back to ancient times. (An example of such traditions would be lifelong training in techniques of plot terracing, weather anticipation and response, fertilization, grafting, seed care, and dedicated gardening.) Plants that are intently monitored often benefit from not only active external protection but also a greater overall vigor. While primitive in the sense of being the most labor-intensive solution by far, where practical or necessary it is more than adequate.

Plant Resistance

Sophisticated agricultural developments now allow growers to choose from among systematically cross-bred species to ensure the greatest hardiness in their crops, as suited for a particular region's pathological profile. Breeding practices have been perfected over centuries, but with the advent of genetic manipulation even finer control of a crop's immunity traits is possible. The engineering of food plants may be less rewarding, however, as higher output is frequently offset by popular suspicion and negative opinion about this "tampering" with nature.

Chemical

Many natural and synthetic compounds can be employed to combat the above threats. This method works by directly eliminating disease-causing organisms or curbing their spread; however, it has been shown to have too broad an effect, typically, to be good for the local ecosystem. From an economic standpoint, all but the simplest natural additives may disqualify a product from "organic" status, potentially reducing the value of the yield.

Biological

Crop rotation may be an effective means to prevent a parasitic population from becoming well-established, as an organism affecting leaves would be starved when the leafy crop is replaced by a tuberous type, etc. Other means to undermine parasites without attacking them directly may exist.

Integrated

The use of two or more of these methods in combination offers a higher chance of effectiveness.

Timeline of Plant Pathology

300–286 BC Theophrastus, father of botany, wrote and studied diseases of trees, cereals and legumes

1665 Robert Hooke illustrates a plant-pathogenic fungal disease, rose rust

1675 Anton van Leeuwenhouek invents the compound microscope, in 1683 describes bacteria seen with the microscope

1729 Pier Antonio Micheli, father of mycology, observes spores for the first time, conducts germination experiments

1755 Tillet reports on treatment of seeds

1802 Lime sulfur first used to control plant disease

1845–1849 Potato late blight epidemic in Ireland

1853 Heinrich Anton de Bary father of modern mycology, establishes that fungi are the cause, not the result, of plant diseases, publishes "Untersuchungen uber die Brandpilze"

1858 Julius Kühn publishes "Die Krankheiten der Kultergewachse"

1865 M. Planchon discovers a new species of *Phylloxera*, which was named *Phylloxera vastatrix.*

1868–1882 Coffee rust epidemic in Sri Lanka

1875 Mikhail Woronin identified the cause of clubroot as a "plasmodiophorous organism" and gave it the name *Plasmodiophora brassicae*

1876 *Fusarium oxysporum* f.sp. *cubense*, responsible for Panama disease, discovered in bananas in Australia

1878–1885 Downy mildew of grape epidemic in France

1879 Robert Koch establishes germ theory: diseases are caused by microorganisms

1882 *Lehrbuch der Baumkrankheiten* (*Textbook of Diseases of Trees*), by Robert Hartig, is published in Berlin, the first textbook of forest pathology.

1885 Bordeaux mixture introduced by Pierre-Marie-Alexis Millardet to control downy mildew on grape

1885 Experimental proof that bacteria can cause plant diseases: "Erwinia amylovora" and fire blight of apple

1886–1898 Recognition of plant viral diseases: Tobacco mosaic virus

1889 Introduction of hot water treatment of seed for disease control by Jensen

1902 First chair of plant pathology established, in Copenhagen

1904 Mendelian inheritance of cereal rust resistance demonstrated

1907 First academic department of plant pathology established, at Cornell University

1908 American Phytopathological Society founded

1910 Panama disease reaches Western Hemisphere

1911 Scientific journal *Phytopathology* founded

1925 Panama disease reaches every banana-growing country in the Western Hemisphere

1951 European and Mediterranean Plant Protection Organization (EPPO) founded

1967 Recognition of plant pathogenic mycoplasma-like organisms

1971 T. O. Diener discovers viroids, organisms smaller than viruses

The historical landmarks in plant pathology are taken from unless otherwise noted.

Plant Genetics

Plant genetics is a very broad term. There are many facets of genetics in general, and of course there are many facets to plants. The definition of genetics is the branch of biology that deals with heredity, especially the mechanisms of hereditary transmission and the variation of inherited characteristics among similar or related organisms. And the definition of a plant is any of various photosynthetic, eukaryotic, multicellular organisms of the kingdom Plantae characteristically producing embryos, containing chloroplasts, having cell walls which contain cellulose, and lacking the power of locomotion. Although there has been a revolution in the biological sciences in the past twenty years, there is still a great deal that remains to be discovered. The completion of the sequencing of the genomes of rice and some agriculturally and scientifically important plants (for example Physcomitrella patens) has increased the possibilities of plant genetic research immeasurably.

Features of Plant Biology

Plant genetics is different from that of animals in a few ways. Like mitochondria, chloroplasts have their own DNA, complicating pedigrees somewhat. Like animals, plants have somatic mutations regularly, but these mutations can contribute to the germ line with ease, since flowers develop at the ends of branches composed of somatic cells. People have known of this for centuries, and mutant branches are called "sports". If the fruit on the sport is economically desirable, a new cultivar may be obtained.

Some plant species are capable of self-fertilization, and some are nearly exclusively self-fertilizers. This means that a plant can be both mother and father to its offspring, a rare occurrence in animals. Scientists and hobbyists attempting to make crosses between different plants must take special measures to prevent the plants from self-fertilizing.

Plants are generally more capable of surviving, and indeed flourishing, as polyploids. Polyploidy, the presence of extra sets of chromosomes, is not usually compatible with life in animals. In plants, polyploid individuals are created frequently by a variety of processes, and once created usually cannot cross back to the parental type. Polyploid individuals, if capable of self-fertilizing, can give rise to a new genetically distinct lineage, which can be the start of a new species. This is often called "instant speciation". Polyploids generally have larger fruit, an economically desirable trait, and many human food crops, including wheat, maize, potatoes, peanuts, strawberries and tobacco, are either accidentally or deliberately created polyploids.

Hybrids between plant species are easy to create by hand-pollination, and may be more successful on average than hybrids between animal species. Often tens of thousands of offspring from a single cross are raised and tested to obtain a single individual with desired characteristics. People create hybrids for economic and aesthetic reasons, especially with orchids.

DNA

Deoxyribonucleic acid (DNA) is a nucleic acid that contains the genetic instructions used in the development and functioning of all known living organisms and some viruses. The main role of DNA molecules is the long-term storage of information. DNA is often compared to a set of blueprints or a recipe, or a code, since it contains the instructions needed to construct other components of cells, such as proteins and RNA molecules. The DNA segments that carry this genetic information are called genes, but other DNA sequences have structural purposes, or are involved in regulating the use of this genetic information. Geneticists, including plant geneticists, use this sequencing of DNA to their advantage as they splice and delete certain genes and regions of the DNA molecule to produce a different or desired genotype and thus, also producing a different phenotype.

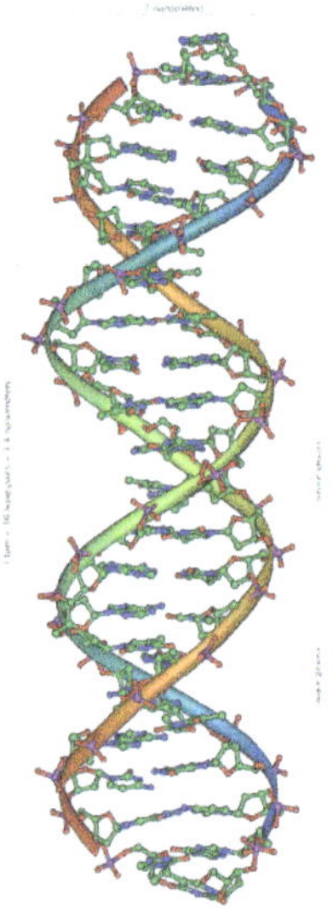

The structure of part of a DNA double helix

Gregor Mendel

Gregor Mendel was an Augustinian priest and scientist born on 20 July 1822 in Austria-Hungary and is well known for discovering genetics. He went to the Abbey of St. Thomas in Brno. He is often called the father of genetics for his study of the inheritance of certain traits in pea plants. Mendel showed that the inheritance of these traits follows particular laws, which were later named after him. The significance of Mendel's work was not recognized until the turn of the 20th century. Its rediscovery prompted the foundation of the discipline of genetics allows geneticists today to accurately predict the outcome of such crosses and in determining the phenotypical effects of the crosses. He died on 6 January 1884 from chronic nephritis.

Modern Ways to Genetically Modify Plants

There are two predominant procedures of transforming genes in organisms: the "Gene gun" method and the *Agrobacterium* method.

"Gene Gun" Method

The "Gene Gun" method is also referred to as "biolistics" (ballistics using biological components). This technique is used for in vivo (within a living organism) transformation and has been especially useful in transforming monocot species like corn and rice. This approach literally shoots genes into plant cells and plant cell chloroplasts. DNA is coated onto small particles of gold or tungsten approximately two micrometres in diameter. The particles are placed in a vacuum chamber and the plant tissue to be engineered is placed below the chamber. The particles are propelled at high velocity using a short pulse of high pressure helium gas, and hit a fine mesh baffle placed above the tissue while the DNA coating continues into any target cell or tissue.

Agrobacterium Method

Transformation via *Agrobacterium* has been successfully practiced in dicots, i.e. broadleaf plants, such as soybeans and tomatoes, for many years. Recently it has been adapted and is now effective in monocots like grasses, including corn and rice. In general, the *Agrobacterium* method is considered preferable to the gene gun, because of a greater frequency of single-site insertions of the foreign DNA, which allows for easier monitoring. In this method, the tumor inducing (Ti) region is removed from the T-DNA (transfer DNA) and replaced with the desired gene and a marker, which is then inserted into the organism. This may involve direct inoculation of the tissue with a culture of transformed Agrobacterium, or inoculation following treatment with micro-projectile bombardment, which wounds the tissue. Wounding of the target tissue causes the release of phenolic compounds by the plant, which induces invasion of the tissue by Agrobacterium. Because of this, microprojectile bombardment often increases the efficiency of infection with Agrobacterium. The marker is used to find the organism which has successfully taken up the desired gene. Tissues of the organism are then transferred to a medium containing an antibiotic or herbicide, depending on which marker was used. The *Agrobacterium* present is also killed by the antibiotic. Only tissues expressing the marker will survive and possess the gene of interest. Thus, subsequent steps in the process will only use these surviving plants. In order to obtain whole plants from these tissues, they are grown under controlled environmental conditions in tissue culture. This is a process of a series of media, each containing nutrients and hormones. Once the plants are grown and produce seed, the process of evaluating the progeny begins. This process entails selection of the seeds with the desired traits and then retesting and growing to make sure that the entire process has been completed successfully with the desired results.

Genetically Engineered Crops

The use of genetically engineered crops has helped many farmers deal with pest problems that reduce their crop production. The impact of pest-resistant crops has led to a much higher yield for farmers in today's world. They can use less pesticides which reduces the chemicals that they put into the ground. Certain engineered crops have led to farmers all over the world and in the United States to increase crop yield exponentially in recent years. Farmers can use a glyphosate herbicide to kill weeds, yet the genetically engineered corn is resistant to the herbicide and is left unaffected. Thus, fields are produced that are virtually weed free. Genetically engineered crops can also benefit farmers when dealing with potentially harmful viruses and bacteria. In the 1990s a mutant strain of virus was decimating the commercial corn fields of the United States. Scientists found a virus resistant strain of maize in the highlands of Mexico and extracted the part of the maize's genome that coded for resistance against the virus and incorporated it into their existing strain of commercial corn. This allowed the commercial strain to produce progeny that were resistant to the virus. Thus, the crops were saved from decimation.

Potential detrimental effects of genetically engineered plants

According to John E. Berringer the outcome of releasing genetically modified organisms into the environment is still not known.

Plant Nutrition

Plant nutrition is the study of the chemical elements and compounds necessary for plant growth, plant metabolism and their external supply. In 1972, E. Epstein defined two criteria for an element to be essential for plant growth:

1. in its absence the plant is unable to complete a normal life cycle.
2. or that the element is part of some essential plant constituent or metabolite.

This is in accordance with Justus von Liebig's law of the minimum. The essential plant nutrients include carbon and oxygen which are absorbed from the air, whereas other nutrients including hydrogen are typically obtained from the soil (exceptions include some parasitic or carnivorous plants).

Plants must obtain the following mineral nutrients from their growing medium:

- the macronutrients: nitrogen (N), phosphorus (P), potassium (K), calcium (Ca), sulfur (S), magnesium (Mg); and
- the micronutrients (or trace minerals): boron (B), chlorine (Cl), manganese (Mn), iron (Fe), zinc (Zn), copper (Cu), molybdenum (Mo), nickel (Ni).

The macronutrients are consumed in larger quantities and are usually present in plant tissue in concentrations of between 0.2% and 4.0% (on a dry matter weight basis). Micronutrients are present in plant tissue in quantities measured in parts per million, ranging from 0.1 to 200 ppm, or less than 0.02% dry weight.

Farmer spreading decomposing manure to improve soil fertility and plant nutrition

Most soil conditions across the world can provide plants adapted to that climate and soil with sufficient nutrition for a complete life cycle, without the addition of nutrients as fertilizer. However, if the soil is cropped it is necessary to artificially modify soil fertility through the addition of fertilizer to promote vigorous growth and increase or sustain yield. This is done because, even with adequate water and light, nutrient deficiency can limit growth and crop yield.

Processes

Plants take up essential elements from the soil through their roots and from the air (mainly consisting of nitrogen and oxygen) through their leaves. Nutrient uptake in the soil is achieved by cation exchange, wherein root hairs pump hydrogen ions (H^+) into the soil through proton pumps. These hydrogen ions displace cations attached to negatively charged soil particles so that the cations are available for uptake by the root. In the leaves, stomata open to take in carbon dioxide and expel oxygen. The carbon dioxide molecules are used as the carbon source in photosynthesis.

The root, especially the root hair, is the essential organ for the uptake of nutrients. The structure and architecture of the root can alter the rate of nutrient uptake. Nutrient ions are transported to the center of the root, the stele in order for the nutrients to reach the conducting tissues, xylem and phloem. The Casparian strip, a cell wall outside the stele but within the root, prevents passive flow of water and nutrients, helping to regulate the uptake of nutrients and water. Xylem moves water and inorganic molecules within the plant and phloem accounts for organic molecule transportation. Water potential plays a key role in a plant's nutrient uptake. If the water potential is more negative within the plant than the surrounding soils, the nutrients will move from the region of higher solute concentration—in the soil—to the area of lower solute concentration - in the plant.

There are three fundamental ways plants uptake nutrients through the root:

1. Simple diffusion occurs when a nonpolar molecule, such as O_2, CO_2, and NH_3 follows a concentration gradient, moving passively through the cell lipid bilayer membrane without the use of transport proteins.

2. Facilitated diffusion is the rapid movement of solutes or ions following a concentration gradient, facilitated by transport proteins.

3. Active transport is the uptake by cells of ions or molecules against a concentration gradient; this requires an energy source, usually ATP, to power molecular pumps that move the ions or molecules through the membrane.

Nutrients can be moved within plants to where they are most needed. For example, a plant will try to supply more nutrients to its younger leaves than to its older ones. When nutrients are mobile within the plant, symptoms of any deficiency become apparent first on the older leaves. However, not all nutrients are equally mobile. Nitrogen, phosphorus, and potassium are mobile nutrients while the others have varying degrees of mobility. When a less-mobile nutrient is deficient, the younger leaves suffer because the nutrient does not move up to them but stays in the older leaves. This phenomenon is helpful in determining which nutrients a plant may be lacking.

Many plants engage in symbiosis with microorganisms. Two important types of these relationship are

1. with bacteria such as rhizobia, that carry out biological nitrogen fixation, in which atmospheric nitrogen (N_2) is converted into ammonium (NH+4); and
2. with mycorrhizal fungi, which through their association with the plant roots help to create a larger effective root surface area. Both of these mutualistic relationships enhance nutrient uptake.

Though nitrogen is plentiful in the Earth's atmosphere, relatively few plants harbour nitrogen-fixing bacteria, so most plants rely on nitrogen compounds present in the soil to support their growth. These can be supplied by mineralization of soil organic matter or added plant residues, nitrogen fixing bacteria, animal waste, through the breaking of triple bonded Nitrogen molecules by lightening strikes or through the application of fertilizers.

Functions of Nutrients

At least 17 elements are known to be essential nutrients for plants. In relatively large amounts, the soil supplies nitrogen, phosphorus, potassium, calcium, magnesium, and sulfur; these are often called the macronutrients. In relatively small amounts, the soil supplies iron, manganese, boron, molybdenum, copper, zinc, chlorine, and cobalt, the so-called micronutrients. Nutrients must be available not only in sufficient amounts but also in appropriate ratios.

Plant nutrition is a difficult subject to understand completely, partially because of the variation between different plants and even between different species or individuals of a given clone. Elements present at low levels may cause deficiency symptoms, and toxicity is possible at levels that are too high. Furthermore, deficiency of one element may present as symptoms of toxicity from another element, and vice versa. An abundance of one nutrient may cause a deficiency of another nutrient. For example, K^+ uptake can be influenced by the amount of NH+ 4 available.

Although nitrogen is plentiful in the Earth's atmosphere, relatively few plants engage in nitrogen fixation (conversion of atmospheric nitrogen to a biologically useful form). Most plants, therefore, require nitrogen compounds to be present in the soil in which they grow.

Carbon and oxygen are absorbed from the air while other nutrients are absorbed from the soil. Green plants obtain their carbohydrate supply from the carbon dioxide in the air by the process of photosynthesis. Each of these nutrients is used in a different place for a different essential function.

Macronutrients (Derived from Air and Water)

Carbon

Carbon forms the backbone of most plant biomolecules, including proteins, starches and cellulose. Carbon is fixed through photosynthesis; this converts carbon dioxide from the air into carbohydrates which are used to store and transport energy within the plant.

Hydrogen

Hydrogen also is necessary for building sugars and building the plant. It is obtained almost entirely from water. Hydrogen ions are imperative for a proton gradient to help drive the electron transport chain in photosynthesis and for respiration.

Oxygen

Oxygen is a component of many organic and inorganic molecules within the plant, and is acquired in many forms. These include: O_2 and CO_2 (mainly from the air via leaves) and H_2O, NO−3, H_2PO−4 and SO2−4 (mainly from the soil water via roots). Plants produce oxygen gas (O_2) along with glucose during photosynthesis but then require O_2 to undergo aerobic cellular respiration and break down this glucose to produce ATP.

Macronutrients (Primary)

Nitrogen

Nitrogen is a major constituent of several of the most important plant substances. For example, nitrogen compounds comprise 40% to 50% of the dry matter of protoplasm, and it is a constituent of amino acids, the building blocks of proteins. It is also an essential constituent of chlorophyll. Nitrogen deficiency most often results in stunted growth, slow growth, and chlorosis. Nitrogen deficient plants will also exhibit a purple appearance on the stems, petioles and underside of leaves from an accumulation of anthocyanin pigments. Most of the nitrogen taken up by plants is from the soil in the forms of NO−3, although in acid environments such as boreal forests where nitrification is less likely to occur, ammonium NH+4 is more likely to be the dominating source of nitrogen. Amino acids and proteins can only be built from NH+4, so NO−3 must be reduced. In many agricultural settings, nitrogen is the limiting nutrient for rapid growth. Nitrogen is transported via the xylem from the roots to the leaf canopy as nitrate ions, or in an organic form, such as amino acids or amides. Nitrogen can also be transported in the phloem sap as amides, amino acids and ureides; it is therefore mobile within the plant, and the older leaves exhibit chlorosis and necrosis earlier than the younger leaves.

There is an abundant supply of nitrogen in the earth's atmosphere — N_2 gas comprises nearly 79% of air. However, N_2 is unavailable for use by most organisms because there is a triple bond between the two nitrogen atoms, making the molecule almost inert. In order for nitrogen to be used for growth it must be "fixed" (combined) in the form of ammonium (NH_4) or nitrate (NO_3) ions. The weathering of rocks releases these ions so slowly that it has a negligible effect on the availability of fixed nitrogen. Therefore, nitrogen is often the limiting factor for growth and biomass production in all environments where there is a suitable climate and availability of water to support life.

Nitrogen enters the plant largely through the roots. A "pool" of soluble nitrogen accumulates. Its composition within a species varies widely depending on several factors, including day length, time of day, night temperatures, nutrient deficiencies, and nutrient imbalance. Short day length promotes asparagine formation, whereas glutamine is produced under long day regimes. Darkness favors protein breakdown accompanied by high asparagine accumulation. Night temperature modifies the effects due to night length, and soluble nitrogen tends to accumulate owing to retarded synthesis and breakdown of proteins. Low night temperature conserves glutamine; high night temperature increases accumulation of asparagine because of breakdown. Deficiency of K accentuates differences between long- and short-day plants. The pool of soluble nitrogen is much smaller than in well-nourished plants when N and P are deficient since uptake of nitrate and further reduction and conversion of N to organic forms is restricted more than is protein synthesis. Deficiencies of Ca, K, and S affect the conversion of organic N to protein more than uptake and reduction. The size of the pool of soluble N is no guide *per se* to growth rate, but the size of the pool in relation to total N might be a useful ratio in this regard. Nitrogen availability in the rooting medium also affects the size and structure of tracheids formed in the long lateral roots of white spruce (Krasowski and Owens 1999).

Microorganisms have a central role in almost all aspects of nitrogen availability, and therefore for life support on earth. Some bacteria can convert N_2 into ammonia by the process termed *nitrogen fixation*; these bacteria are either free-living or form symbiotic associations with plants or other organisms (e.g., termites, protozoa), while other bacteria bring about transformations of ammonia to nitrate, and of nitrate to N_2 or other nitrogen gases. Many bacteria and fungi degrade organic matter, releasing fixed nitrogen for reuse by other organisms. All these processes contribute to the nitrogen cycle.

Phosphorus

Like nitrogen, phosphorus is involved with many vital plant processes. Within a plant, it is present mainly as a structural component of the nucleic acids, deoxyribonucleic nucleic acid (DNA) and ribose nucleic acid (RNA), and as a constituent of fatty phospholipids, of importance in membrane development and function. It is present in both organic and inorganic forms, both of which are readily translocated within the plant. All energy transfers in the cell are critically dependent on phosphorus. As with all living things, phosphorus is part of the Adenosine triphosphate (ATP), which is of immediate use in all processes that require energy with the cells. Phosphorus can also be used to modify the activity of various enzymes by phosphorylation, and is used for cell signaling. Phosphorus is concentrated at the most actively growing points of a plant and stored within seeds in anticipation of their germination. Phosphorus is most commonly found in the soil in the form of polyprotic phosphoric acid (H_3PO_4), but is taken up most readily in the form of H_2PO−4. Phosphorus is available to plants in limited quantities in most soils because it is released very slowly from insoluble phosphates and is rapidly fixed once again. Under most environmental conditions it is the element that limits growth because of this constriction and due to its high demand by plants and microorganisms. Plants can increase phosphorus uptake by a mutualism with mycorrhiza. A Phosphorus deficiency in plants is characterized by an intense green coloration or reddening in leaves due to lack of chlorophyll. If the plant is experiencing high phosphorus deficiencies the leaves may become denatured and show signs of death. Occasionally the leaves may appear purple from an accumulation of anthocyanin. Because phosphorus is a mobile nutrient, older leaves will show the first signs of deficiency.

On some soils, the phosphorus nutrition of some conifers, including the spruces, depends on the ability of mycorrhizae to take up, and make soil phosphorus available to the tree, hitherto unobtainable to the non-mycorrhizal root. Seedling white spruce, greenhouse-grown in sand testing negative for phosphorus, were very small and purple for many months until spontaneous mycorrhizal inoculation, the effect of which was manifested by a greening of foliage and the development of vigorous shoot growth.

Phosphorus deficiency can produce symptoms similar to those of nitrogen deficiency, but as noted by Russel: "Phosphate deficiency differs from nitrogen deficiency in being extremely difficult to diagnose, and crops can be suffering from extreme starvation without there being any obvious signs that lack of phosphate is the cause". Russell's observation applies to at least some coniferous seedlings, but Benzian found that although response to phosphorus in very acid forest tree nurseries in England was consistently high, no species (including Sitka spruce) showed any visible symptom of deficiency other than a slight lack of lustre. Phosphorus levels have to be exceedingly low before visible symptoms appear in such seedlings. In sand culture at 0 ppm phosphorus, white spruce seedlings were very small and tinted deep purple; at 0.62 ppm, only the smallest seedlings were deep purple; at 6.2 ppm, the seedlings were of good size and color.

It is useful to apply a high phosphorus content fertilizer, such as bone meal, to perennials to help with successful root formation.

Potassium

Unlike other major elements, potassium does not enter into the composition of any of the important plant constituents involved in metabolism, but it does occur in all parts of plants in substantial amounts. It seems to be of particular importance in leaves and at growing points. Potassium is outstanding among the nutrient elements for its mobility and solubility within plant tissues. Processes involving potassium include the formation of carbohydrates and proteins, the regulation of internal plant moisture, as a catalyst and condensing agent of complex substances, as an accelerator of enzyme action, and as contributor to photosynthesis, especially under low light intensity.

When soil-potassium levels are high, plants take up more potassium than needed for healthy growth. The term *luxury consumption* has been applied to this. When potassium is moderately deficient, the effects first appear in the older tissues, and from there progress towards the growing points. Acute deficiency severely affects growing points, and die-back commonly occurs. Symptoms of potassium deficiency in white spruce include: browning and death of needles (chlorosis); reduced growth in height and diameter; impaired retention of needles; and reduced needle length. A relationship between potassium nutrition and cold resistance has been found in several tree species, including two species of spruce.

Potassium regulates the opening and closing of the stomata by a potassium ion pump. Since stomata are important in water regulation, potassium regulates water loss from the leaves and increases drought tolerance. Potassium deficiency may cause necrosis or interveinal chlorosis. The potassium ion (K^+) is highly mobile and can aid in balancing the anion (negative) charges within the plant. Potassium helps in fruit coloration, shape and also increases its brix. Hence, quality fruits are produced in potassium-rich soils. Potassium serves as an activator of enzymes used in photosynthesis and respiration. Potassium is used to build cellulose and aids in photosynthesis by the formation of a chlorophyll precursor. Potassium deficiency may result in higher risk of pathogens, wilting, chlorosis, brown spotting, and higher chances of damage from frost and heat.

Macronutrients (Secondary and Tertiary)

Sulphur

Sulphur is a structural component of some amino acids and vitamins, and is essential in the manufacturing of chloroplasts. Sulphur is also found in the iron-sulphur complexes of the electron transport chains in photosynthesis. It is immobile and deficiency, therefore, affects younger tissues first. Symptoms of deficiency include yellowing of leaves and stunted growth.

Calcium

Calcium regulates transport of other nutrients into the plant and is also involved in the activation of certain plant enzymes. Calcium deficiency results in stunting. This nutrient is involved in photosynthesis and plant structure. Blossom end rot is also a result of inadequate calcium.

Calcium in plants occurs chiefly in the leaves, with lower concentrations in seeds, fruits, and roots. A major function is as a constituent of cell walls. When coupled with certain acidic compounds of the jelly-like pectins of the middle lamella, calcium forms an insoluble salt. It is also intimately involved in meristems, and is particularly important in root development, with roles in cell division, cell elongation, and the detoxification of hydrogen ions. Other functions attributed to calcium are; the neutralization of organic acids; inhibition of some potassium-activated ions; and a role in nitrogen absorption. A notable feature of calcium-deficient plants is a defective root system. Roots are usually affected before above-ground parts.

Magnesium

The outstanding role of magnesium in plant nutrition is as a constituent of the chlorophyll molecule. As a carrier, it is also involved in numerous enzyme reactions as an effective activator, in which it is closely associated with energy-supplying phosphorus compounds. Magnesium is very mobile in plants, and, like potassium, when deficient is translocated from older to younger tissues, so that signs of deficiency appear first on the oldest tissues and then spread progressively to younger tissues.

Micro-nutrients

Some elements are directly involved in plant metabolism (Arnon and Stout, 1939). However, this principle does not account for the so-called beneficial elements, whose presence, while not required, has clear positive effects on plant growth. Mineral elements that either stimulate growth but are not essential, or that are essential only for certain plant species, or under given conditions, are usually defined as beneficial elements.

Plants are able sufficiently to accumulate most trace elements. Some plants are sensitive indicators of the chemical environment in which they grow (Dunn 1991), and some plants have barrier mechanisms that exclude or limit the uptake of a particular element or ion species, e.g., alder twigs commonly accumulate molybdenum but not arsenic, whereas the reverse is true of spruce bark (Dunn 1991). Otherwise, a plant can integrate the geochemical signature of the soil mass permeated by its root system together with the contained groundwaters. Sampling is facilitated by the tendency of many elements to accumulate in tissues at the plant's extremities.

Iron

Iron is necessary for photosynthesis and is present as an enzyme cofactor in plants. Iron deficiency can result in interveinal chlorosis and necrosis. Iron is not a structural part of chlorophyll but very much essential for its synthesis. Copper deficiency can be responsible for promoting an iron deficiency.

Molybdenum

Molybdenum is a cofactor to enzymes important in building amino acids and is involved in nitrogen metabolism. Molybdenum is part of the nitrate reductase enzyme (needed for the reduction of nitrate) and the nitrogenase enzyme (required for biological nitrogen fixation).

Boron

Boron is found in the highly insoluble mineral, tourmaline. It is absorbed by plants in the form of the anion BO3−3. It is available to plants in moderately soluble mineral forms of Ca, Mg and Na borates and the highly soluble form of organic compounds. Concentration in soil must, in general, be below 5 ppm in the soil water solution, above that toxicity results. Its availability in soils ranges from 20 to 200 pounds per acre in the first eight inches, of which less than 5% is available. It is available to plants over a range of pH, from 5.0 to 7.5. It is mobile in the soil, hence, it is prone to leaching. Leaching removes substantial amounts of boron in sandy soil, but little in fine silt or clay soil. Boron's fixation to those minerals at high pH can render boron unavailable, while low pH frees the fixed boron, leaving it prone to leaching in wet climates. It precipitates with other minerals in the form of borax in which form it was first used over 400 years ago as a soil supplement. Decomposition of organic material causes boron to be deposited in the topmost soil layer; organic forms of boron are more soluble than their mineral form, hence are more available in the top few inches. When soil dries it can cause a precipitous drop in the availability of boron to plants as the plants cannot draw nutrients from that desiccated layer. Hence, boron deficiency diseases appear in dry weather.

Boron has many functions within a plant: it affects flowering and fruiting, pollen germination, cell division, and active salt absorption. The metabolism of amino acids and proteins, carbohydrates, calcium, and water are strongly affected by boron. Many of those listed functions may be embodied by its function in moving the highly polar sugars through cell membranes by reducing their polarity and hence the energy needed to pass the sugar. If sugar cannot pass to the fastest growing parts rapidly enough, those parts die. Boron is relatively immobile within a plant suggesting that the molecule is fixed to the points in the membrane where they facilitate sugar transport.

Boron is not relocatable in the plant via the phloem. It must be supplied to the growing parts via the xylem. Foliar sprays affect only those parts sprayed, which may be insufficient for the fastest growing parts, and is very temporary.

Boron is essential for the proper forming and strengthening of cell walls. Lack of boron results in short thick cells producing stunted fruiting bodies and roots. Calcium to boron ratio must be maintained in a narrow range for normal plant growth. For alfalfa, that calcium to boron ratio must be from 80:1 to 600:1. Boron deficiency appears at 800:1 and higher. For alfalfa, similar ratios exist for magnesium, copper, nitrogen and potassium. Boron levels within plants differ with plant species and range from 2.3 p.p.m for barley to 94.7 p.p.m for poppy . Lack of boron causes failure of calcium metabolism which produces hollow heart in beets and peanuts.

Inadequate amounts of boron affect many agricultural crops, legume forage crops most strongly. Of the micronutrients, boron deficiencies are second most common after zinc. Deficiencies of boron when soil is cropped are common and require the application of mineral supplement; one ton of alfalfa hay carries with it one ounce of boron, 100 bushels of peaches 4 ounces. Deficiency results in the death of the terminal growing points. Symptoms first appear as stunted growth, then to cellular changes, which leads to physical changes, and finally death of the plant.

Boron supplements derive from dry lake bed deposits such as those in Death Valley, USA, in the form of sodium tetraborate, from which less soluble calcium borate is made. Foliar sprays are used on fruit crop trees in soils of high alkalinity. Boron is often applied to fields as a contaminant in other soil amendments but is not generally adequate to make up the rate of loss by cropping. The rates of application of borate to produce an adequate alfalfa crop range from 15 pounds per acre for a sandy-silt, acidic soil of low organic matter, to 60 pounds per acre for a soil with high organic matter, high cation exchange capacity and high pH.

Boron concentration in soil water solution higher than one ppm is toxic to most plants. Toxic concentrations within plants are 10 to 50 ppm for small grains and 200 ppm in boron-tolerant crops such as sugar beets, rutabaga, cucumbers, and conifers. Toxic soil conditions are generally limited to arid regions or can be caused by underground borax deposits in contact with water or volcanic gases dissolved in percolating water. Application rates should be limited to a few pounds per acre in a test plot to determine if boron is needed generally. Otherwise, testing for boron levels in plant material is required to determine remedies. Excess boron can be removed by irrigation and assisted by application of elemental sulfur to lower the pH and increase boron's solubility. Application of calcium will increase soil alkalinity, causing boron to fix on the mineral soil component and remove some fraction, thereby reducing boron toxicity.

Boron deficiencies must be detected by analysis of plant material to apply a correction before the obvious symptoms appear, after which it is too late to prevent crop loss. Strawberries deficient in boron will produce lumpy fruit; apricots will not blossom or, if they do, will not fruit or will drop their fruit depending on the level of boron deficit. Broadcast of boron supplements is effective and long term; a foliar spray is immediate but must be repeated.

Boron is an essential element for the health of animals which derive their boron from plant material.

Copper

Copper is important for photosynthesis. Symptoms for copper deficiency include chlorosis.It is involved in many enzyme processes; necessary for proper photosynthesis; involved in the manufacture of lignin (cell walls) and involved in grain production. It is also hard to find in some soil conditions.

Manganese

Manganese is necessary for photosynthesis, including the building of chloroplasts. Manganese deficiency may result in coloration abnormalities, such as discolored spots on the foliage.

Sodium

Sodium is involved in the regeneration of phosphoenolpyruvate in CAM and C4 plants. Sodium can potentially replace potassium's regulation of stomatal opening and closing.

Essentiality of sodium:

- Essential for C4 plants rather C3
- Substitution of K by Na: Plants can be classified into four groups:

1. Group A—a high proportion of K can be replaced by Na and stimulate the growth, which cannot be achieved by the application of K
2. Group B—specific growth responses to Na are observed but they are much less distinct
3. Group C—Only minor substitution is possible and Na has no effect
4. Group D—No substitution occurs

- Stimulate the growth—increase leaf area and stomata. Improves the water balance
- Na functions in metabolism

1. C4 metabolism
2. Impair the conversion of pyruvate to phosphoenol-pyruvate
3. Reduce the photosystem II activity and ultrastructural changes in mesophyll chloroplast

- Replacing K functions

1. Internal osmoticum
2. Stomatal function
3. Photosynthesis
4. Counteraction in long distance transport
5. Enzyme activation

- Improves the crop quality e.g. improves the taste of carrots by increasing sucrose

Zinc

Zinc is required in a large number of enzymes and plays an essential role in DNA transcription. A typical symptom of zinc deficiency is the stunted growth of leaves, commonly known as "little leaf" and is caused by the oxidative degradation of the growth hormone auxin.

Nickel

In higher plants, nickel is absorbed by plants in the form of Ni^{2+} ion. Nickel is essential for activation of urease, an enzyme involved with nitrogen metabolism that is required to process urea. Without nickel, toxic levels of urea accumulate, leading to the formation of necrotic lesions. In lower plants, nickel activates several enzymes involved in a variety of processes, and can substitute for zinc and iron as a cofactor in some enzymes.

Chlorine

Chlorine, as compounded chloride, is necessary for osmosis and ionic balance; it also plays a role in photosynthesis.

Cobalt

Cobalt has proven to be beneficial to at least some plants although it does not appear to be essential for most species. It has, however, been shown to be essential for nitrogen fixation by the nitrogen-fixing bacteria associated with legumes and other plants.

Aluminium

- Tea has a high tolerance for aluminum (Al) toxicity and the growth is stimulated by Al application. The possible reason is the prevention of Cu, Mn or P toxicity effects.
- There have been reports that Al may serve as a fungicide against certain types of root rot.

Silicon

Silicon is not considered an essential element for plant growth and development. It is always found in abundance in the environment and hence if needed it is available. It is found in the structures of plants and improves the health of plants.

In plants, silicon has been shown in experiments to strengthen cell walls, improve plant strength, health, and productivity. There have been studies showing evidence of silicon improving drought and frost resistance, decreasing lodging potential and boosting the plant's natural pest and disease fighting systems. Silicon has also been shown to improve plant vigor and physiology by improving root mass and density, and increasing above ground plant biomass and crop yields. Silicon is currently under consideration by the Association of American Plant Food Control Officials (AAPFCO) for elevation to the status of a "plant beneficial substance".

Higher plants differ characteristically in their capacity to take up silicon. Depending on their SiO_2 content they can be divided into three major groups:

- Wetland graminae-wetland rice, horsetail (10–15%)
- Dryland graminae-sugar cane, most of the cereal species and few dicotyledons species (1–3%)
- Most of dicotyledons especially legumes (<0.5%)
- The long distance transport of Si in plants is confined to the xylem. Its distribution within the shoot organ is therefore determined by transpiration rate in the organs

- The epidermal cell walls are impregnated with a film layer of silicon and effective barrier against water loss, cuticular transpiration rate in the organs.

Vanadium

Vanadium may be required by some plants, but at very low concentrations. It may also be substituting for molybdenum.

Selenium

Selenium is probably not essential for flowering plants, but it can be beneficial; it can stimulate plant growth, improve tolerance of oxidative stress, and increase resistance to pathogens and herbivory.

Selenium is, however, an essential mineral element for animal (including human) nutrition and selenium deficiencies are known to occur when food or animal feed is grown on selenium-deficient soils. The use of inorganic selenium fertilizers can increase selenium concentrations in edible crops and animal diets thereby improving animal health.

Nutrient Deficiency

The effect of a nutrient deficiency can vary from a subtle depression of growth rate to obvious stunting, deformity, discoloration, distress, and even death. Visual symptoms distinctive enough to be useful in identifying a deficiency are rare. Most deficiencies are multiple and moderate. However, while a deficiency is seldom that of a single nutrient, nitrogen is commonly the nutrient in shortest supply.

Chlorosis of foliage is not always due to mineral nutrient deficiency. Solarization can produce superficially similar effects, though mineral deficiency tends to cause premature defoliation, whereas solarization does not, nor does solarization depress nitrogen concentration.

Nutrient Status of Plants

Nutrient status (mineral nutrient and trace element composition, also called ionome and nutrient profile) of plants are commonly portrayed by tissue elementary analysis. Interpretation of the results of such studies, however, has been controversial. During the last decades the nearly two-century-old "law of minimum" or "Liebig's law" (that states that plant growth is controlled not by the total amount of resources available, but by the scarcest resource) has been replaced by several mathematical approaches that use different models in order to take the interactions between the individual nutrients into account. The latest developments in this field are based on the fact that the nutrient elements (and compounds) do not act independently from each other; Baxter, 2015, because there may be direct chemical interactions between them or they may influence each other's uptake, translocation, and biological action via a number of mechanisms as exemplified for the case of ammonia.

Plant Nutrition in Agricultural Systems

Hydroponics

Hydroponics is a method for growing plants in a water-nutrient solution without the use of nutrient-rich soil. It allows researchers and home gardeners to grow their plants in a controlled environment. The most common solution is the Hoagland solution, developed by D. R. Hoagland in 1933. The solution consists of all the essential nutrients in the correct proportions necessary for most plant growth. An aerator is used to prevent an anoxic event or hypoxia. Hypoxia can affect nutrient uptake of a plant because, without oxygen present, respiration becomes inhibited within the root cells. The nutrient film technique is a hydroponic technique in which the roots are not fully submerged. This allows for adequate aeration of the roots, while a "film" thin layer of nutrient-rich water is pumped through the system to provide nutrients and water to the plant.

Paleoethnobotany

Pal(a)eoethnobotany or Archaeobotany, "is the study of remains of plants cultivated or used by man in ancient times, which have survived in archaeological contexts". Paleoethnobotany is the archaeological sub-field that studies plant remains from archaeological sites. Basing on the recovery and identification of plant remains and the ecological and cultural information available for modern plants, the major research themes are the use of wild plants, the origins of agriculture and domestication, and the co-evolution of human-plant interactions.

Preservation

Plant macrofossils are preserved through four main modes of preservation at archaeological sites. First, plant remains, usually cereal grains, chaff, seeds and charcoal are largely reduced to elemental carbon (charred) when they are heated in a reducing atmosphere. These are referred to as 'charred' or 'carbonised' plant remains. This mode of preservation is biased towards plant remains that come into contact with fire, through cooking or fuel use, and those which are less fragile, such as cereal grains and nutshell. Second, plant remains deposited in permanently waterlogged anoxic conditions are preserved as the absence of oxygen prohibits microbial activity. This mode of preservation occurs in deep archaeological features such as wells, in urban settlements where organic refuse is rapidly deposited, and at settlements adjacent to lakes or rivers. A wide range of plant remains are usually preserved, including seeds, fruit stones, nutshells, leaves, straw and other vegetative material. Third, calcium-phosphate mineralisation of plant remains occurs usually in latrine pits and in middens, as plant remains are completely replaced by calcium-phosphate. In latrine pits, plant remains which have been consumed by humans are the most common items, such as seeds of flavourings, fruit pips and fruit stones. Finally, plant remains are preserved by desiccation in arid environments, where the absence of water limits decomposition. Delicate vegetative plant remains are preserved, such as onion skin and artichoke bracts, alongside fruit stones, cereal chaff and seeds of wild plants.

Recovery Methods

Paleoethnobotanists use a variety of methods to recover and identify plant remains. Charred plant remains are usually recovered by flotation. The matrix (the soil from a suspected archaeological feature) is slowly added to agitated water. The soil, sand, and other heavy material, known as *heavy fraction*, will sink to the bottom. The less dense organic material such as charred seeds, grains and charcoal will tend to float to the surface. The material that floats to the top, called *light fraction*, is poured into a sieve (usually 250-500 μm). The light fraction is then dried and later examined under a low power microscope. Samples of the heavy fraction are also gathered for later analysis. Flotation can be undertaken manually with buckets, or by machine-assisted flotation where water is circulated through a series of tanks by a pump. Waterlogged plant remains are separated from the matrix by a combination of wet-sieving and/or small-scale flotation in a laboratory. Desiccated plant remains are usually recovered by dry-sieving, using a stack of different sieves to separate larger items such as cereal straw and fruit stones from smaller items such as weed seeds.

Identification

Identification of macroremains is then usually carried out under a stereomicroscope, using morphological features such as shape and surface features in the case of seeds, or microanatomy in the case of wood or charcoal. Identification literature as well as a comparative collection of modern plant materials are crucial for reliable results. Depending on the type of material, and its condition, also other methods such as thin sections or SEM are applied.

Paleoethnobotanists also recover and analyze microremains (such as phytoliths and pollen), human and animal excrements (paleofeces, sometimes called coprolites), or plant impressions in ceramic sherds and clay (such as in daub).

Palynology is a mature and distinct scientific discipline that studies pollen, typically in the context of reconstructing past environments.

Dendrochronology, the study of growth rings on trees relating to study of past environments, is another scientific discipline useful to paleoethnobotanical study.

Research

The work done in paleoethnobotany can be divided into field work, collections management, systematic description of species, and theories into the origins of human and plant interaction. Some examples of this analysis:

A paleoethnobotanist may find discrete concentrations of burned or dried remnants of seeds in an area of discolored soil (a possible hearth feature). If later analyses indicates that the remnants were of only mature wild seeds of a type of plant that grows locally, it could be inferred that the site was only visited seasonally. Such an inference could be supported by a lack of other features that would suggest that no permanent shelters were built at the site.

Alternatively, a paleoethnobotanist may find that a fire pit feature contains concentrated remnants of a wide variety of edible wild plants that mature throughout the year. An archaeologist may find features at the site that indicate some sort of semi-permanent dwellings (such as post holes and middens). The middens may have concentrations of animal remains, identified by a zooarchaeologist as those of wild game, with a variety of species-specific maturity levels. In that case, a more permanent settlement may be inferred, perhaps to the level of a village. Such an analysis of the archaeological features could suggest a society of hunter-gatherers who inhabited the site on a more-or-less year-round basis.

A paleoethnobotanist may also find concentrated remains of plants that typically are only grown through active cultivation (such as cereals, legumes, and oilseeds). At the same site, an archaeologist might identify features such as stone walls surrounding enclosures arrayed in a pattern, and deep, layered middens with concentrations of domesticated animal remains such as goats or pigs. An analysis of the site, set within the context of the archaeological features and animal and plant remains, would suggest a settled agrarian community.

Phytosociology

Phytosociology is the branch of science which deals with plant communities, their composition and development, and the relationships between the species within them. A phytosociological system is a system for classifying these communities. The term "phytosociology" was coined in 1896 by Józef Paczoski.

Overview

The aim of phytosociology is to achieve a sufficient empirical model of vegetation using plant *taxa* combinations that characterize univocally vegetation units. Vegetation units as understood by phytosociologists may express largely abstract vegetation concepts (e.g. the set of all hard-leaved evergreen forests of western Mediterranean area) or actual readily recognizable vegetation types (e.g. cork-oak oceanic forests on Pleistocene dunes with dense canopy in SW-Iberian Peninsula). Such conceptual units are called "syntaxa" (singular "syntaxon") and can be set in a hierarchy system called "synsystem" or syntaxonomic system. The act of creation, amelioration or adjusting the synsystem is called "syntaxonomy". Therefore, the syntaxonomic system is putatively a sufficient empirical representation of vegetation of a given territory. An *International Code of Phytosociological Nomenclature*, issuing the rules for naming "syntaxa" exists and its use has increased among vegetation scientists.

Association Model

The basic unit of syntaxonomy is the "association". The association is a conceptual model of a concrete phytocoenosis (the plant component of a biocoenosis). The association is defined by its characteristic combination of plant taxa, habitat features, physiognomy, biogeographical area, role in ecological succession, historical (e.g. history of use by humans) and paleo-biogeographical relationships. Associations with floristic and territorial affinities can be grouped in larger ecological conceptual units (i.e. syntaxa) called "alliances". Similar alliances may be grouped in "orders" and orders in vegetation "classes". The setting of syntaxa in such a hierarchy makes up the syntaxonomical system, or the reference model of the given vegetation and territory.

In spite of early attempts (e.g. Charles Flauhault in the late 19th century), this science started in Europe, with the Swiss botanist and ecologist Josias Braun-Blanquet (1884–1980).

Vegetation Complexes

Nowadays, phytosociologists try to include higher levels of complexity in the perception of vegetation, namely by describing whole successional units (vegetation series) or, in general, vegetation complexes. These lie in the scope of Landscape Phytosociology. Other developments include the use of multivariate statistics for the definition of "syntaxa" and their environmental interpretation.

On the one hand, some authors consider phytosociology to be in the scope of contemporary vegetation science, a successful approach because of its highly descriptive and predictive powers, and its usefulness in nature management issues. On the other hand, there are numerous critics who have focused on several methodological limitations: the absence of statistical approaches, the complexity and non-stability of the nomenclatural system, the mistakes in the predictive models, and certain basic assumptions.

Habitat-type Classification

Even if in continental Europe, a complete synsystem describing vegetation types has been developed and it is a basis for habitat-type classification (e.g. NATURA 2000 typology and habitat network), there are numerous scientific experts who do not have a positive opinion about the suitability for phytosociology to be the main geobotanical approach for managing vegetation systems. An important point of disagreement is the floristic-phytosociological assumption that the forest patches of the Mediterranean species of pines mainly derived from afforestations, non-stables and incidentals.

Data Collections

Phytosociological data contain information collected in relevés (or plots) listing each species cover-abundance values and the measured environmental variables. This data is conveniently databanked in a program like TURBOVEG allowing for editing, storage and export to other applications.

Data is usually classified and sorted using TWINSPAN in host programs like JUICE to create realistic species-relevé associations. Further patterns are investigated using clustering and resemblance methods, and ordination techniques available in software packages like CANOCO or the R-package vegan.

Phenology

Phenology is the study of periodic plant and animal life cycle events and how these are influenced by seasonal and interannual variations in climate, as well as habitat factors (such as elevation). The word is derived from the Greek φαίνω (*phainō*), "to show, to bring to light, make to appear" + λόγος (*logos*), amongst others "study, discourse, reasoning" and indicates that phenology has been principally concerned with the dates of first occurrence of biological events in their annual cycle. Examples include the date of emergence of leaves and flowers, the first flight of butterflies and the first appearance of migratory birds, the date of leaf colouring and fall in deciduous trees, the dates of egg-laying of birds and amphibia, or the timing of the developmental cycles of temperate-zone honey bee colonies. In the scientific literature on ecology, the term is used more generally to indicate the time frame for any seasonal biological phenomena, including the dates of last appearance (e.g., the seasonal phenology of a species may be from April through September).

Phenological development of Olive flowering, following BBCH standard scale. a-50, b-51, c-54, d-57, (<15% open flowers); f-65, (>15% open flowers); g-67, (<15% open flowers); h-68 (Oteros et al., 2013)

Because many such phenomena are very sensitive to small variations in climate, especially to temperature, phenological records can be a useful proxy for temperature in historical climatology, especially in the study of climate change and global warming. For example, viticultural records of grape harvests in Europe have been used to reconstruct a record of summer growing season temperatures going back more than 500 years. In addition to providing a longer historical baseline than instrumental measurements, phenological observations provide high temporal resolution of ongoing changes related to global warming.

Past Records

Observations of phenological events have provided indications of the progress of the natural calendar since ancient agricultural times. Many cultures have traditional phenological proverbs and sayings which indicate a time for action: "When the sloe tree is white as a sheet, sow your barley whether it be dry or wet" or attempt to forecast future climate: "If oak's before ash, you're in for a splash. If ash before oak, you're in for a soak". But the indications can be pretty unreliable, as an alternative version of the rhyme shows: "If the oak is out before the ash, 'Twill be a summer of wet and splash; If the ash is out before the oak,'Twill be a summer of fire and smoke." Theoretically, though, these are not mutually exclusive, as one forecasts immediate conditions and one forecasts future conditions.

The North American Bird Phenology Program at USGS Patuxent Wildlife Research Center (PWRC) is in possession of a collection of millions of bird arrival and departure date records for over 870 species across North America, dating between 1880 and 1970. This program, originally started by Wells W. Cooke, involved over 3,000 observers including many notable naturalists of the time. The program ran for 90 years and came to a close in 1970 when other programs starting up at PWRC took precedence. The program was again started in 2009 to digitize the collection of records and now with the help of citizens worldwide, each record is being transcribed into a database which will be publicly accessible for use.

The English naturalists Gilbert White and William Markwick reported the seasonal events of more than 400 plant and animal species, Gilbert White in Selborne, Hampshire and William Markwick in Battle, Sussex over a 25-year period between 1768 and 1793. The data, reported in White's *Natural History and Antiquities of Selborne* are reported as the earliest and latest dates for each event over 25 years; so annual changes cannot therefore be determined.

In Japan and China the time of blossoming of cherry and peach trees is associated with ancient festivals and some of these dates can be traced back to the eighth century. Such historical records may, in principle, be capable of providing estimates of climate at dates before instrumental records became available. For example, records of the harvest dates of the pinot noir grape in Burgundy have been used in an attempt to reconstruct spring–summer temperatures from 1370 to 2003; the reconstructed values during 1787–2000 have a correlation with Paris instrumental data of about 0.75.

Modern Records

Robert Marsham is the founding father of modern phenological recording. Marsham was a wealthy landowner who kept systematic records of "Indications of spring" on his estate at Stratton Strawless, Norfolk, from 1736. These were in the form of dates of the first occurrence of events such as flowering, bud burst, emergence or flight of an insect. Consistent records of the same events or "phenophases" were maintained by generations of the same family over unprecedentedly long periods of time, eventually ending with the death of Mary Marsham in 1958, so that trends can be observed and related to long-term climate records. The data show significant variation in dates which broadly correspond with warm and cold years. Between 1850 and 1950 a long-term trend of gradual climate warming is observable, and during this same period the Marsham record of oak leafing dates tended to become earlier.

After 1960 the rate of warming accelerated, and this is mirrored by increasing earliness of oak leafing, recorded in the data collected by Jean Combes in Surrey. Over the past 250 years, the first leafing date of oak appears to have advanced by about 8 days, corresponding to overall warming on the order of 1.5 °C in the same period.

Towards the end of the 19th century the recording of the appearance and development of plants and animals became a national pastime, and between 1891 and 1948 a programme of phenological recording was organised across the British Isles by the Royal Meteorological Society (RMS). Up to 600 observers submitted returns in some years, with numbers averaging a few hundred. During this period 11 main plant phenophases were consistently recorded over the 58 years from 1891–1948, and a further 14 phenophases were recorded for the 20 years between 1929 and 1948. The returns were summarised each year in the Quarterly Journal of the RMS as *The Phenological Reports*. The 58-year data have been summarised by Jeffree (1960), and show that flowering dates could be as many as 21 days early and as many as 34 days late, with extreme earliness greatest in summer flowering species, and extreme lateness in spring flowering species. In all 25 species, the timings of all phenological events are significantly related to temperature, indicating that phenological events are likely to get earlier as climate warms.

The Phenological Reports ended suddenly in 1948 after 58 years, and Britain was without a national recording scheme for almost 50 years, just at a time when climate change was becoming evident. During this period, important contributions were made by individual dedicated observers. The naturalist and author Richard Fitter recorded the First Flowering Date (FFD) of 557 species of British flowering plants in Oxfordshire between about 1954 and 1990. Writing in Science in 2002, Richard Fitter and his son Alistair Fitter found that "the average FFD of 385 British plant species has advanced by 4.5 days during the past decade compared with the previous four decades." They note that FFD is sensitive to temperature, as is generally agreed, that "150 to 200 species may be flowering on average 15 days earlier in Britain now than in the very recent past" and that these earlier FFDs will have "profound ecosystem and evolutionary consequences".

In the last decade, national recording in Britain has been resumed by the UK Phenology network , run by Woodland Trust and the Centre for Ecology and Hydrology and the BBC Springwatch survey. There is a USA National Phenology Network in which both professional scientists and lay recorders participate, a European Phenology Network that has monitoring, research and educational remits and many other countries such as Canada (Alberta Plantwatch and Saskatchewan PlantWatch), China and Australia have phenological programs.

In eastern North America, almanacs are traditionally used for information on action phenology (in agriculture), taking into account the astronomical positions at the time. William Felker has studied phenology in Ohio, USA since 1973 and now publishes "Poor Will's Almanack", a phenological almanac for farmers (not to be confused with a late 18th century almanac by the same name).

Airborne Sensors

Recent technological advances in studying the earth from space have resulted in a new field of phenological research that is concerned with observing the phenology of whole ecosystems and stands of vegetation on a global scale using proxy approaches. These methods complement the traditional phenological methods which recorded the first occurrences of individual species and phenophases.

The most successful of these approaches is based on tracking the temporal change of a Vegetation Index (like Normalized Difference Vegetation Index(NDVI)). NDVI makes use of the vegetation's typical low reflection in the red (red energy is mostly absorbed by growing plants for Photosynthesis) and strong reflection in the Near Infrared (Infrared energy is mostly reflected by plants due to their cellular structure). Due to its robustness and simplicity, NDVI has become one of the most popular remote sensing based products. Typically, a vegetation index is constructed in such a way that the attenuated reflected sunlight energy (1% to 30% of incident sunlight) is amplified by ratio-ing red and NIR following this equation:

$$\mathrm{NDVI} = \frac{\mathrm{NIR} - \mathrm{red}}{\mathrm{NIR} + \mathrm{red}}$$

The evolution of the vegetation index through time, depicted by the graph above, exhibits a strong correlation with the typical green vegetation growth stages (emergence, vigor/growth, maturity, and harvest/senescence). These temporal curves are analyzed to extract useful parameters about the vegetation growing season (start of season, end of season, length of growing season, etc.). Other growing season parameters could potentially be extracted, and global maps of any of these growing season parameters could then be constructed and used in all sorts of climatic change studies.

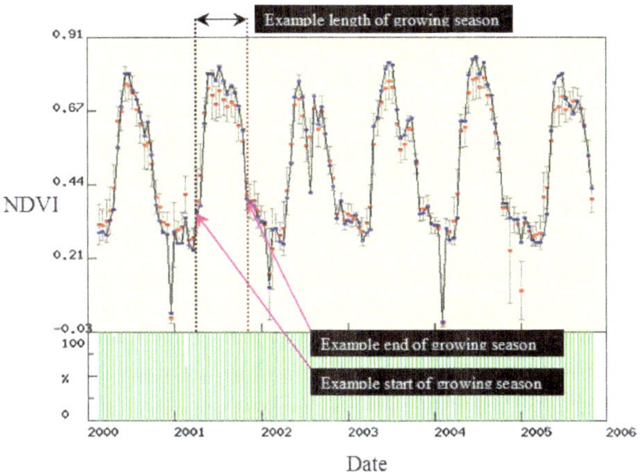

NDVI temporal profile for a typical patch of coniferous forest over a period of six years. This temporal profile depicts the growing season every year as well as changes in this profile from year to year due to climatic and other constraints. Data and graph are based on the MODIS sensor standard public vegetation index product. Data archived at the ORNL DAAC , courtesy of Dr. Robert Cook.

A noteworthy example of the use of remote sensing based phenology is the work of Ranga Myneni from Boston University. This work showed an apparent increase in vegetation productivity that most likely resulted from the increase in temperature and lengthening of the growing season in the boreal forest. Another example based on the MODIS enhanced vegetation index (EVI) reported by Alfredo Huete at the University of Arizona and colleagues showed that the Amazon Rainforest, as opposed to the long held view of a monotonous growing season or growth only during the wet rainy season, does in fact exhibit growth spurts during the dry season.

However, these phenological parameters are only an approximation of the true biological growth stages. This is mainly due to the limitation of current space-based remote sensing, especially the spatial resolution, and the nature of vegetation index. A pixel in an image does not contain a pure target (like a tree, a shrub, etc.) but contains a mixture of whatever intersected the sensor's field of view.

Palynology

Palynology is the "study of dust" (from Greek παλύνω, *palunō*, "strew, sprinkle" and -logy) or "particles that are strewn". A classic palynologist analyses particulate samples collected from the air, from water, or from deposits including sediments of any age. The condition and identification of those particles, organic and inorganic, give the palynologist clues to the life, environment, and energetic conditions that produced them.

The term is sometimes narrowly used to refer to a subset of the discipline, which is defined as "the study of microscopic objects of macromolecular organic composition (i.e., compounds of carbon, hydrogen, nitrogen and oxygen), not capable of dissolution in hydrochloric or hydrofluoric acids". It is the science that studies contemporary and fossil palynomorphs, including pollen, spores, orbicules, dinocysts, acritarchs, chitinozoans and scolecodonts, together with particulate organic matter (POM) and kerogen found in sedimentary rocks and sediments. Palynology does not include diatoms, foraminiferans or other organisms with siliceous or calcareous exoskeletons.

Palynology is an interdisciplinary science and is a branch of earth science (geology or geological science) and biological science (biology), particularly plant science (botany). Stratigraphical palynology is a branch of micropalaeontology and paleobotany, which studies fossil palynomorphs from the Precambrian to the Holocene.

A History of Palynology

Early History

The earliest reported observations of pollen under a microscope are likely to have been in the 1640s by the English botanist Nehemiah Grew, who described pollen and the stamen, and correctly predicted that pollen is required for sexual reproduction in flowering plants.

By the late 1870s, as optical microscopes improved and the principles of stratigraphy were worked out, Robert Kidston and P. Reinsch were able to examine the presence of fossil spores in the Devonian and Carboniferous coal seams and make comparisons between the living spores and the ancient fossil spores. Early investigators include Christian Gottfried Ehrenberg (radiolarians, diatoms and dinoflagellate cysts), Gideon Mantell (desmids) and Henry Hopley White (dinoflagellate cysts).

Palynology 1890s to 1940s

Quantitative analysis of pollen began with Lennart von Post's published work. Although he published in the Swedish language, his methodology gained a wide audience through his lectures. In particular, his Kristiania lecture of 1916 was important in gaining a wider audience. Because the early investigations published in the Nordic languages or Scandinavian languages the field of pollen analysis was confined to those countries. The isolation ended with the German publication of Gunnar Erdtman's 1921 thesis. The methodology of pollen analysis became widespread throughout Europe and North America and revolutionized Quaternary vegetation and climate change research.

Earlier pollen researchers were Früh (1885), who enumerated many common tree pollen types, and a considerable number of spores and herb pollen grains. There is a study of pollen samples taken from sediments of Swedish lakes by Trybom (1888), Pinus (Pine) and Picea (Spruce) pollen was found in such profusion that he considered them to be serviceable as "index fossils". Georg F. L. Sarauw studied fossil pollen of middle Pleistocene age (Cromerian) from the harbour of Copenhagen. Lagerheim (in Witte 1905) and C. A.Weber (in H. A. Weber 1918) appear to be among the first to undertake 'percentage frequency' calculations.

Palynology 1940s to 1989

The term *palynology* was introduced by Hyde and Williams in 1944, following correspondence with the Swedish geologist Antevs, in the pages of the Pollen Analysis Circular (one of the first journals devoted to pollen analysis, produced by Paul Sears in North America). Hyde and Williams chose *palynology* on the basis of the Greek words *paluno* meaning 'to sprinkle' and *pale* meaning 'dust' (and thus similar to the Latin word *pollen*).

Palynology 1990s to 21st Century

Pollen analysis advanced rapidly in this period due to advances in optics and computers. Much of the science was revised by Johannes Iversen and Knut Fægri in their textbook on the subject.

Methods of Study

Palynomorphs

Palynomorphs are broadly defined as organic-walled microfossils between 5 and 500 micrometres in size. They are extracted from sedimentary rocks and sediment cores both physically, by ultrasonic treatment and wet sieving, and chemically, by chemical digestion to remove the non-organic fraction. Palynomorphs may be composed of organic material such as chitin, pseudochitin and sporopollenin. Palynomorphs that have a taxonomy description are sometimes referred to as palynotaxa.

Palynomorphs form a geological record of importance in determining the type of prehistoric life that existed at the time the sedimentary formation was laid down. As a result, these microfossils give important clues to the prevailing climatic conditions of the time. Their paleontological utility derives from an abundance numbering in millions of cells per gram in organic marine deposits, even when such deposits are generally not fossiliferous. Palynomorphs, however, generally have been destroyed in metamorphic or recrystallized rocks.

Typically, palynomorphs are dinoflagellate cysts, acritarchs, spores, pollen, fungi, scolecodonts (scleroprotein teeth, jaws and associated features of polychaete annelid worms), arthropod organs (such as insect mouthparts), chitinozoans and microforams. Palynomorph microscopic structures that are abundant in most sediments are resistant to routine pollen extraction including strong acids and bases, and acetolysis, or density separation.

Chemical Preparation

Chemical digestion follows a number of steps. Initially the only chemical treatment used by researchers was treatment with potassium hydroxide (KOH) to remove humic substances; deflocculation was accomplished through surface treatment or ultra-sonic treatment, although sonification may cause the pollen exine to rupture. The use of hydrofluoric acid (HF) to digest silicate minerals was introduced by Assarson and Granlund in 1924, greatly reducing the amount of time required to scan slides for palynomorphs. Palynological studies using peats presented a particular challenge because of the presence of well-preserved organic material, including fine rootlets, moss leaflets and organic litter. This was the last major challenge in the chemical preparation of materials for palynological study. Acetolysis was developed by Gunnar Erdtman and his brother to remove these fine cellulose materials by dissolving them. In acetolysis the specimen is treated with acetic anhydride and sulfuric acid, dissolving cellulistic materials and thus providing better visibility for palynomorphs.

Some steps of the chemical treatments require special care for safety reasons, in particular the use of HF which diffuses very fast through the skin and, causes severe chemical burns, and can be fatal.

Another treatment includes kerosene flotation for chitinous materials.

Analysis

Once samples have been prepared chemically, they are mounted on microscope slides using silicon oil, glycerol or glycerol-jelly and examined using light microscopy or mounted on a stub for scanning electron microscopy.

Researchers will often study either modern samples from a number of unique sites within a given area, or samples from a single site with a record through time, such as samples obtained from peat or lake sediments. More recent studies have used the modern analog technique in which paleo-samples are compared to modern samples for which the parent vegetation is known

When the slides are observed under a microscope, the researcher counts the number of grains of each pollen taxon. This record is next used to produce a pollen diagram. These data can be used to detect anthropogenic effects, such as logging, traditional patterns of land use or long term changes in regional climate

Palynology can be applied to problems in many fields including geology, botany, paleontology, archaeology, pedology (soil study), and physical geography.

Applications

Palynology is used for a diverse range of applications, related to many scientific disciplines:

- Biostratigraphy and geochronology. Geologists use palynological studies in biostratigraphy to correlate strata and determine the relative age of a given bed, horizon, formation or stratigraphical sequence.
- Palaeoecology and climate change. Palynology can be used to reconstruct past vegetation (land plants) and marine and freshwater phytoplankton communities, and so infer past environmental (palaeoenvironmental) and palaeoclimatic conditions.
- Organic palynofacies studies, which examine the preservation of the particulate organic matter and palynomorphs provides information on the depositional environment of sediments and depositional palaeoenvironments of sedimentary rocks.
- Geothermal alteration studies examine the colour of palynomorphs extracted from rocks to give the thermal alteration and maturation of sedimentary sequences, which provides estimates of maximum palaeotemperatures.
- Limnology studies. Freshwater palynomorphs and animal and plant fragments, including the prasinophytes and desmids (green algae) can be used to study past lake levels and long term climate change.
- Taxonomy and evolutionary studies.
- Forensic palynology: the study of pollen and other palynomorphs for evidence at a crime scene.
- Allergy studies. Studies of the geographic distribution and seasonal production of pollen, can help sufferers of allergies such as hay fever.

- Melissopalynology: the study of pollen and spores found in honey.
- Archaeological palynology examines human uses of plants in the past. This can help determine seasonality of site occupation, presence or absence of agricultural practices or products, and 'plant-related activity areas' within an archaeological context. Bonfire Shelter is one such example of this application.

Because the distribution of acritarchs, chitinozoans, dinoflagellate cysts, pollen and spores provides evidence of stratigraphical correlation through biostratigraphy and palaeoenvironmental reconstruction, one common and lucrative application of palynology is in oil and gas exploration.

Palynology also allows scientists to infer the climatic conditions from the vegetation present in an area thousands or millions of years ago. This is a fundamental part of research into climate change.

Economic Botany

Economic botany is the study of the relationship between people (individuals and cultures) and plants. Economic botany contributes significantly to many fields, including established disciplines such as agronomy, anthropology, archaeology, chemistry, economics, ethnobotany, ethnology, forestry, genetic resources, geography, geology, horticulture, medicine, microbiology, nutrition, pharmacognosy, and pharmacology. This link between botany and anthropology explores the ways humans use plants for food, shelter, medicines, textiles, and more.

History of Economic Botany

In a 1958 essay at the conference that founded the Society for Economic Botany, David J. Rogers wrote, "A current viewpoint is that economic botany should concern itself with basic botanical, phytochemical and ethnological studies of plants known to be useful or those which may have potential uses so far underdeveloped. Economic botany is, then, a composite of those sciences working specifically with plants of importance to [people]." Closely allied with economic botany is ethnobotany, which emphasizes plants in the context of anthropology.

Botany itself came about through medicine and the development of herbal remedies. Thus at its advent, botany was economic as well as systematic. As plants became useful for herbals and curatives, their economic value increased. An early set of instructions drawn up by a cosmographer of Charles the fifth instructed explorers to

"determine what are the items of sustenance of the land and which ones are generally used, whether fruits or seeds, and all manner of spices, drugs, or whatever other scents, and find out the time in which one can reproduce the trees, plants, herbs, and fruits that these parts offer, and if the natives use them for medicines, as we do."

Teosinte and rice are two examples of plants modified so that their economic values would increase.

Teosinte

The teosintes are grasses of the genus *Zea*. Native Americans bred and selected teosinte for the traits we see in corn today (large ears, multiple rows of kernels). The first ears of maize were very short, with only 8 rows of kernels. Modern corn is the result of several thousand generations of selective breeding. Modern corn is incapable of reproducing without human help; the kernels will stay firmly attached to the cob and rot. This doesn't represent a useful adaptation for the species, but is excellent for harvesting and transporting corn.

Rice

Rice was first domesticated approximately 5,000 years ago, in Southeast Asia. Rice and American wild rice are believed to have been domesticated separately. Rice variants have been adapted to the tropics where they provide a grain staple, but rice can be grown almost anywhere. The introduction of dwarf rice variants made several rice-producing countries self-sufficient. Rice is suited to countries with high rainfall.

Economically Important Food Plants

Plants that humans use for food are of high economic importance. Research into food plants generally involves increasing the size of the edible plant organ in question, or increasing the areas where the plant can be grown, and less frequently, finding new crop species. Results of such research are often published in the journal *Economic Botany*. The New Zealand-based Plant & Food Research publishes its own journal on cultivar development and sustainable production systems for high quality produce, and the design and development of new and novel functional foods.

Florida Oranges

Citrus has been a major commercial product in Florida since the 19th century. Florida produces over 70% of the U.S. citrus supply. The color of oranges is not related to ripening, but is a serious component for sales. The orange color only develops in areas with cool nighttime temperatures. In tropical climates, growers often expose the fruit to ethylene, to promote the loss of chlorophyll and expose the beta-carotenes (the orange color).

North American Apples

Apples are not native to North America, but today the North American continent boasts the greatest diversity of apples in the world. Part of this is due to "Johnny Appleseed," real name John Chapman. Chapman spent 48 years travelling all along the American northwest spreading apple seeds and planting trees. While apples come in literally thousands of varieties, the majority of the apple market is based on three: Red Delicious, Golden Delicious, and Granny Smith. The Red Delicious is the ideal apple, for marketing purposes, for the apple industry. Its large, intensely red, and instantly recognizable.

Economically Valuable Medicinal Plants

Medical research in the U.S. alone has a budget of $95 billion. A large portion of that money is spent on research into plants and plant extracts. Several key medical discoveries have been made by studying plants and the compounds they produce, to see the effect they have on humans.

Ephedrine

The *Ephedra* plant is the natural source of ephedrine, the plant's principal alkaloid. Ephedrine is actually a very interesting case of economic botany in medicine. Ephedrine imitates epinephrine in its effect on the human body. While it has medicinal uses, ephedrine can be highly toxic. Because of this fact, medical researchers studied the compound and produced pseudoephedrine, which is used in over-the-counter medications and in the illegal manufacture of methamphetamine.

Echinacea

One of many herbal remedies out there, Echinacea represents a sizable industry. Many people take echinacea for cold and flu-like symptoms, but studies show that the plant has had mixed success fighting these viruses. However, those same studies show the plant possibly being useful for the treatment of upper respiratory infections. NCCAM is currently studying echinacea for the treatment of upper respiratory infections as well as its effect on the immune system.

Ornamental Plants

Ornamental plants can be found in almost any store, and many people have at least one in their home. However, ornamental plants are not limited to houseplants. Landscaping agencies make heavy use of ornamental plants, usually with an accompanying high cost. Trees, shrubs, flowers, and grasses, all of these are planted by professional landscaping agencies regularly, with a large economic effect.

References

- Fosket, Donald E. (1994). Plant Growth and Development: A Molecular Approach. San Diego: Academic Press. pp. 498–509. ISBN 0-12-262430-0.
- Adrian Charles Slack; Jane Gate (1980). Carnivorous Plants. Cambridge, Mass. : MIT Press. p. 160. ISBN 978-0-262-19186-9.
- Kingsley Rowland Stern; Shelley Jansky (1991). Introductory Plant Biology. WCB/McGraw-Hill. p. 309. ISBN 978-0-697-09948-8.
- Raven, P. H., R. F. Evert, & S. E. Eichhorn. Biology of Plants, 7th ed., page 9. (New York: W. H. Freeman, 2005). ISBN 0-7167-1007-2.
- Harold C. Bold, C. J. Alexopoulos, and T. Delevoryas. Morphology of Plants and Fungi, 5th ed., page 3. (New York: Harper-Collins, 1987). ISBN 0-06-040839-1.
- Michael A Dirr; Charles W Heuser, jr. (2006). "2". The Reference Manual of Woody Plant Propagation (Second ed.). Varsity Press Inc. pp. 26, 28, 29. ISBN 0942375092.
- Jackson RW (editor). (2009). Plant Pathogenic Bacteria: Genomics and Molecular Biology. Caister Academic Press. ISBN 978-1-904455-37-0.
- Marschner, Petra, ed. (2012). Marschner's mineral nutrition of higher plants (3rd ed.). Amsterdam: Elsevier/Academic Press. ISBN 9780123849052.
- Roy, R.N.; Finck, A.; Blair, G.J.; Tandon, H.L.S. (2006). "Chapter 3: Plant nutrients and basics of plant nutrition". Plant nutrition for food security: a guide for integrated nutrient management (PDF). Rome: Food and

Agriculture Organization of the United Nations. pp. 25–42. ISBN 92-5-105490-8. Retrieved 20 June 2016.

- Norman P. A. Huner; William Hopkins. "3 & 4". Introduction to Plant Physiology 4th Edition. John Wiley & Sons, Inc. ISBN 978-0-470-24766-2.
- Barker, AV; Pilbeam, DJ (2015). Handbook of Plant Nutrition. (2nd ed.). CRC Press. ISBN 9781439881972. Retrieved 5 June 2016.
- White, Philip J. (2016). "Selenium accumulation by plants". Annals of Botany. 117: 217–235. doi:10.1093/aob/mcv180. Retrieved 5 June 2016.
- Jones, Cynthia S. (1999-11-01). "An Essay on Juvenility, Phase Change, and Heteroblasty in Seed Plants". International Journal of Plant Sciences. 160 (S6): –105–S111. doi:10.1086/314215. ISSN 1058-5893. Retrieved 2016-10-16.
- "Plasmopara viticola, the Cause of Downy Mildew of Grapes". The Origin of Plant Pathology and The Potato Famine, and Other Stories of Plant Diseases. Retrieved 4 February 2015.
- "Fusarium oxysporum : The End of the Banana Industry?". The Origin of Plant Pathology and The Potato Famine, and Other Stories of Plant Diseases. Retrieved 4 February 2015.
- Lindsey, Rebecca; Robert Simmon (June 30, 2006). "Defying Dry: Amazon Greener in Dry Season than Wet". The Earth Observatory. EOS Project Science Office, NASA Goddard. Retrieved 29 August 2013
- "1st large-scale map of a plant's protein network addresses evolution, disease process". Dana-Farber Cancer Institute. July 29, 2011. Retrieved 24 July 2012.

Key Elements of Botany

3

The key elements of botany discussed within the text are flowers, seed, bud, cell wall, tree and forest. Each section is an integral whole as well as part of a composite evolutionary process. The section strategically encompasses and incorporates the major components and key concepts of botany, providing a complete understanding.

Seed

A seed is an embryonic plant enclosed in a protective outer covering. The formation of the seed is part of the process of reproduction in seed plants, the spermatophytes, including the gymnosperm and angiosperm plants.

Brown flax seeds

Seeds are the product of the ripened ovule, after fertilization by pollen and some growth within the mother plant. The embryo is developed from the zygote and the seed coat from the integuments of the ovule.

Seeds have been an important development in the reproduction and success of gymnosperms and angiosperms plants, relative to more primitive plants such as ferns, mosses and liverworts, which do not have seeds and use other means to propagate themselves. Seed plants now dominate biological niches on land, from forests to grasslands both in hot and cold climates.

The term “seed” also has a general meaning that antedates the above—anything that can be sown, e.g. “seed” potatoes, “seeds” of corn or sunflower “seeds”. In the case of sunflower and corn “seeds”, what is sown is the seed enclosed in a shell or husk, whereas the potato is a tuber.

Many structures commonly referred to as “seeds” are actually dry fruits. Plants producing berries are called baccate. Sunflower seeds are sometimes sold commercially while still enclosed within the hard wall of the fruit, which must be split open to reach the seed. Different groups of plants have other modifications, the so-called stone fruits (such as the peach) have a hardened fruit layer (the endocarp) fused to and surrounding the actual seed. Nuts are the one-seeded, hard-shelled fruit of some plants with an indehiscent seed, such as an acorn or hazelnut.

Production

Seeds are produced in several related groups of plants, and their manner of production distinguishes the angiosperms ("enclosed seeds") from the gymnosperms ("naked seeds"). Angiosperm seeds are produced in a hard or fleshy structure called a fruit that encloses the seeds, hence the name. (Some fruits have layers of both hard and fleshy material). In gymnosperms, no special structure develops to enclose the seeds, which begin their development "naked" on the bracts of cones. However, the seeds do become covered by the cone scales as they develop in some species of conifer.

Seed production in natural plant populations varies widely from year-to-year in response to weather variables, insects and diseases, and internal cycles within the plants themselves. Over a 20-year period, for example, forests composed of loblolly pine and shortleaf pine produced from 0 to nearly 5 million sound pine seeds per hectare. Over this period, there were six bumper, five poor, and nine good seed crops, when evaluated in regard to producing adequate seedlings for natural forest reproduction.

Development

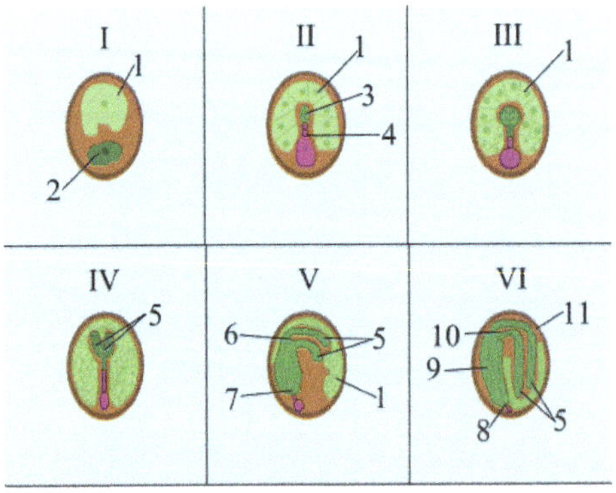

Stages of seed development:

I Zygote	IV Heart
II Proembryo	V Torpedo
III Globular	VI Mature Embryo

Key: *1. Endosperm 2. Zygote 3. Embryo 4. Suspensor 5. Cotyledons 6. Shoot Apical Meristem 7. Root Apical Meristem 8. Radicle 9. Hypocotyl 10. Epicotyl 11. Seed Coat*

Angiosperm (flowering plants) seeds consist of three genetically distinct constituents: (1) the embryo formed from the zygote, (2) the endosperm, which is normally triploid, (3) the seed coat from tissue derived from the maternal tissue of the ovule. In angiosperms, the process of seed development begins with double fertilization, which involves the fusion of two male gametes with the egg cell and the central cell to form the primary endosperm and the zygote. Right after fertilization, the zygote is mostly inactive, but the primary endosperm divides rapidly to form the endosperm tissue. This tissue becomes the food the young plant will consume until the roots have developed after germination.

Ovule

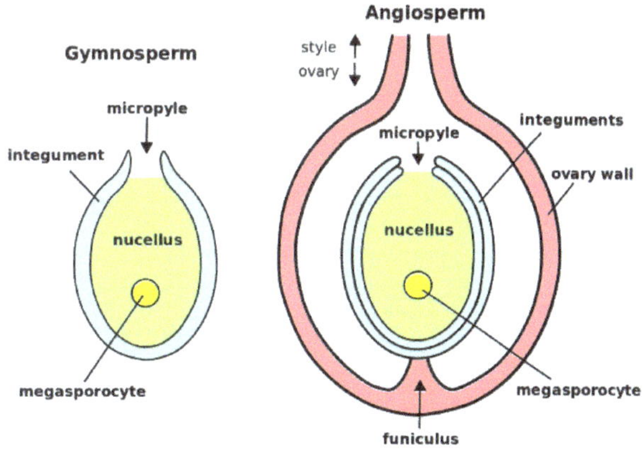

Plant ovules: Gymnosperm ovule on left, angiosperm ovule (inside ovary) on right

After fertilization the ovules develop into the seeds. The ovule consists of a number of components:

- The funicle (*funiculus, funiculi*) or seed stalk which attaches the ovule to the placenta and hence ovary or fruit wall, at the pericarp.
- The nucellus, the remnant of the megasporangium and main region of the ovule where the megagametophyte develops.
- The micropyle, a small pore or opening in the apex of the integument of the ovule where the pollen tube usually enters during the process of fertilization.
- The chalaza, the base of the ovule opposite the micropyle, where integument and nucellus are joined together).

The shape of the ovules as they develop often affects the final shape of the seeds. Plants generally produce ovules of four shapes: the most common shape is called anatropous, with a curved shape. Orthotropous ovules are straight with all the parts of the ovule lined up in a long row producing an uncurved seed. Campylotropous ovules have a curved megagametophyte often giving the seed a tight "C" shape. The last ovule shape is called amphitropous, where the ovule is partly inverted and turned back 90 degrees on its stalk (the funicle or funiculus).

In the majority of flowering plants, the zygote's first division is transversely oriented in regards to the long axis, and this establishes the polarity of the embryo. The upper or chalazal pole becomes the main area of growth of the embryo, while the lower or micropylar pole produces the stalk-like suspensor that attaches to the micropyle. The suspensor absorbs and manufacturers nutrients from the endosperm that are used during the embryo's growth.

Embryo

The inside of a *Ginkgo* seed, showing a well-developed embryo, nutritive tissue (megagametophyte), and a bit of the surrounding seed coat

The main components of the embryo are:

- The cotyledons, the seed leaves, attached to the embryonic axis. There may be one (Monocotyledons), or two (Dicotyledons). The cotyledons are also the source of nutrients in the non-endospermic dicotyledons, in which case they replace the endosperm, and are thick and leathery. In endospermic seeds the cotyledons are thin and papery. Dicotyledons have the point of attachment opposite one another on the axis.
- The epicotyl, the embryonic axis above the point of attachment of the cotyledon(s).
- The plumule, the tip of the epicotyl, and has a feathery appearance due to the presence of young leaf primordia at the apex, and will become the shoot upon germination.
- The hypocotyl, the embryonic axis below the point of attachment of the cotyledon(s), connecting the epicotyle and the radicle, being the stem-root transition zone.
- The radicle, the basal tip of the hypocotyl, grows into the primary root.

Monocotyledonous plants have two additional structures in the form of sheaths. The plumule is covered with a coleoptile that forms the first leaf while the radicle is covered with a coleorhiza that connects to the primary root and adventitious roots form from the sides. Here the hypocotyl is a rudimentary axis between radicle and plumule. The seeds of corn are constructed with these structures; pericarp, scutellum (single large cotyledon) that absorbs nutrients from the endosperm, plumule, radicle, coleoptile and coleorhiza—these last two structures are sheath-like and enclose the plumule and radicle, acting as a protective covering.

Seed Coat

The maturing ovule undergoes marked changes in the integuments, generally a reduction and disorganisation but occasionally a thickening. The seed coat forms from the two integuments or outer layers of cells of the ovule, which derive from tissue from the mother plant, the inner integument forms the tegmen and the outer forms the testa. (The seed coats of some mononocotyledon plants, such as the grasses, are not distinct structures, but are fused with the fruit wall to form a pericarp.) The testae of both monocots and dicots are often marked with patterns and textured markings, or have wings or tufts of hair. When the seed coat forms from only one layer, it is also called the testa, though not all such testae are homologous from one species to the next. The funiculus abscises (detaches at fixed point – abscission zone), the scar forming an oval depression, the hilum. Anatropous ovules have a portion of the funiculus that is adnate (fused to the seed coat), and which forms a longitudinal ridge, or raphe, just above the hilum. In bitegmic ovules (e.g. *Gossypium* described here) both inner and outer integuments contribute to the seed coat formation. With continuing maturation the cells enlarge in the outer integument. While the inner epidermis may remain a single layer, it may also divide to produce two to three layers and accumulates starch, and is referred to as the colourless layer. By contrast the outer epidermis becomes tanniferous. The inner integument may consist of eight to fifteen layers. (Kozlowski 1972)

As the cells enlarge, and starch is deposited in the outer layers of the pigmented zone below the outer epidermis, this zone begins to lignify, while the cells of the outer epidermis enlarge radially and their walls thicken, with nucleus and cytoplasm compressed into the outer layer. these cells which are broader on their inner surface are called palisade cells. In the inner epidermis the cells also enlarge radially with plate like thickening of the walls. The mature inner integument has a palisade layer, a pigmented zone with 15-20 layers, while the innermost layer is known as the fringe layer. (Kozlowski 1972)

Gymnosperms

In gymnosperms, which do not form ovaries, the ovules and hence the seeds are exposed. This is the basis for their nomenclature – naked seeded plants (gymnosperms). Two sperm cells transferred from the pollen do not develop seed by double fertilization, but one sperm nucleus unites with the egg nucleus and the other sperm is not used. Sometimes each sperm fertilizes an egg cell and one zygote is then aborted or absorbed during early development. The seed is composed of the embryo (the result of fertilization) and tissue from the mother plant, which also form a cone around the seed in coniferous plants such as pine and spruce.

Shape and Appearance

A large number of terms are used to describe seed shapes, many of which are largely self-explanatory such as *Bean-shaped* (reniform) – resembling a kidney, with lobed ends on either side of the hilum, *Square* or *Oblong* – angular with all sides more or less equal or longer than wide, *Triangular* – three sided, broadest below middle, *Elliptic* or *Ovate* or *Obovate* – rounded at both ends, or egg shaped (ovate or obovate, broader at one end), being rounded but either symmetrical about the middle or broader below the middle or broader above the middle.

Other less obvious terms include discoid (resembling a disc or plate, having both thickness and parallel faces and with a rounded margin), ellipsoid, globose (spherical), or subglobose (Inflated, but less than spherical), lenticular, oblong, ovoid, reniform and sectoroid. Striate seeds are striped with parallel, longitudinal lines or ridges. The commonest colours are brown and black, other colours are infrequent. The surface varies from highly polished to considerably roughened. The surface may have a variety of appendages (see Seed coat). A seed coat with the consistency of cork is referred to as suberose. Other terms include crustaceous (hard, thin or brittle).

Structure

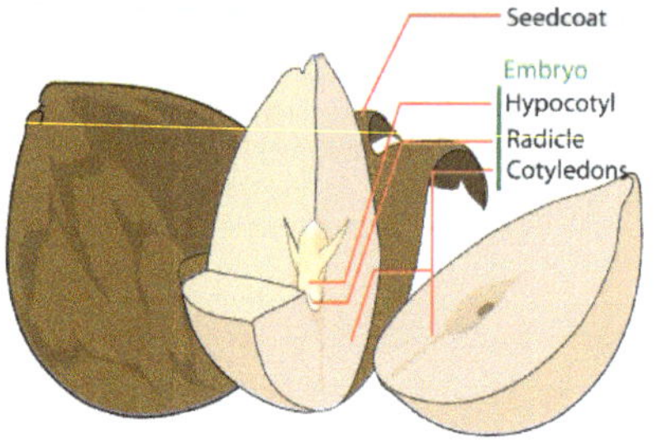

The parts of an avocado seed (a dicot), showing the seed coat, endosperm, and embryo

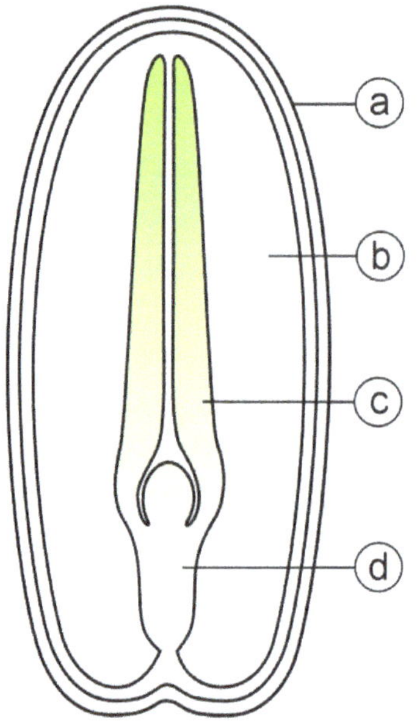

Diagram of the internal structure of a dicot seed and embryo:
(a) seed coat, (b) endosperm, (c) cotyledon, (d) hypocotyl

A typical seed includes two basic parts:

1. an embryo;
2. a seed coat.

In addition, the endosperm forms a supply of nutrients for the embryo in most monocotyledons and the endospermic dicotyledons.

Seed Types

Seeds have been considered to occur in many structurally different types (Martin 1946). These are based on a number of criteria, of which the dominant one is the embryo-to-seed size ratio. This reflects the degree to which the developing cotyledons absorb the nutrients of the endosperm, and thus obliterate it.

Six types occur amongst the monocotyledons, ten in the dicotyledons, and two in the gymnosperms (linear and spatulate). This classification is based on three characteristics: embryo morphology, amount of endosperm and the position of the embryo relative to the endosperm.

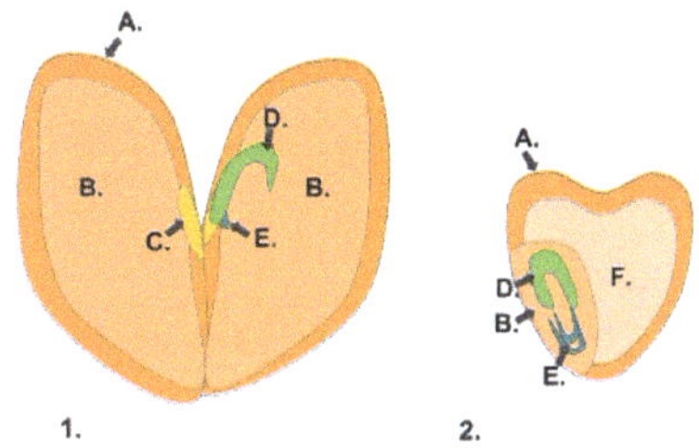

Diagram of a generalized dicot seed (1) versus a generalized monocot seed (2). A. Seed Coat B. Cotyledon C. Hilum D. Plumule E. Radicle F. Endosperm

Embryo

In endospermic seeds, there are two distinct regions inside the seed coat, an upper and larger endosperm and a lower smaller embryo. The embryo is the fertilised ovule, an immature plant from which a new plant will grow under proper conditions. The embryo has one cotyledon or seed leaf in monocotyledons, two cotyledons in almost all dicotyledons and two or more in gymnosperms. In the fruit of grains (caryopses) the single monocotyledon is shield shaped and hence called a scutellum. The scutellum is pressed closely against the endosperm from which it absorbs food, and passes it to the growing parts. Embryo descriptors include small, straight, bent, curved and curled.

Nutrient Storage

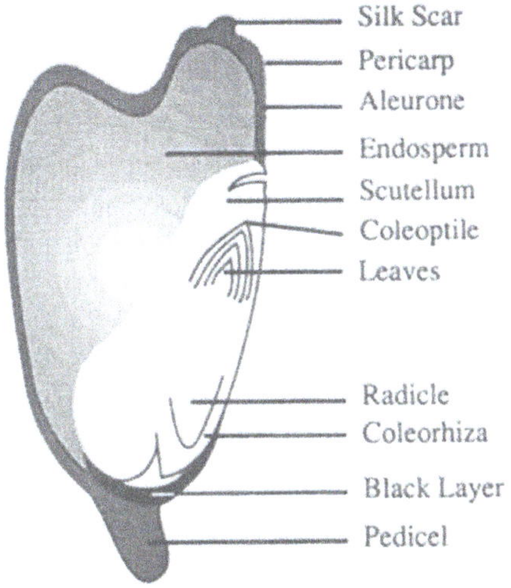

Layers within an endospermic maize seed

Within the seed, there usually is a store of nutrients for the seedling that will grow from the embryo. The form of the stored nutrition varies depending on the kind of plant. In angiosperms, the stored food begins as a tissue called the endosperm, which is derived from the mother plant and the pollen via double fertilization. It is usually triploid, and is rich in oil or starch, and protein. In gymnosperms, such as conifers, the food storage tissue (also called endosperm) is part of the female gametophyte, a haploid tissue. The endosperm is surrounded by the aleurone layer (peripheral endosperm), filled with proteinaceous aleurone grains.

Originally, by analogy with the animal ovum, the outer nucellus layer (perisperm) was referred to as albumen, and the inner endosperm layer as vitellus. Although misleading, the term began to be applied to all the nutrient matter. This terminology persists in referring to endospermic seeds as 'albuminous'. The nature of this material is used in both describing and classifying seeds, in addition to the embryo to endosperm size ratio. The endosperm may be considered to be farinaceous (or mealy) in which the cells are filled with starch, as for instance cereal grains, or not (non-farinaceous). The endosperm may also be referred to as 'fleshy' or 'cartilaginous' with thicker soft cells such as coconut, but may also be oily as in *Ricinus* (castor oil), *Croton* and Poppy. The endosperm is called 'horny' when the cell walls are thicker such as date and coffee, or 'ruminated' if mottled, as in nutmeg, palms and Annonaceae.

In most monocotyledons (such as grasses and palms) and some (endospermic or albuminous) dicotyledons (such as castor beans) the embryo is embedded in the endosperm (and nucellus, which the seedling will use upon germination. In the non-endospermic dicotyledons the endosperm is absorbed by the embryo as the latter grows within the developing seed, and the cotyledons of the embryo become filled with stored food. At maturity, seeds of these species have no endosperm and are also referred to as exalbuminous seeds. The exalbuminous seeds include the legumes (such as beans and peas), trees such as the oak and walnut, vegetables such as squash and radish, and sunflowers. According to Bewley and Black (1978), Brazil nut storage is in hypocotyl, this place of storage is uncommon among seeds. All gymnosperm seeds are albuminous.

Seed Coat

The seed coat develops from the maternal tissue, the integuments, originally surrounding the ovule. The seed coat in the mature seed can be a paper-thin layer (e.g. peanut) or something more substantial (e.g. thick and hard in honey locust and coconut), or fleshy as in the sarcotesta of pomegranate. The seed coat helps protect the embryo from mechanical injury, predators and drying out. Depending on its development, the seed coat is either bitegmic or unitegmic. Bitegmic seeds form a testa from the outer integument and a tegmen from the inner integument while unitegmic seeds have only one integument. Usually parts of the testa or tegmen form a hard protective mechanical layer. The mechanical layer may prevent water penetration and germination. Amongst the barriers may be the presence of lignified sclereids.

The outer integument has a number of layers, generally between four and eight organised into three layers: (a) outer epidermis, (b) outer pigmented zone of two to five layers containing tannin and starch, and (c) inner epidermis. The endotegmen is derived from the inner epidermis of the inner integument, the exotegmen from the outer surface of the inner integument. The endotesta is derived from the inner epidermis of the outer integument, and the outer layer of the testa from the outer surface of the outer integument is referred to as the exotesta. If the exotesta is also the mechanical layer, this is called an exotestal seed, but if the mechanical layer is the endotegmen, then the seed is endotestal. The exotesta may consist of one or more rows of cells that are elongated and pallisade like (e.g. Fabaceae), hence 'palisade exotesta'.

In addition to the three basic seed parts, some seeds have an appendage, an aril, a fleshy outgrowth of the funicle (funiculus), (as in yew and nutmeg) or an oily appendage, an elaiosome (as in *Corydalis*), or hairs (trichomes). In the latter example these hairs are the source of the textile crop cotton. Other seed appendages include the raphe (a ridge), wings, caruncles (a soft spongy outgrowth from the outer integument in the vicinity of the micropyle), spines, or tubercles.

A scar also may remain on the seed coat, called the hilum, where the seed was attached to the ovary wall by the funicle. Just below it is a small pore, representing the micropyle of the ovule.

Size and Seed Set

A collection of various vegetable and herb seeds

Seeds are very diverse in size. The dust-like orchid seeds are the smallest, with about one million seeds per gram; they are often embryonic seeds with immature embryos and no significant energy reserves. Orchids and a few other groups of plants are mycoheterotrophs which depend on mycorrhizal fungi for nutrition during germination and the early growth of the seedling. Some terrestrial orchid seedlings, in fact, spend the first few years of their lives deriving energy from the fungi and do not produce green leaves. At over 20 kg, the largest seed is the *coco de mer*. Plants that produce smaller seeds can generate many more seeds per flower, while plants with larger seeds invest more resources into those seeds and normally produce fewer seeds. Small seeds are quicker to ripen and can be dispersed sooner, so fall blooming plants often have small seeds. Many annual plants produce great quantities of smaller seeds; this helps to ensure at least a few will end in a favorable place for growth. Herbaceous perennials and woody plants often have larger seeds; they can produce seeds over many years, and larger seeds have more energy reserves for germination and seedling growth and produce larger, more established seedlings after germination.

Functions

Seeds serve several functions for the plants that produce them. Key among these functions are nourishment of the embryo, dispersal to a new location, and dormancy during unfavorable conditions. Seeds fundamentally are means of reproduction, and most seeds are the product of sexual reproduction which produces a remixing of genetic material and phenotype variability on which natural selection acts.

Embryo Nourishment

Seeds protect and nourish the embryo or young plant. They usually give a seedling a faster start than a sporeling from a spore, because of the larger food reserves in the seed and the multicellularity of the enclosed embryo.

Dispersal

Unlike animals, plants are limited in their ability to seek out favorable conditions for life and growth. As a result, plants have evolved many ways to disperse their offspring by dispersing their seeds. A seed must somehow "arrive" at a location and be there at a time favorable for germination and growth. When the fruits open and release their seeds in a regular way, it is called dehiscent, which is often distinctive for related groups of plants; these fruits include capsules, follicles, legumes, silicles and siliques. When fruits do not open and re-lease their seeds in a regular fashion, they are called indehiscent, which include the fruits achenes, caryopsis, nuts, samaras, and utricles.

By Wind (Anemochory)

Dandelion seeds are contained within achenes, which can be carried long distances by the wind.

The seed pod of milkweed (*Asclepias syriaca*)

- Some seeds (e.g., pine) have a wing that aids in wind dispersal.

- The dustlike seeds of orchids are carried efficiently by the wind.
- Some seeds (e.g. milkweed, poplar) have hairs that aid in wind dispersal.

Other seeds are enclosed in fruit structures that aid wind dispersal in similar ways:

- Dandelion achenes have hairs.
- Maple samaras have two wings.

By Water (Hydrochory)

- Some plants, such as *Mucuna* and *Dioclea*, produce buoyant seeds termed sea-beans or drift seeds because they float in rivers to the oceans and wash up on beaches.

By Animals (Zoochory)

- Seeds (burrs) with barbs or hooks (e.g. acaena, burdock, dock) which attach to animal fur or feathers, and then drop off later.
- Seeds with a fleshy covering (e.g. apple, cherry, juniper) are eaten by animals (birds, mammals, reptiles, fish) which then disperse these seeds in their droppings.
- Seeds (nuts) are attractive long-term storable food resources for animals (e.g. acorns, hazelnut, walnut); the seeds are stored some distance from the parent plant, and some escape being eaten if the animal forgets them.

Myrmecochory is the dispersal of seeds by ants. Foraging ants disperse seeds which have appendages called elaiosomes (e.g. bloodroot, trilliums, acacias, and many species of Proteaceae). Elaiosomes are soft, fleshy structures that contain nutrients for animals that eat them. The ants carry such seeds back to their nest, where the elaiosomes are eaten. The remainder of the seed, which is hard and inedible to the ants, then germinates either within the nest or at a removal site where the seed has been discarded by the ants. This dispersal relationship is an example of mutualism, since the plants depend upon the ants to disperse seeds, while the ants depend upon the plants seeds for food. As a result, a drop in numbers of one partner can reduce success of the other. In South Africa, the Argentine ant (*Linepithema humile*) has invaded and displaced native species of ants. Unlike the native ant species, Argentine ants do not collect the seeds of *Mimetes cucullatus* or eat the elaiosomes. In areas where these ants have invaded, the numbers of *Mimetes* seedlings have dropped.

Dormancy

Seed dormancy has two main functions: the first is synchronizing germination with the optimal conditions for survival of the resulting seedling; the second is spreading germination of a batch of seeds over time so a catastrophe (e.g. late frosts, drought, herbivory) does not result in the death of all offspring of a plant (bet-hedging). Seed dormancy is defined as a seed failing to germinate under environmental conditions optimal for germination, normally when the environment is at a suitable temperature with proper soil moisture. This true dormancy or innate dormancy is therefore caused by conditions within the seed that prevent germination. Thus dormancy is a state of the seed, not of the environment. Induced dormancy, enforced dormancy or seed quiescence occurs when a seed fails to germinate because the external environmental conditions are inappropriate for germination, mostly in response to conditions being too dark or light, too cold or hot, or too dry.

Seed dormancy is not the same as seed persistence in the soil or on the plant, though even in scientific publications dormancy and persistence are often confused or used as synonyms.

Often, seed dormancy is divided into four major categories: exogenous; endogenous; combinational; and secondary. A more recent system distinguishes five classes: morphological, physiological, morphophysiological, physical and combinational dormancy.

Exogenous dormancy is caused by conditions outside the embryo, including:

- Physical dormancy or hard seed coats occurs when seeds are impermeable to water. At dormancy break, a specialized structure, the 'water gap', is disrupted in response to environmental cues, especially temperature, so water can enter the seed and germination can occur. Plant families where physical dormancy occurs include Anacardiaceae, Cannaceae, Convulvulaceae, Fabaceae and Malvaceae.
- Chemical dormancy considers species that lack physiological dormancy, but where a chemical prevents germination. This chemical can be leached out of the seed by rainwater or snow melt or be deactivated somehow. Leaching of chemical inhibitors from the seed by rain water is often cited as an important cause of dormancy release in seeds of desert plants, but little evidence exists to support this claim.

Endogenous dormancy is caused by conditions within the embryo itself, including:

- In morphological dormancy, germination is prevented due to morphological characteristics of the embryo. In some species, the embryo is just a mass of cells when seeds are dispersed; it is not differentiated. Before germination can take place, both differentiation and growth of the embryo have to occur. In other species, the embryo is differentiated but not fully grown (underdeveloped) at dispersal, and embryo growth up to a species specific length is required before germination can occur. Examples of plant families where morphological dormancy occurs are Apiaceae, Cycadaceae, Liliaceae, Magnoliaceae and Ranunculaceae.
- Morphophysiological dormancy includes seeds with underdeveloped embryos, and also have physiological components to dormancy. These seeds, therefore, require a dormancy-breaking treatments, as well as a period of time to develop fully grown embryos. Plant families where morphophysiological dormancy occurs include Apiaceae, Aquifoliaceae, Liliaceae, Magnoliaceae, Papaveraceae and Ranunculaceae. Some plants with morphophysiological dormancy, such as *Asarum* or *Trillium* species, have multiple types of dormancy, one affects radicle (root) growth, while the other affects plumule (shoot) growth. The terms "double dormancy" and "two-year seeds" are used for species whose seeds need two years to complete germination or at least two winters and one summer. Dormancy of the radicle (seedling root)is broken during the first winter after dispersal while dormancy of the shoot bud is broken during the second winter.
- Physiological dormancy means the embryo, due to physiological causes, cannot generate enough power to break through the seed coat, endosperm or other covering structures.

Dormancy is typically broken at cool wet, warm wet or warm dry conditions. Abscisic acid is usually the growth inhibitor in seeds, and its production can be affected by light.

- o Drying, in some plants, including a number of grasses and those from seasonally arid regions, is needed before they will germinate. The seeds are released, but need to have a lower moisture content before germination can begin. If the seeds remain moist after dispersal, germination can be delayed for many months or even years. Many herbaceous plants from temperate climate zones have physiological dormancy that disappears with drying of the seeds. Other species will germinate after dispersal only under very narrow temperature ranges, but as the seeds dry, they are able to germinate over a wider temperature range.

- In seeds with combinational dormancy, the seed or fruit coat is impermeable to water and the embryo has physiological dormancy. Depending on the species, physical dormancy can be broken before or after physiological dormancy is broken.

- Secondary dormancy* is caused by conditions after the seed has been dispersed and occurs in some seeds when nondormant seed is exposed to conditions that are not favorable to germination, very often high temperatures. The mechanisms of secondary dormancy are not yet fully understood, but might involve the loss of sensitivity in receptors in the plasma membrane.

The following types of seed dormancy do not involve seed dormancy, strictly speaking, as lack of germination is prevented by the environment, not by characteristics of the seed itself (see Germination):

- Photodormancy or light sensitivity affects germination of some seeds. These photoblastic seeds need a period of darkness or light to germinate. In species with thin seed coats, light may be able to penetrate into the dormant embryo. The presence of light or the absence of light may trigger the germination process, inhibiting germination in some seeds buried too deeply or in others not buried in the soil.

- Thermodormancy is seed sensitivity to heat or cold. Some seeds, including cocklebur and amaranth, germinate only at high temperatures (30 °C or 86 °F); many plants that have seeds that germinate in early to midsummer have thermodormancy, so germinate only when the soil temperature is warm. Other seeds need cool soils to germinate, while others, such as celery, are inhibited when soil temperatures are too warm. Often, thermodormancy requirements disappear as the seed ages or dries.

Not all seeds undergo a period of dormancy. Seeds of some mangroves are viviparous; they begin to germinate while still attached to the parent. The large, heavy root allows the seed to penetrate into the ground when it falls. Many garden plant seeds will germinate readily as soon as they have water and are warm enough; though their wild ancestors may have had dormancy, these cultivated plants lack it. After many generations of selective pressure by plant breeders and gardeners, dormancy has been selected out.

For annuals, seeds are a way for the species to survive dry or cold seasons. Ephemeral plants are usually annuals that can go from seed to seed in as few as six weeks.

Persistence and Seed Banks

Germination

Germinating sunflower seedlings

Seed germination is a process by which a seed embryo develops into a seedling. It involves the reactivation of the metabolic pathways that lead to growth and the emergence of the radicle or seed root and plumule or shoot. The emergence of the seedling above the soil surface is the next phase of the plant's growth and is called seedling establishment.

Three fundamental conditions must exist before germination can occur. (1) The embryo must be alive, called seed viability. (2) Any dormancy requirements that prevent germination must be overcome. (3) The proper environmental conditions must exist for germination.

Seed viability is the ability of the embryo to germinate and is affected by a number of different conditions. Some plants do not produce seeds that have functional complete embryos, or the seed may have no embryo at all, often called empty seeds. Predators and pathogens can damage or kill the seed while it is still in the fruit or after it is dispersed. Environmental conditions like flooding or heat can kill the seed before or during germination. The age of the seed affects its health and germination ability: since the seed has a living embryo, over time cells die and cannot be replaced. Some seeds can live for a long time before germination, while others can only survive for a short period after dispersal before they die.

Seed vigor is a measure of the quality of seed, and involves the viability of the seed, the germination percentage, germination rate and the strength of the seedlings produced.

The germination percentage is simply the proportion of seeds that germinate from all seeds subject to the right conditions for growth. The germination rate is the length of time it takes for the seeds to germinate. Germination percentages and rates are affected by seed viability, dormancy and environmental effects that impact on the seed and seedling. In agriculture and horticulture quality seeds have high viability, measured by germination percentage plus the rate of germination. This is given as a percent of germination over a certain amount of time, 90% germination in 20 days, for example. 'Dormancy' is covered above; many plants produce seeds with varying degrees of dormancy, and different seeds from the same fruit can have different degrees of dormancy. It's possible to have seeds with no dormancy if they are dispersed right away and do not dry (if the seeds dry they go into physiological dormancy). There is great variation amongst plants and a dormant seed is still a viable seed even though the germination rate might be very low.

Environmental conditions affecting seed germination include; water, oxygen, temperature and light.

Three distinct phases of seed germination occur: water imbibition; lag phase; and radicle emergence.

In order for the seed coat to split, the embryo must imbibe (soak up water), which causes it to swell, splitting the seed coat. However, the nature of the seed coat determines how rapidly water can penetrate and subsequently initiate germination. The rate of imbibition is dependent on the permeability of the seed coat, amount of water in the environment and the area of contact the seed has to the source of water. For some seeds, imbibing too much water too quickly can kill the seed. For some seeds, once water is imbibed the germination process cannot be stopped, and drying then becomes fatal. Other seeds can imbibe and lose water a few times without causing ill effects, but drying can cause secondary dormancy.

Repair of DNA Damage

During seed dormancy, often associated with unpredictable and stressful environments, DNA damages accumulate as the seeds age. In rye seeds, the reduction of DNA integrity due to damage is associated with loss of seed viability during storage. Upon germination, seeds of *Vicia faba* undergo DNA repair. A plant DNA ligase that is involved in repair of single- and double-strand breaks during seed germination is an important determinant of seed longevity. Also, in Arabidopsis seeds, the activities of the DNA repair enzymes Poly ADP ribose polymerases (PARP) are likely needed for successful germination. Thus DNA damages that accumulate during dormancy appear to be a problem for seed survival, and the enzymatic repair of DNA damages during germination appears to be important for seed viability.

Inducing Germination

A number of different strategies are used by gardeners and horticulturists to break seed dormancy.

Scarification allows water and gases to penetrate into the seed; it includes methods to physically break the hard seed coats or soften them by chemicals, such as soaking in hot water or poking holes in the seed with a pin or rubbing them on sandpaper or cracking with a press or hammer. Sometimes fruits are harvested while the seeds are still immature and the seed coat is not fully developed and sown right away before the seed coat become impermeable. Under natural conditions, seed coats are worn down by rodents chewing on the seed, the seeds rubbing against rocks (seeds are moved by the wind or water currents), by undergoing freezing and thawing of surface water, or passing through an animal's digestive tract. In the latter case, the seed coat protects the seed from digestion, while often weakening the seed coat such that the embryo is ready to sprout when it is deposited, along with a bit of fecal matter that acts as fertilizer, far from the parent plant. Microorganisms are often effective in breaking down hard seed coats and are sometimes used by people as a treatment; the seeds are stored in a moist warm sandy medium for several months under nonsterile conditions.

Stratification, also called moist-chilling, breaks down physiological dormancy, and involves the addition of moisture to the seeds so they absorb water, and they are then subjected to a period of moist chilling to after-ripen the embryo. Sowing in late summer and fall and allowing to overwinter under cool conditions is an effective way to stratify seeds; some seeds respond more favorably to periods of oscillating temperatures which are a part of the natural environment.

Leaching or the soaking in water removes chemical inhibitors in some seeds that prevent germination. Rain and melting snow naturally accomplish this task. For seeds planted in gardens, running water is best—if soaked in a container, 12 to 24 hours of soaking is sufficient. Soaking longer, especially in stagnant water, can result in oxygen starvation and seed death. Seeds with hard seed coats can be soaked in hot water to break open the impermeable cell layers that prevent water intake.

Other methods used to assist in the germination of seeds that have dormancy include prechilling, predrying, daily alternation of temperature, light exposure, potassium nitrate, the use of plant growth regulators, such as gibberellins, cytokinins, ethylene, thiourea, sodium hypochlorite, and others. Some seeds germinate best after a fire. For some seeds, fire cracks hard seed coats, while in others, chemical dormancy is broken in reaction to the presence of smoke. Liquid smoke is often used by gardeners to assist in the germination of these species.

Sterile Seeds

Seeds may be sterile for few reasons: they may have been irradiated, unpollinated, cells lived past expectancy, or bred for the purpose.

Evolution and Origin of Seeds

The origin of seed plants is a problem that still remains unsolved. However, more and more data tends to place this origin in the middle Devonian. The description in 2004 of the proto-seed *Runcaria heinzelinii* in the Givetian of Belgium is an indication of that ancient origin of seed-plants. As with modern ferns, most land plants before this time reproduced by sending spores into the air, that would land and become whole new plants.

The first "true" seeds are described from the upper Devonian, which is probably the theater of their true first evolutionary radiation. The seed plants progressively became one of the major elements of nearly all ecosystems.

Economic Importance

Phaseolus vulgaris seeds are diverse in size, shape, and color.

Edible Seeds

Many seeds are edible and the majority of human calories comes from seeds, especially from cereals, legumes and nuts. Seeds also provide most cooking oils, many beverages and spices and some important food additives. In different seeds the seed embryo or the endosperm dominates and provides most of the nutrients. The storage proteins of the embryo and endosperm differ in their amino acid content and physical properties. For example, the gluten of wheat, important in providing the elastic property to bread dough is strictly an endosperm protein.

Seeds are used to propagate many crops such as cereals, legumes, forest trees, turfgrasses and pasture grasses. Particularly in developing countries, a major constraint faced is the inadequacy of the marketing channels to get the seed to poor farmers. Thus the use of farmer-retained seed remains quite common.

Seeds are also eaten by animals, and are fed to livestock. Many seeds are used as birdseed.

Poison and Food Safety

While some seeds are edible, others are harmful, poisonous or deadly. Plants and seeds often contain chemical compounds to discourage herbivores and seed predators. In some cases, these compounds simply taste bad (such as in mustard), but other compounds are toxic or break down into toxic compounds within the digestive system. Children, being smaller than adults, are more susceptible to poisoning by plants and seeds.

A deadly poison, ricin, comes from seeds of the castor bean. Reported lethal doses are anywhere from two to eight seeds, though only a few deaths have been reported when castor beans have been ingested by animals.

In addition, seeds containing amygdalin—apple, apricot, bitter almond, peach, plum, cherry, quince, and others—when consumed in sufficient amounts, may cause cyanide poisoning. Other seeds that contain poisons include annona, cotton, custard apple, datura, uncooked durian, golden chain, horse-chestnut, larkspur, locoweed, lychee, nectarine, rambutan, rosary pea, sour sop, sugar apple, wisteria, and yew. The seeds of the strychnine tree are also poisonous, containing the poison strychnine.

The seeds of many legumes, including the common bean (*Phaseolus vulgaris*), contain proteins called lectins which can cause gastric distress if the beans are eaten without cooking. The common bean and many others, including the soybean, also contain trypsin inhibitors which interfere with the action of the digestive enzyme trypsin. Normal cooking processes degrade lectins and trypsin inhibitors to harmless forms.

Other uses

Cotton fiber grows attached to cotton plant seeds. Other seed fibers are from kapok and milkweed.

Many important nonfood oils are extracted from seeds. Linseed oil is used in paints. Oil from jojoba and crambe are similar to whale oil.

Seeds are the source of some medicines including castor oil, tea tree oil and the cancer drug, Laetrile.

Many seeds have been used as beads in necklaces and rosaries including Job's tears, Chinaberry, rosary pea, and castor bean. However, the latter three are also poisonous.

Other seed uses include:

- Seeds once used as weights for balances.
- Seeds used as toys by children, such as for the game Conkers.
- Resin from *Clusia rosea* seeds used to caulk boats.
- Nematicide from milkweed seeds.
- Cottonseed meal used as animal feed and fertilizer.

Seed Records

The massive fruit of the coco de mer

- The oldest viable carbon-14-dated seed that has grown into a plant was a Judean date palm seed about 2,000 years old, recovered from excavations at Herod the Great's palace on Masada in Israel. It was germinated in 2005. (A reported regeneration of *Silene stenophylla* (narrow-leafed campion) from material preserved for 31,800 years in the Siberian permafrost was achieved using fruit tissue, not seed.)
- The largest seed is produced by the coco de mer, or "double coconut palm", *Lodoicea maldivica*. The entire fruit may weigh up to 23 kilograms (50 pounds) and usually contains a single seed.
- The smallest seeds are produced by epiphytic orchids. They are only 85 micrometers long, and weigh 0.81 micrograms. They have no endosperm and contain underdeveloped embryos.

- The earliest fossil seeds are around 365 million years old from the Late Devonian of West Virginia. The seeds are preserved immature ovules of the plant *Elkinsia polymorpha*.

In Religion

The Book of Genesis in the Old Testament begins with an explanation of how all plant forms began:

And God said, Let the earth bring forth grass, the herb yielding seed, and the fruit tree yielding fruit after his kind, whose seed is in itself, upon the earth: and it was so. And the earth brought forth grass, and herb yielding seed after its kind, and the tree yielding fruit, whose seed was in itself, after its kind: and God saw that it was good. And the evening and the morning were the third day.

The Quran speaks about seed germination:

It is Allah Who causeth the seed-grain and the date-stone to split and sprout. He causeth the living to issue from the dead, and He is the one to cause the dead to issue from the living. That is Allah: then how are ye deluded away from the truth?

Bud

European beech (*Fagus sylvatica*) bud

In botany, a bud is an undeveloped or embryonic shoot and normally occurs in the axil of a leaf or at the tip of a stem. Once formed, a bud may remain for some time in a dormant condition, or it may form a shoot immediately. Buds may be specialized to develop flowers or short shoots, or may have the potential for general shoot development. The term bud is also used in zoology, where it refers to an outgrowth from the body which can develop into a new individual.

Overview

The buds of many woody plants, especially in temperate or cold climates, are protected by a covering of modified leaves called *scales* which tightly enclose the more delicate parts of the bud. Many bud scales are covered by a gummy substance which serves as added protection. When the bud develops, the scales may enlarge somewhat but usually just drop off, leaving on the surface of the growing stem a series of horizontally-elongated scars. By means of these scars one can determine the age of any young branch, since each year's growth ends in the formation of a bud, the formation of which produces an additional group of bud scale scars. Continued growth of the branch causes these scars to be obliterated after a few years so that the total age of older branches cannot be determined by this means.

Inflorescence bud scales in *Halesia carolina*

In many plants scales do not form over the bud, and the bud is then called a naked bud. The minute underdeveloped leaves in such buds are often excessively hairy. Naked buds are found in some shrubs, like some species of the Sumac and Viburnums (*Viburnum alnifolium* and *V. lantana*) and in herbaceous plants. In many of the latter, buds are even more reduced, often consisting of undifferentiated masses of cells in the axils of leaves. A terminal bud occurs on the end of a stem and lateral buds are found on the side. A head of cabbage (see Brassica) is an exceptionally large terminal bud, while Brussels sprouts are large lateral buds.

Since buds are formed in the axils of leaves, their distribution on the stem is the same as that of leaves. There are alternate, opposite, and whorled buds, as well as the terminal bud at the tip of the stem. In many plants buds appear in unexpected places: these are known as adventitious buds.

Often it is possible to find a bud in a remarkable series of gradations of bud scales. In the buckeye, for example, one may see a complete gradation from the small brown outer scale through larger scales which on unfolding become somewhat green to the inner scales of the bud, which are remarkably leaf-like. Such a series suggests that the scales of the bud are in truth leaves, modified to protect the more delicate parts of the plant during unfavorable periods.

Types of Buds

Buds are often useful in the identification of plants, especially for woody plants in winter when leaves have fallen. Buds may be classified and described according to different criteria: location, status, morphology, and function.

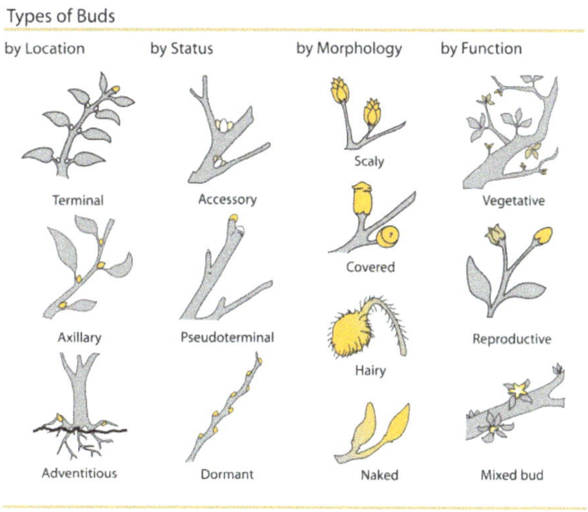

Plant buds classification

Terminal, vegetative bud of *Ficus carica*

Botanists commonly use the following terms:

- for location:
    - terminal, when located at the tip of a stem (apical is equivalent but rather reserved for the one at the top of the plant);
    - axillary, when located in the axil of a leaf (lateral is the equivalent but some adventitious buds may be lateral too);
    - adventitious, when occurring elsewhere, for example on trunk or on roots (some adventitious buds may be former axillary ones reduced and hidden under the bark, other adventitious buds are completely new formed ones).
- for status:
    - accessory, for secondary buds formed besides a principal bud (axillary or terminal);
    - resting, for buds that form at the end of a growth season, which will lie dormant until onset of the next growth season;

- dormant or latent, for buds whose growth has been delayed for a rather long time. The term is usable as a synonym of *resting*, but is better employed for buds waiting undeveloped for years, for example epicormic buds;
- pseudoterminal, for an axillary bud taking over the function of a terminal bud (characteristic of species whose growth is sympodial: terminal bud dies and is replaced by the closer axillary bud, for examples beech, persimmon, *Platanus* have sympodial growth).

- for morphology:
  - scaly or covered (perulate), when scales, also referred to as a perule (lat. perula, perulaei) (which are in fact transformed and reduced leaves) cover and protect the embryonic parts;
  - naked, when not covered by scales;
  - hairy, when also protected by hairs (it may apply either to scaly or to naked buds).
- for function:
  - vegetative, if only containing vegetative pieces: embryonic shoot with leaves (a leaf bud is the same);
  - reproductive, if containing embryonic flower(s) (a flower bud is the same);
  - mixed, if containing both embryonic leaves and flowers.

Inflorescence bud of a sunflower

Within Zoology

The term bud (as in budding) is used by analogy within zoology as well, where it refers to an outgrowth from the body which develops into a new individual. It is a form of asexual reproduction limited to animals or plants of relatively simple structure. In this process a portion of the wall of the parent cell softens and pushes out. The protuberance thus formed enlarges rapidly while at this time the nucleus of the parent cell divides (see: mitosis, meiosis). One of the resulting nuclei passes into the bud, and then the bud is cut off from its parent cell and the process is repeated. Often the daughter cell will begin to bud before it becomes separated from the parent, so that whole colonies of adhering cells may be formed. Eventually cross walls cut off the bud from the original cell.

Flower

A flower, sometimes known as a bloom or blossom, is the reproductive structure found in plants that are floral (plants of the division Magnoliophyta, also called angiosperms). The biological function of a flower is to effect reproduction, usually by providing a mechanism for the union of sperm with eggs. Flowers may facilitate outcrossing (fusion of sperm and eggs from different individuals in a population) or allow selfing (fusion of sperm and egg from the same flower). Some flowers produce diaspores without fertilization (parthenocarpy). Flowers contain sporangia and are the site where gametophytes develop. Many flowers have evolved to be attractive to animals, so as to cause them to be vectors for the transfer of pollen. After fertilization, the ovary of the flower develops into fruit containing seeds.

A poster with flowers or clusters of flowers produced by twelve species of flowering plants from different families

Various flowers from different families.

In addition to facilitating the reproduction of flowering plants, flowers have long been admired and used by humans to beautify their environment, and also as objects of romance, ritual, religion, medicine and as a source of food.

Morphology

Main parts of a mature flower (Ranunculus glaberrimus)

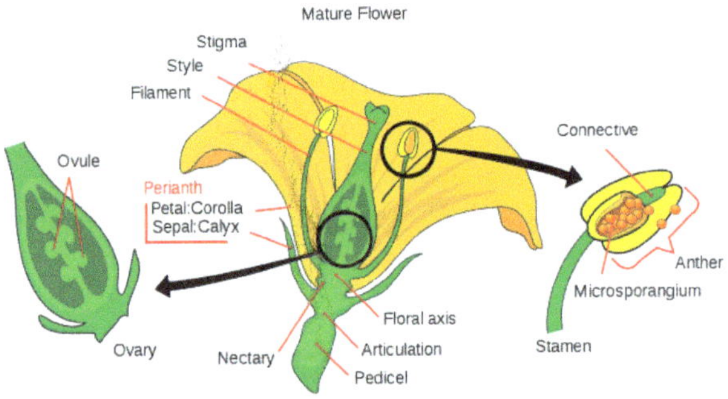

Diagram of flower parts

Floral Parts

The essential parts of a flower can be considered in two parts: the vegetative part, consisting of petals and associated structures in the perianth, and the reproductive or sexual parts. A stereotypical flower consists of four kinds of structures attached to the tip of a short stalk. Each of these kinds of parts is arranged in a whorl on the receptacle. The four main whorls (starting from the base of the flower or lowest node and working upwards) are as follows:

Vegetative (Perianth)

Collectively the calyx and corolla form the perianth (see diagram).

- *Calyx*: the outermost whorl consisting of units called *sepals*; these are typically green and enclose the rest of the flower in the bud stage, however, they can be absent or prominent and petal-like in some species.
- *Corolla*: the next whorl toward the apex, composed of units called *petals*, which are typically thin, soft and colored to attract animals that help the process of pollination.

Reproductive

Reproductive parts of Easter Lily (*Lilium longiflorum*). 1. Stigma, 2. Style, 3. Stamens, 4. Filament, 5. Petal

- *Androecium* (from Greek *andros oikia*: man's house): the next whorl (sometimes multiplied into several whorls), consisting of units called stamens. Stamens consist of two parts: a stalk called a filament, topped by an anther where pollen is produced by meiosis and eventually dispersed.

- *Gynoecium* (from Greek *gynaikos oikia*: woman's house): the innermost whorl of a flower, consisting of one or more units called carpels. The carpel or multiple fused carpels form a hollow structure called an ovary, which produces ovules internally. Ovules are megasporangia and they in turn produce megaspores by meiosis which develop into female gametophytes. These give rise to egg cells. The gynoecium of a flower is also described using an alternative terminology wherein the structure one sees in the innermost whorl (consisting of an ovary, style and stigma) is called a pistil. A pistil may consist of a single carpel or a number of carpels fused together. The sticky tip of the pistil, the stigma, is the receptor of pollen. The supportive stalk, the style, becomes the pathway for pollen tubes to grow from pollen grains adhering to the stigma. The relationship to the gynoecium on the receptacle is described as hypogynous (beneath a superior ovary), perigynous (surrounding a superior ovary), or epigynous (above inferior ovary).

Structure

Although the arrangement described above is considered "typical", plant species show a wide variation in floral structure. These modifications have significance in the evolution of flowering plants and are used extensively by botanists to establish relationships among plant species.

The four main parts of a flower are generally defined by their positions on the receptacle and not by their function. Many flowers lack some parts or parts may be modified into other functions and/or look like what is typically another part. In some families, like Ranunculaceae, the petals are greatly reduced and in many species the sepals are colorful and petal-like. Other flowers have modified stamens that are petal-like; the double flowers of Peonies and Roses are mostly petaloid stamens. Flowers show great variation and plant scientists describe this variation in a systematic way to identify and distinguish species.

Specific terminology is used to describe flowers and their parts. Many flower parts are fused together; fused parts originating from the same whorl are connate, while fused parts originating from different whorls are adnate; parts that are not fused are free. When petals are fused into a tube or ring that falls away as a single unit, they are sympetalous (also called gamopetalous). Connate petals may have distinctive regions: the cylindrical base is the tube, the expanding region is the throat and the flaring outer region is the limb. A sympetalous flower, with bilateral symmetry with an upper and lower lip, is bilabiate. Flowers with connate petals or sepals may have various shaped corolla or calyx, including campanulate, funnelform, tubular, urceolate, salverform or rotate.

Referring to "fusion," as it is commonly done, appears questionable because at least some of the processes involved may be non-fusion processes. For example, the addition of intercalary growth at or below the base of the primordia of floral appendages such as sepals, petals, stamens and carpels may lead to a common base that is not the result of fusion.

Left: A normal zygomorphic Streptocarpus flower. Right: An aberrant peloric Streptocarpus flower. Both of these flowers appeared on the Streptocarpus hybrid 'Anderson's Crows' Wings'.

Many flowers have a symmetry. When the perianth is bisected through the central axis from any point, symmetrical halves are produced, forming a radial symmetry. These flowers are also known to be actinomorphic or regular, e.g. rose or trillium. When flowers are bisected and produce only one line that produces symmetrical halves the flower is said to be irregular or zygomorphic, e.g. snapdragon or most orchids.

Flowers may be directly attached to the plant at their base (sessile—the supporting stalk or stem is highly reduced or absent). The stem or stalk subtending a flower is called a peduncle. If a peduncle supports more than one flower, the stems connecting each flower to the main axis are called pedicels. The apex of a flowering stem forms a terminal swelling which is called the *torus* or receptacle.

Inflorescence

In those species that have more than one flower on an axis, the collective cluster of flowers is termed an *inflorescence*. Some inflorescences are composed of many small flowers arranged in a formation that resembles a single flower. The common example of this is most members of the very large composite (Asteraceae) group. A single daisy or sunflower, for example, is not a flower but a flower *head*—an inflorescence composed of numerous flowers (or florets). An inflorescence may include specialized stems and modified leaves known as bracts.

The familiar calla lily is not a single flower. It is actually an inflorescence of tiny flowers pressed together on a central stalk that is surrounded by a large petal-like bract.

Floral Diagrams and Floral Formulae

A *floral formula* is a way to represent the structure of a flower using specific letters, numbers and symbols, presenting substantial information about the flower in a compact form. It can represent a taxon, usually giving ranges of the numbers of different organs, or particular species. Floral formulae have been developed in the early 19th century and their use has declined since. Prenner *et al.* (2010) devised an extension of the existing model to broaden the descriptive capability of the formula. The format of floral formulae differs in different parts of the world, yet they convey the same information.

The structure of a flower can also be expressed by the means of *floral diagrams*. The use of schematic diagrams can replace long descriptions or complicated drawings as a tool for understanding both floral structure and evolution. Such diagrams may show important features of flowers, including the relative positions of the various organs, including the presence of fusion and symmetry, as well as structural details.

Development

A flower develops on a modified shoot or *axis* from a determinate apical meristem (*determinate* meaning the axis grows to a set size). It has compressed internodes, bearing structures that in classical plant morphology are interpreted as highly modified leaves. Detailed developmental studies, however, have shown that stamens are often initiated more or less like modified stems (caulomes) that in some cases may even resemble branchlets. Taking into account the whole diversity in the development of the androecium of flowering plants, we find a continuum between modified leaves (phyllomes), modified stems (caulomes), and modified branchlets (shoots).

Flowering Transition

The transition to flowering is one of the major phase changes that a plant makes during its life cycle. The transition must take place at a time that is favorable for fertilization and the formation of seeds, hence ensuring maximal reproductive success. To meet these needs a plant is able to interpret important endogenous and environmental cues such as changes in levels of plant hormones and seasonable temperature and photoperiod changes. Many perennial and most biennial plants require vernalization to flower. The molecular interpretation of these signals is through the transmission of a complex signal known as florigen, which involves a variety of genes, including CONSTANS, FLOWERING LOCUS C and FLOWERING LOCUS T. Florigen is produced in the leaves in reproductively favorable conditions and acts in buds and growing tips to induce a number of different physiological and morphological changes.

The first step of the transition is the transformation of the vegetative stem primordia into floral primordia. This occurs as biochemical changes take place to change cellular differentiation of leaf, bud and stem tissues into tissue that will grow into the reproductive organs. Growth of the central part of the stem tip stops or flattens out and the sides develop protuberances in a whorled or spiral fashion around the outside of the stem end. These protuberances develop into the sepals, petals, stamens, and carpels. Once this process begins, in most plants, it cannot be reversed and the stems develop flowers, even if the initial start of the flower formation event was dependent of some environmental cue. Once the process begins, even if that cue is removed the stem will continue to develop a flower.

Organ Development

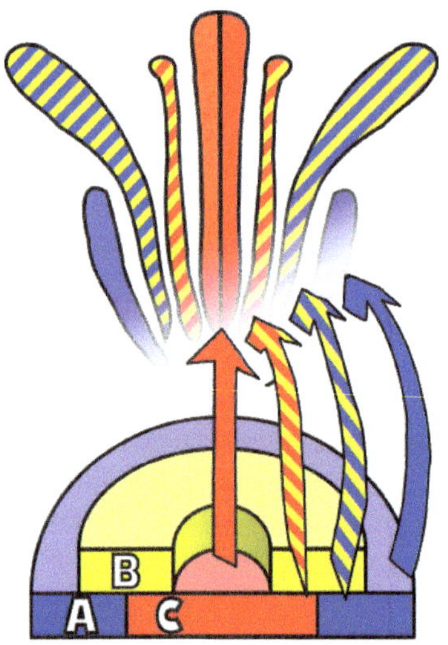

The ABC model of flower development

The molecular control of floral organ identity determination appears to be fairly well understood in some species. In a simple model, three gene activities interact in a combinatorial manner to determine the developmental identities of the organ primordia within the floral meristem. These gene functions are called A, B and C-gene functions. In the first floral whorl only A-genes are expressed, leading to the formation of sepals. In the second whorl both A- and B-genes are expressed, leading to the formation of petals. In the third whorl, B and C genes interact to form stamens and in the center of the flower C-genes alone give rise to carpels. The model is based upon studies of homeotic mutants in *Arabidopsis thaliana* and snapdragon, *Antirrhinum majus*. For example, when there is a loss of B-gene function, mutant flowers are produced with sepals in the first whorl as usual, but also in the second whorl instead of the normal petal formation. In the third whorl the lack of B function but presence of C-function mimics the fourth whorl, leading to the formation of carpels also in the third whorl.

Most genes central in this model belong to the MADS-box genes and are transcription factors that regulate the expression of the genes specific for each floral organ.

Floral Function

An example of a "perfect flower", this *Crateva religiosa* flower has both stamens (outer ring) and a pistil (center).

The principal purpose of a flower is the reproduction of the individual and the species. All flowering plants are *heterosporous*, producing two types of spores. Microspores are produced by meiosis inside anthers while megaspores are produced inside ovules, inside an ovary. In fact, anthers typically consist of four microsporangia and an ovule is an integumented megasporangium. Both types of spores develop into gametophytes inside sporangia. As with all heterosporous plants, the gametophytes also develop inside the spores (are endosporic).

In the majority of species, individual flowers have both functional carpels and stamens. Botanists describe these flowers as being *perfect* or *bisexual* and the species as *hermaphroditic*. Some flowers lack one or the other reproductive organ and called *imperfect* or *unisexual*. If unisex flowers are found on the same individual plant but in different locations, the species is said to be *monoecious*. If each type of unisex flower is found only on separate individuals, the plant is *dioecious*.

Flower Specialization and Pollination

Flowering plants usually face selective pressure to optimize the transfer of their pollen, and this is typically reflected in the morphology of the flowers and the behaviour of the plants. Pollen may be transferred between plants via a number of 'vectors'. Some plants make use of abiotic vectors — namely wind (anemophily) or, much less commonly, water (hydrophily). Others use biotic vectors including insects (entomophily), birds (ornithophily), bats (chiropterophily) or other animals. Some plants make use of multiple vectors, but many are highly specialised.

Cleistogamous flowers are self-pollinated, after which they may or may not open. Many Viola and some Salvia species are known to have these types of flowers.

The flowers of plants that make use of biotic pollen vectors commonly have glands called nectaries that act as an incentive for animals to visit the flower. Some flowers have patterns, called nectar guides, that show pollinators where to look for nectar. Flowers also attract pollinators by scent and color. Still other flowers use mimicry to attract pollinators. Some species of orchids, for example, produce flowers resembling female bees in color, shape, and scent. Flowers are also specialized in shape and have an arrangement of the stamens that ensures that pollen grains are transferred to the bodies of the pollinator when it lands in search of its attractant (such as nectar, pollen, or a mate). In pursuing this attractant from many flowers of the same species, the pollinator transfers pollen to the stigmas—arranged with equally pointed precision—of all of the flowers it visits.

Anemophilous flowers use the wind to move pollen from one flower to the next. Examples include grasses, birch trees, ragweed and maples. They have no need to attract pollinators and therefore tend not to be "showy" flowers. Male and female reproductive organs are generally found in separate flowers, the male flowers having a number of long filaments terminating in exposed stamens, and the female flowers having long, feather-like stigmas. Whereas the pollen of animal-pollinated flowers tends to be large-grained, sticky, and rich in protein (another "reward" for pollinators), anemophilous flower pollen is usually small-grained, very light, and of little nutritional value to animals.

Pollination

Grains of pollen sticking to this bee will be transferred to the next flower it visits

The primary purpose of a flower is reproduction. Since the flowers are the reproductive organs of plant, they mediate the joining of the sperm, contained within pollen, to the ovules — contained in the ovary. Pollination is the movement of pollen from the anthers to the stigma. The joining of the sperm to the ovules is called fertilization. Normally pollen is moved from one plant to another, but many plants are able to self pollinate. The fertilized ovules produce seeds that are the next generation. Sexual reproduction produces genetically unique offspring, allowing for adaptation. Flowers have specific designs which encourages the transfer of pollen from one plant to another of the same species. Many plants are dependent upon external factors for pollination, including: wind and animals, and especially insects. Even large animals such as birds, bats, and pygmy possums can be employed. The period of time during which this process can take place (the flower is fully expanded and functional) is called *anthesis*. The study of pollination by insects is called *anthecology*.

Pollen

The types of pollen that most commonly cause allergic reactions are produced by the plain-looking plants (trees, grasses, and weeds) that do not have showy flowers. These plants make small, light, dry pollen grains that are custom-made for wind transport.

The type of allergens in the pollen is the main factor that determines whether the pollen is likely to cause hay fever. For example, pine tree pollen is produced in large amounts by a common tree, which would make it a good candidate for causing allergy. It is, however, a relatively rare cause of allergy because the types of allergens in pine pollen appear to make it less allergenic.

Among North American plants, weeds are the most prolific producers of allergenic pollen. Ragweed is the major culprit, but other important sources are sagebrush, redroot pigweed, lamb's quarters, Russian thistle (tumbleweed), and English plantain.

There is much confusion about the role of flowers in allergies. For example, the showy and entomophilous goldenrod (*Solidago*) is frequently blamed for respiratory allergies, of which it is innocent, since its pollen cannot be airborne. Instead the allergen is usually the pollen of the contemporary bloom of anemophilous ragweed (*Ambrosia*), which can drift for many kilometers.

Scientists have collected samples of ragweed pollen 400 miles out at sea and 2 miles high in the air. A single ragweed plant can generate a million grains of pollen per day.

It is common to hear people say they are allergic to colorful or scented flowers like roses. In fact, only florists, gardeners, and others who have prolonged, close contact with flowers are likely to be sensitive to pollen from these plants. Most people have little contact with the large, heavy, waxy pollen grains of such flowering plants because this type of pollen is not carried by wind but by insects such as butterflies and bees.

Attraction Methods

A Bee orchid has evolved over many generations to better mimic a female bee to attract male bees as pollinators.

Plants cannot move from one location to another, thus many flowers have evolved to attract animals to transfer pollen between individuals in dispersed populations. Flowers that are insect-pollinated are called *entomophilous*; literally "insect-loving" in Greek. They can be highly modified along with the pollinating insects by co-evolution. Flowers commonly have glands called *nectaries* on various parts that attract animals looking for nutritious nectar. Birds and bees have color vision, enabling them to seek out "colorful" flowers.

Some flowers have patterns, called nectar guides, that show pollinators where to look for nectar; they may be visible only under ultraviolet light, which is visible to bees and some other insects. Flowers also attract pollinators by scent and some of those scents are pleasant to our sense of smell. Not all flower scents are appealing to humans; a number of flowers are pollinated by insects that are attracted to rotten flesh and have flowers that smell like dead animals, often called Carrion flowers, including *Rafflesia*, the titan arum, and the North American pawpaw (*Asimina triloba*). Flowers pollinated by night visitors, including bats and moths, are likely to concentrate on scent to attract pollinators and most such flowers are white.

Other flowers use mimicry to attract pollinators. Some species of orchids, for example, produce flowers resembling female bees in color, shape, and scent. Male bees move from one such flower to another in search of a mate.

Pollination Mechanism

The pollination mechanism employed by a plant depends on what method of pollination is utilized.

Most flowers can be divided between two broad groups of pollination methods:

Entomophilous: flowers attract and use insects, bats, birds or other animals to transfer pollen from one flower to the next. Often they are specialized in shape and have an arrangement of the stamens that ensures that pollen grains are transferred to the bodies of the pollinator when it lands in search of its attractant (such as nectar, pollen, or a mate). In pursuing this attractant from many flowers of the same species, the pollinator transfers pollen to the stigmas—arranged with equally pointed precision—of all of the flowers it visits. Many flowers rely on simple proximity between flower parts to ensure pollination. Others, such as the *Sarracenia* or lady-slipper orchids, have elaborate designs to ensure pollination while preventing self-pollination.

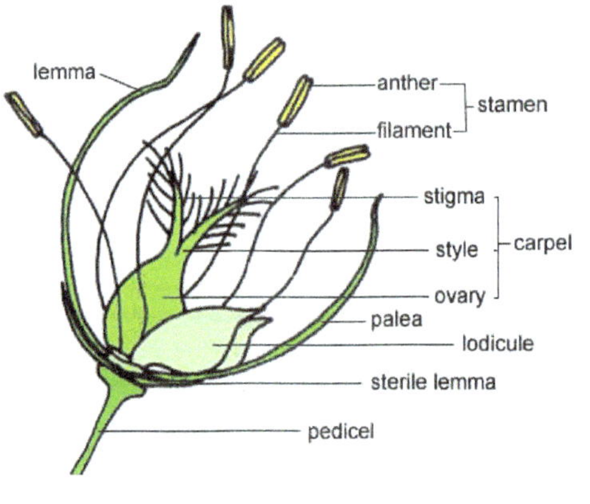

Grass flower with vestigial perianth or lodicules

Anemophilous: flowers use the wind to move pollen from one flower to the next, examples include the grasses, Birch trees, Ragweed and Maples. They have no need to attract pollinators and therefore tend not to grow large blossoms. Whereas the pollen of entomophilous flowers tends to be large-grained, sticky, and rich in protein (another "reward" for pollinators), anemophilous flower pollen is usually small-grained, very light, and of little nutritional value to insects, though it may still be gathered in times of dearth. Honeybees and bumblebees actively gather anemophilous corn (maize) pollen, though it is of little value to them.

Some flowers are self-pollinated and use flowers that never open or are self-pollinated before the flowers open, these flowers are called cleistogamous. Many Viola species and some Salvia have these types of flowers.

Flower-pollinator Relationships

Many flowers have close relationships with one or a few specific pollinating organisms. Many flowers, for example, attract only one specific species of insect, and therefore rely on that insect for successful reproduction. This close relationship is often given as an example of coevolution, as the flower and pollinator are thought to have developed together over a long period of time to match each other's needs.

This close relationship compounds the negative effects of extinction. The extinction of either member in such a relationship would mean almost certain extinction of the other member as well. Some endangered plant species are so because of shrinking pollinator populations.

Fertilization and Dispersal

Some flowers with both stamens and a pistil are capable of self-fertilization, which does increase the chance of producing seeds but limits genetic variation. The extreme case of self-fertilization occurs in flowers that always self-fertilize, such as many dandelions. Conversely, many species of plants have ways of preventing self-fertilization. Unisexual male and female flowers on the same plant may not appear or mature at the same time, or pollen from the same plant may be incapable of fertilizing its ovules. The latter flower types, which have chemical barriers to their own pollen, are referred to as self-sterile or self-incompatible.

Evolution

While land plants have existed for about 425 million years, the first ones reproduced by a simple adaptation of their aquatic counterparts: spores. In the sea, plants—and some animals—can simply scatter out genetic clones of themselves to float away and grow elsewhere. This is how early plants reproduced. But plants soon evolved methods of protecting these copies to deal with drying out and other abuse which is even more likely on land than in the sea. The protection became the seed, though it had not yet evolved the flower. Early seed-bearing plants include the ginkgo and conifers. The earliest fossil of a flowering plant, *Archaefructus liaoningensis*, is dated about 125 million years old.

Several groups of extinct gymnosperms, particularly seed ferns, have been proposed as the ancestors of flowering plants but there is no continuous fossil evidence showing exactly how flowers evolved. The apparently sudden appearance of relatively modern flowers in the fossil record posed such a problem for the theory of evolution that it was called an "abominable mystery" by Charles Darwin. Recently discovered angiosperm fossils such as *Archaefructus*, along with further discoveries of fossil gymnosperms, suggest how angiosperm characteristics may have been acquired in a series of steps.

Archaefructus liaoningensis, one of the earliest known flowering plants

Recent DNA analysis (molecular systematics) shows that *Amborella trichopoda*, found on the Pacific island of New Caledonia, is the sister group to the rest of the flowering plants, and morphological studies suggest that it has features which may have been characteristic of the earliest flowering plants.

The general assumption is that the function of flowers, from the start, was to involve animals in the reproduction process. Pollen can be scattered without bright colors and obvious shapes, which would therefore be a liability, using the plant's resources, unless they provide some other benefit. One proposed reason for the sudden, fully developed appearance of flowers is that they evolved in an isolated setting like an island, or chain of islands, where the plants bearing them were able to develop a highly specialized relationship with some specific animal (a wasp, for example), the way many island species develop today. This symbiotic relationship, with a hypothetical wasp bearing pollen from one plant to another much the way fig wasps do today, could have eventually resulted in both the plant(s) and their partners developing a high degree of specialization. Island genetics is believed to be a common source of speciation, especially when it comes to radical adaptations which seem to have required inferior transitional forms. Note that the wasp example is not incidental; bees, apparently evolved specifically for symbiotic plant relationships, are descended from wasps.

Likewise, most fruit used in plant reproduction comes from the enlargement of parts of the flower. This fruit is frequently a tool which depends upon animals wishing to eat it, and thus scattering the seeds it contains.

While many such symbiotic relationships remain too fragile to survive competition with mainland organisms, flowers proved to be an unusually effective means of production, spreading (whatever their actual origin) to become the dominant form of land plant life.

While there is only hard proof of such flowers existing about 130 million years ago, there is some circumstantial evidence that they did exist up to 250 million years ago. A chemical used by plants to defend their flowers, oleanane, has been detected in fossil plants that old, including gigantopterids, which evolved at that time and bear many of the traits of modern, flowering plants, though they are not known to be flowering plants themselves, because only their stems and prickles have been found preserved in detail; one of the earliest examples of petrification.

Amborella trichopoda, the sister group to the rest of the flowering plants

The similarity in leaf and stem structure can be very important, because flowers are genetically just an adaptation of normal leaf and stem components on plants, a combination of genes normally responsible for forming new shoots. The most primitive flowers are thought to have had a variable number of flower parts, often separate from (but in contact with) each other. The flowers would have tended to grow in a spiral pattern, to be bisexual (in plants, this means both male and female parts on the same flower), and to be dominated by the ovary (female part). As flowers grew more advanced, some variations developed parts fused together, with a much more specific number and design, and with either specific sexes per flower or plant, or at least "ovary inferior".

Flower evolution continues to the present day; modern flowers have been so profoundly influenced by humans that many of them cannot be pollinated in nature. Many modern, domesticated flowers used to be simple weeds, which only sprouted when the ground was disturbed. Some of them tended to grow with human crops, and the prettiest did not get plucked because of their beauty, developing a dependence upon and special adaptation to human affection.

Color

Many flowering plants reflect as much light as possible within the range of visible wavelengths of the pollinator the plant intends to attract. Flowers that reflect the full range of visible light are generally perceived as *white* by a human observer. An important feature of white flowers is that they reflect equally across the visible spectrum. While many flowering plants use white to attract pollinators, the use of color is also widespread (even within the same species). Color allows a flowering plant to be more specific about the pollinator it seeks to attract. The color model used by human color reproduction technology (CMYK) relies on the modulation of pigments that divide the spectrum into broad areas of absorption. Flowering plants by contrast are able to shift the transition point wavelength between absorption and reflection. If it is assumed that the visual systems of most pollinators view the visible spectrum as circular then it may be said that flowering plants produce color by absorbing the light in one region of the spectrum and reflecting the light in the other region. With CMYK, color is produced as a function of the amplitude of the broad regions of absorption. Flowering plants by contrast produce color by modifying the frequency (or rather wavelength) of the light reflected. Most flowers absorb light in the blue to yellow region of the spectrum and reflect light from the green to red region of the spectrum. For many species of flowering plant, it is the transition point that characterizes the color that they produce. Color may be modulated by shifting the transition point between absorption and reflection and in this way a flowering plant may specify which pollinator it seeks to attract.

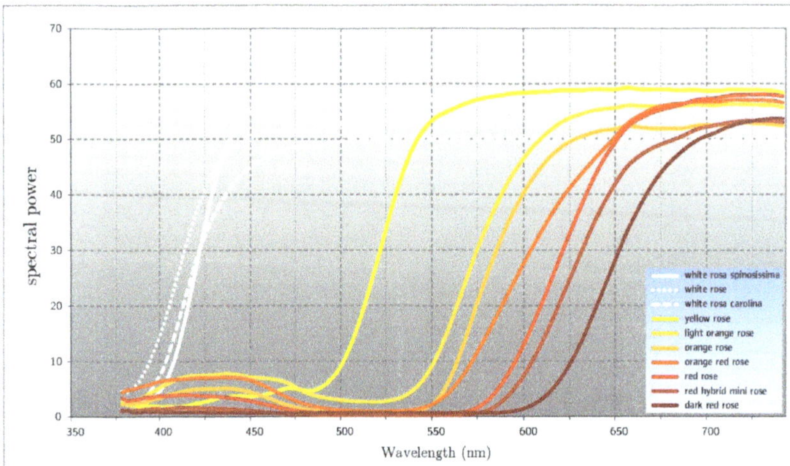

Spectrum of the flowers of the Rose family of plants. A red rose absorbs about 99.7% of light across a broad area below the red wavelengths of the spectrum, leading to an exceptionally *pure* red. A yellow rose will reflect about 5% of blue light, producing an unsaturated yellow (a yellow with a degree of white in it).

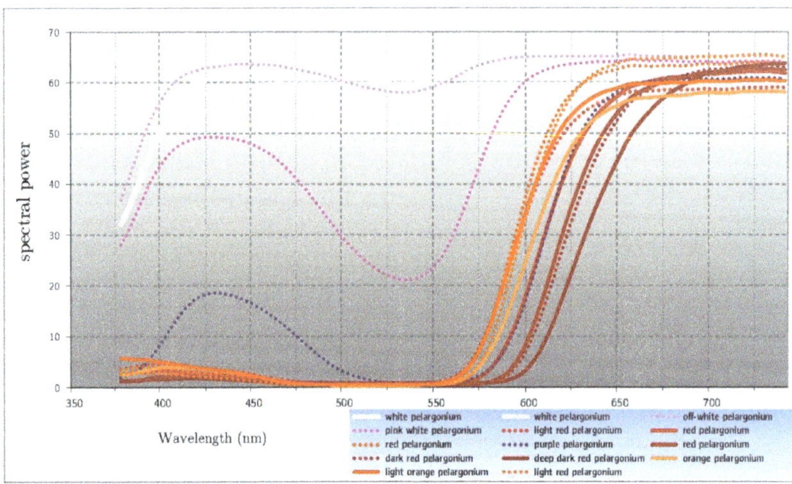

Spectrum of the flowers of the Pelargonium family of plants. Pelargonium produces exceptionally saturated colours in the orange to red range, with absorption of 99.6% across a broad area of the spectrum. Pelargonium is able to modulate its absorption producing pink with multiple levels of white.

Some flowering plants also have a limited ability to modulate areas of absorption. This is typically not as precise as control over wavelength. Humans observers will perceive this as degrees of saturation (the amount of *white* in the color).

Symbolism

Red Rose

Lilies are often used to denote life or resurrection

Flowers are common subjects of still life paintings, such as this one by Ambrosius Bosschaert the Elder

Many flowers have important symbolic meanings in Western culture. The practice of assigning meanings to flowers is known as floriography. Some of the more common examples include:

- Red roses are given as a symbol of love, beauty, and passion.
- Poppies are a symbol of consolation in time of death. In the United Kingdom, New Zealand, Australia and Canada, red poppies are worn to commemorate soldiers who have died in times of war.

- Irises/Lily are used in burials as a symbol referring to "resurrection/life". It is also associated with stars (sun) and its petals blooming/shining.
- Daisies are a symbol of innocence.

Flowers within art are also representative of the female genitalia, as seen in the works of artists such as Georgia O'Keeffe, Imogen Cunningham, Veronica Ruiz de Velasco, and Judy Chicago, and in fact in Asian and western classical art. Many cultures around the world have a marked tendency to associate flowers with femininity.

The great variety of delicate and beautiful flowers has inspired the works of numerous poets, especially from the 18th-19th century Romantic era. Famous examples include William Wordsworth's *I Wandered Lonely as a Cloud* and William Blake's *Ah! Sun-Flower*.

Because of their varied and colorful appearance, flowers have long been a favorite subject of visual artists as well. Some of the most celebrated paintings from well-known painters are of flowers, such as Van Gogh's sunflowers series or Monet's water lilies. Flowers are also dried, freeze dried and pressed in order to create permanent, three-dimensional pieces of flower art.

Their symbolism in dreams has also been discussed, with possible interpretations including "blossoming potential".

The Roman goddess of flowers, gardens, and the season of Spring is Flora. The Greek goddess of spring, flowers and nature is Chloris.

In Hindu mythology, flowers have a significant status. Vishnu, one of the three major gods in the Hindu system, is often depicted standing straight on a lotus flower. Apart from the association with Vishnu, the Hindu tradition also considers the lotus to have spiritual significance. For example, it figures in the Hindu stories of creation.

Usage

Flower market - Detroit's Eastern Market

A woman spreading flowers over a lingam in a temple in Varanasi

In modern times people have sought ways to cultivate, buy, wear, or otherwise be around flowers and blooming plants, partly because of their agreeable appearance and smell. Around the world, people use flowers for a wide range of events and functions that, cumulatively, encompass one's lifetime:

- For new births or christenings
- As a corsage or boutonniere worn at social functions or for holidays
- As tokens of love or esteem
- For wedding flowers for the bridal party, and for decorations for the hall
- As brightening decorations within the home
- As a gift of remembrance for *bon voyage* parties, welcome-home parties, and "thinking of you" gifts
- For funeral flowers and expressions of sympathy for the grieving
- For worshiping goddesses. In Hindu culture adherents commonly bring flowers as a gift to temples

People therefore grow flowers around their homes, dedicate entire parts of their living space to flower gardens, pick wildflowers, or buy flowers from florists who depend on an entire network of commercial growers and shippers to support their trade.

Flowers provide less food than other major plants parts (seeds, fruits, roots, stems and leaves) but they provide several important foods and spices. Flower vegetables include broccoli, cauliflower and artichoke. The most expensive spice, saffron, consists of dried stigmas of a crocus. Other flower spices are cloves and capers. Hops flowers are used to flavor beer. Marigold flowers are fed to chickens to give their egg yolks a golden yellow color, which consumers find more desirable; dried and ground marigold flowers are also used as a spice and colouring agent in Georgian cuisine. Flowers of the dandelion and elder are often made into wine. Bee pollen, pollen collected from bees, is considered a health food by some people. Honey consists of bee-processed flower nectar and is often named for the type of flower, e.g. orange blossom honey, clover honey and tupelo honey.

Hundreds of fresh flowers are edible but few are widely marketed as food. They are often used to add color and flavor to salads. Squash flowers are dipped in breadcrumbs and fried. Edible flowers include nasturtium, chrysanthemum, carnation, cattail, honeysuckle, chicory, cornflower, canna, and sunflower. Some edible flowers are sometimes candied such as daisy, rose, and violet (one may also come across a candied pansy).

Flowers can also be made into herbal teas. Dried flowers such as chrysanthemum, rose, jasmine, camomile are infused into tea both for their fragrance and medical properties. Sometimes, they are also mixed with tea leaves for the added fragrance.

Flowers have been used since as far back as 50,000 years in funeral rituals. Many cultures do draw a connection between flowers and life and death, and because of their seasonal return flowers also suggest rebirth, which may explain why many people place flowers upon graves. In ancient times the Greeks would place a crown of flowers on the head of the deceased as well as cover tombs with wreaths and flower petals. Rich and powerful women in ancient Egypt would wear floral head-dresses and necklaces upon their death as representations of renewal and a joyful afterlife, and the Mexicans to this day use flowers prominently in their Day of the Dead celebrations in the same way that their Aztec ancestors did.

Eight Flowers, a painting by artist Qian Xuan, 13th century, Palace Museum, Beijing.

Fruit

In botany, a fruit is the seed-bearing structure in flowering plants (also known as angiosperms) formed from the ovary after flowering.

Culinary fruits

Several culinary fruits

Mixed fruit

The Medici citrus collection by Bartolomeo Bimbi, 1715

Fruits are the means by which angiosperms disseminate seeds. Edible fruits, in particular, have propagated with the movements of humans and animals in a symbiotic relationship as a means for seed dispersal and nutrition; in fact, humans and many animals have become dependent on fruits as a source of food. Accordingly, fruits account for a substantial fraction of the world's agricultural output, and some (such as the apple and the pomegranate) have acquired extensive cultural and symbolic meanings.

In common language usage, "fruit" normally means the fleshy seed-associated structures of a plant that are sweet or sour, and edible in the raw state, such as apples, bananas, grapes, lemons, oranges, and strawberries. On the other hand, in botanical usage, "fruit" includes many structures that are not commonly called "fruits", such as bean pods, corn kernels, tomatoes, and wheat grains. The section of a fungus that produces spores is also called a fruiting body.

Botanic Fruit and Culinary Fruit

Many common terms for seeds and fruit do not correspond to the botanical classifications. In culinary terminology, a *fruit* is usually any sweet-tasting plant part, especially a botanical fruit; a *nut* is any hard, oily, and shelled plant product; and a *vegetable* is any savory or less sweet plant product. However, in botany, a *fruit* is the ripened ovary or carpel that contains seeds, a *nut* is a type of fruit and not a seed, and a *seed* is a ripened ovule.

Venn diagram representing the relationship between (culinary) vegetables and botanical fruits

Examples of culinary "vegetables" and nuts that are botanically fruit include corn, cucurbits (e.g., cucumber, pumpkin, and squash), eggplant, legumes (beans, peanuts, and peas), sweet pepper, and tomato. In addition, some spices, such as allspice and chili pepper, are fruits, botanically speaking. In contrast, rhubarb is often referred to as a fruit, because it is used to make sweet desserts such as pies, though only the petiole (leaf stalk) of the rhubarb plant is edible, and edible gymnosperm seeds are often given fruit names, e.g., ginkgo nuts and pine nuts.

Botanically, a cereal grain, such as corn, rice, or wheat, is also a kind of fruit, termed a caryopsis. However, the fruit wall is very thin and is fused to the seed coat, so almost all of the edible grain is actually a seed.

Fruit Structure

The outer, often edible layer, is the *pericarp*, formed from the ovary and surrounding the seeds, although in some species other tissues contribute to or form the edible portion. The pericarp may be described in three layers from outer to inner, the *epicarp*, *mesocarp* and *endocarp*.

Fruit that bears a prominent pointed terminal projection is said to be *beaked*.

Fruit Development

A fruit results from maturation of one or more flowers, and the gynoecium of the flower(s) forms all or part of the fruit.

Inside the ovary/ovaries are one or more ovules where the megagametophyte contains the egg cell. After double fertilization, these ovules will become seeds. The ovules are fertilized in a process that starts with pollination, which involves the movement of pollen from the stamens to the stigma of flowers. After pollination, a tube grows from the pollen through the stigma into the ovary to the ovule and two sperm are transferred from the pollen to the megagametophyte. Within the megagametophyte one of the two sperm unites with the egg, forming a zygote, and the second sperm enters the central cell forming the endosperm mother cell, which completes the double fertilization process. Later the zygote will give rise to the embryo of the seed, and the endosperm mother cell will give rise to endosperm, a nutritive tissue used by the embryo.

The development sequence of a typical drupe, the nectarine (*Prunus persica*) over a 7.5 month period, from bud formation in early winter to fruit ripening in midsummer

As the ovules develop into seeds, the ovary begins to ripen and the ovary wall, the *pericarp*, may become fleshy (as in berries or drupes), or form a hard outer covering (as in nuts). In some multiseeded fruits, the extent to which the flesh develops is proportional to the number of fertilized ovules. The pericarp is often differentiated into two or three distinct layers called the *exocarp* (outer layer, also called epicarp), *mesocarp* (middle layer), and *endocarp* (inner layer). In some fruits, especially simple fruits derived from an inferior ovary, other parts of the flower (such as the floral tube, including the petals, sepals, and stamens), fuse with the ovary and ripen with it. In other cases, the sepals, petals and/or stamens and style of the flower fall off. When such other floral parts are a significant part of the fruit, it is called an *accessory fruit*. Since other parts of the flower may contribute to the structure of the fruit, it is important to study flower structure to understand how a particular fruit forms.

There are three general modes of fruit development:

- Apocarpous fruits develop from a single flower having one or more separate carpels, and they are the simplest fruits.
- Syncarpous fruits develop from a single gynoecium having two or more carpels fused together.
- Multiple fruits form from many different flowers.

Plant scientists have grouped fruits into three main groups, simple fruits, aggregate fruits, and composite or multiple fruits. The groupings are not evolutionarily relevant, since many diverse plant taxa may be in the same group, but reflect how the flower organs are arranged and how the fruits develop.

Simple Fruit

Epigynous berries are simple fleshy fruit. Clockwise from top right: cranberries, lingonberries, blueberries, red huckleberries

Simple fruits can be either dry or fleshy, and result from the ripening of a simple or compound ovary in a flower with only one pistil. Dry fruits may be either dehiscent (they open to discharge seeds), or indehiscent (they do not open to discharge seeds). Types of dry, simple fruits, and examples of each, include:

- achene – Most commonly seen in aggregate fruits (e.g., strawberry)
- capsule – (e.g., Brazil nut)
- caryopsis – (e.g., wheat)
- cypsela – an achene-like fruit derived from the individual florets in a capitulum (e.g., dandelion).
- fibrous drupe – (e.g., coconut, walnut)
- follicle – is formed from a single carpel, opens by one suture (e.g., milkweed), commonly seen in aggregate fruits (e.g., magnolia)
- legume – (e.g., bean, pea, peanut)
- loment – a type of indehiscent legume
- nut – (e.g., beech, hazelnut, oak acorn)
- samara – (e.g., ash, elm, maple key)
- schizocarp – (e.g., carrot seed)
- silique – (e.g., radish seed)
- silicle – (e.g., shepherd's purse)
- utricle – (e.g., beet)

Lilium unripe capsule fruit

Fruits in which part or all of the *pericarp* (fruit wall) is fleshy at maturity are *simple fleshy fruits.* Types of simple, fleshy, fruits (with examples) include:

- berry – (e.g., cranberry, gooseberry, redcurrant, tomato)
- stone fruit or drupe (e.g., apricot, cherry, olive, peach, plum)

Dewberry flowers. Note the multiple pistils, each of which will produce a drupelet. Each flower will become a blackberry-like aggregate fruit.

An aggregate fruit, or *etaerio*, develops from a single flower with numerous simple pistils.

- Magnolia and peony, collection of follicles developing from one flower.
- Sweet gum, collection of capsules.
- Sycamore, collection of achenes.
- Teasel, collection of cypsellas

- Tuliptree, collection of samaras.

The pome fruits of the family Rosaceae, (including apples, pears, rosehips, and saskatoon berry) are a syncarpous fleshy fruit, a simple fruit, developing from a half-inferior ovary.

Schizocarp fruits form from a syncarpous ovary and do not really dehisce, but rather split into segments with one or more seeds; they include a number of different forms from a wide range of families. Carrot seed is an example.

Aggregate Fruit

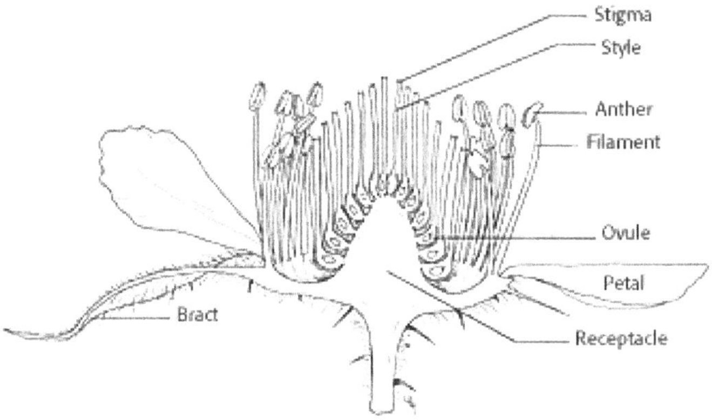

Figure 167. - Longitudinal section of 'Willamett' raspberry flower, x10.

Detail of raspberry flower

Aggregate fruits form from single flowers that have multiple carpels which are not joined together, i.e. each pistil contains one carpel. Each pistil forms a fruitlet, and collectively the fruitlets are called an etaerio. Four types of aggregate fruits include etaerios of achenes, follicles, drupelets, and berries. Ranunculaceae species, including *Clematis* and *Ranunculus* have an etaerio of achenes, *Calotropis* has an etaerio of follicles, and *Rubus* species like raspberry, have an etaerio of drupelets. *Annona* have an etaerio of berries.

The raspberry, whose pistils are termed *drupelets* because each is like a small drupe attached to the receptacle. In some bramble fruits (such as blackberry) the receptacle is elongated and part of the ripe fruit, making the blackberry an *aggregate-accessory* fruit. The strawberry is also an aggregate-accessory fruit, only one in which the seeds are contained in achenes. In all these examples, the fruit develops from a single flower with numerous pistils.

Multiple Fruits

A multiple fruit is one formed from a cluster of flowers (called an *inflorescence*). Each flower produces a fruit, but these mature into a single mass. Examples are the pineapple, fig, mulberry, osage-orange, and breadfruit.

In the photograph on the right, stages of flowering and fruit development in the noni or Indian mulberry (*Morinda citrifolia*) can be observed on a single branch. First an inflorescence of white flowers called a head is produced. After fertilization, each flower develops into a drupe, and as the drupes expand, they become *connate* (merge) into a *multiple fleshy fruit* called a *syncarp*.

In some plants, such as this noni, flowers are produced regularly along the stem and it is possible to see together examples of flowering, fruit development, and fruit ripening.

Berries

Berries are another type of fleshy fruit; they are simple fruit created from a single ovary. The ovary may be compound, with several carpels. Types include (examples follow in the table below):

- Pepo – berries whose skin is hardened, cucurbits
- Hesperidium – berries with a rind and a juicy interior, like most citrus fruit

Accessory Fruit

The fruit of a pineapple includes tissue from the sepals as well as the pistils of many flowers. It is an accessory fruit and a multiple fruit.

Some or all of the edible part of accessory fruit is not generated by the ovary. Accessory fruit can be simple, aggregate, or multiple, i.e., they can include one or more pistils and other parts from the same flower, or the pistils and other parts of many flowers.

Table of fruit examples

Types of fleshy fruits					
True berry	**Pepo**	**Hesperidium**	**Aggregate fruit**	**Multiple fruit**	**Accessory fruit**
Blackcurrant, Blueberry, Chili pepper, Cranberry, Eggplant, Gooseberry, Grape, Guava, Kiwifruit, Lucuma, Pomegranate, Red-currant, Tomato	Cucumber, Gourd, Mel-on, Pumpkin	Grapefruit, Lemon, Lime, Orange	Blackberry, Boysenberry, Raspberry	Fig, Hedge apple, Mulberry, Pineapple	Apple, Pineap-ple, Rose hip, Stone fruit, Strawberry

Seedless Fruits

An arrangement of fruits commonly thought of as vegetables, including tomatoes and various squash

Seedlessness is an important feature of some fruits of commerce. Commercial cultivars of bananas and pineapples are examples of seedless fruits. Some cultivars of citrus fruits (especially grapefruit, mandarin oranges, navel oranges), satsumas, table grapes, and watermelons are valued for their seedlessness. In some species, seedlessness is the result of *parthenocarpy*, where fruits set without fertilization. Parthenocarpic fruit set may or may not require pollination, but most seedless citrus fruits require a stimulus from pollination to produce fruit.

Seedless bananas and grapes are triploids, and seedlessness results from the abortion of the embryonic plant that is produced by fertilization, a phenomenon known as *stenospermocarpy*, which requires normal pollination and fertilization.

Seed Dissemination

Variations in fruit structures largely depend on their seeds' mode of dispersal. This dispersal can be achieved by animals, explosive dehiscence, water, or wind.

Some fruits have coats covered with spikes or hooked burrs, either to prevent themselves from being eaten by animals, or to stick to the feathers, hairs, or legs of animals, using them as dispersal agents. Examples include cocklebur and unicorn plant.

The sweet flesh of many fruits is "deliberately" appealing to animals, so that the seeds held within are eaten and "unwittingly" carried away and deposited (i.e., defecated) at a distance from the parent. Likewise, the nutritious, oily kernels of nuts are appealing to rodents (such as squirrels), which hoard them in the soil to avoid starving during the winter, thus giving those seeds that remain uneaten the chance to germinate and grow into a new plant away from their parent.

Grapes and Mangoes

Other fruits are elongated and flattened out naturally, and so become thin, like wings or helicopter blades, e.g., elm, maple, and tuliptree. This is an evolutionary mechanism to increase dispersal distance away from the parent, via wind. Other wind-dispersed fruit have tiny "parachutes", e.g., dandelion, milkweed, salsify.

Coconut fruits can float thousands of miles in the ocean to spread seeds. Some other fruits that can disperse via water are nipa palm and screw pine.

Some fruits fling seeds substantial distances (up to 100 m in sandbox tree) via explosive dehiscence or other mechanisms, e.g., impatiens and squirting cucumber.

Uses

Nectarines are one of many fruits that can be easily stewed.

Oranges, bananas, pears, apples, and a watermelon

Many hundreds of fruits, including fleshy fruits (like apple, kiwifruit, mango, peach, pear, and watermelon) are commercially valuable as human food, eaten both fresh and as jams, marmalade and other preserves. Fruits are also used in manufactured foods (e.g., cakes, cookies, ice cream, muffins, or yogurt) or beverages, such as fruit juices (e.g., apple juice, grape juice, or orange juice) or alcoholic beverages (e.g., brandy, fruit beer, or wine), Fruits are also used for gift giving, e.g., in the form of Fruit Baskets and Fruit Bouquets.

Many "vegetables" in culinary *parlance* are botanical fruits, including bell pepper, cucumber, eggplant, green bean, okra, pumpkin, squash, tomato, and zucchini. Olive fruit is pressed for olive oil. Spices like allspice, black pepper, paprika, and vanilla are derived from berries.

Nutritional Value

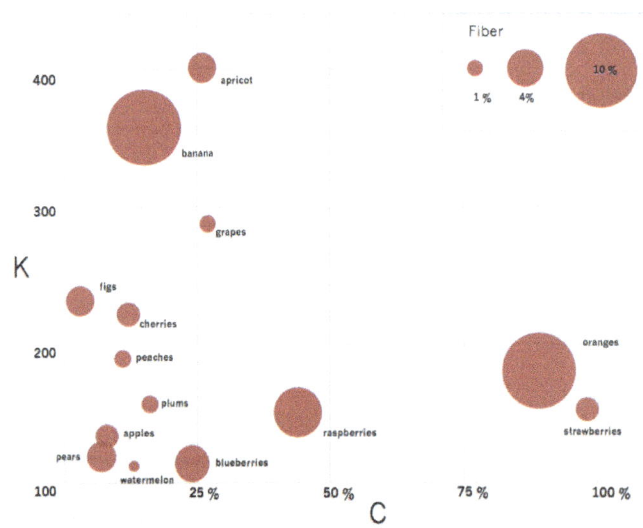

Each point refers to a 100 g serving of the fresh fruit, the daily recommended allowance of vitamin C is on the X axis and mg of Potassium (K) on the Y (offset by 100 mg which every fruit has) and the size of the disk represents amount of fiber (key in upper right). Watermelon, which has almost no fiber, and low levels of vitamin C and potassium, comes in last place.

Fresh fruits are generally high in fiber, vitamin C, and water.

Regular consumption of fruit is generally associated with reduced risks of several diseases and functional declines associated with aging.

Nonfood Uses

Because fruits have been such a major part of the human diet, various cultures have developed many different uses for fruits they do not depend on for food. For example:

- Bayberry fruits provide a wax often used to make candles;
- Many dry fruits are used as decorations or in dried flower arrangements (e.g., annual honesty, cotoneaster, lotus, milkweed, unicorn plant, and wheat). Ornamental trees and shrubs are often cultivated for their colorful fruits, including beautyberry, cotoneaster, holly, pyracantha, skimmia, and viburnum.
- Fruits of opium poppy are the source of opium, which contains the drugs codeine and morphine, as well as the biologically inactive chemical theabaine from which the drug oxycodone is synthesized.
- Osage orange fruits are used to repel cockroaches.
- Many fruits provide natural dyes (e.g., cherry, mulberry, sumac, and walnut).
- Dried gourds are used as bird houses, cups, decorations, dishes, musical instruments, and water jugs.
- Pumpkins are carved into Jack-o'-lanterns for Halloween.
- The spiny fruit of burdock or cocklebur inspired the invention of Velcro.
- Coir fiber from coconut shells is used for brushes, doormats, floor tiles, insulation, mattresses, sacking, and as a growing medium for container plants. The shell of the coconut fruit is used to make bird houses, bowls, cups, musical instruments, and souvenir heads.
- Fruit is often a subject of still life paintings.

Safety

For food safety, the CDC recommends proper fruit handling and preparation to reduce the risk of food contamination and foodborne illness. Fresh fruits and vegetables should be carefully selected; at the store, they should not be damaged or bruised; and pre-cut pieces should be refrigerated or surrounded by ice.

All fruits and vegetables should be rinsed before eating. This recommendation also applies to produce with rinds or skins that are not eaten. It should be done just before preparing or eating to avoid premature spoilage.

Fruits and vegetables should be kept separate from raw foods like meat, poultry, and seafood, as well as from utensils that have come in contact with raw foods. Fruits and vegetables that are not going to be cooked should be thrown away if they have touched raw meat, poultry, seafood, or eggs.

All cut, peeled, or cooked fruits and vegetables should be refrigerated within two hours. After a certain time, harmful bacteria may grow on them and increase the risk of foodborne illness.

Allergies

Fruit allergies make up about 10 percent of all food related allergies

Storage

All fruits benefit from proper post harvest care, and in many fruits, the plant hormone ethylene causes ripening. Therefore, maintaining most fruits in an efficient cold chain is optimal for post harvest storage, with the aim of extending and ensuring shelf life.

Cell Wall

A cell wall is a structural layer that surrounds most types of cells, situated outside the cell membrane. It can be tough, flexible, and sometimes rigid. It provides the cell with both structural support and protection, and also acts as a filtering mechanism. Cell walls are present in most prokaryotes, and in plants, fungi and other eukaryotes, where a major function is to act as pressure vessels, preventing over-expansion of the cell when water enters. Cell walls are absent from *Mycoplasma* bacteria, animal, and human cells.

The composition of cell walls varies between species and may depend on cell type and developmental stage. The primary cell wall of land plants is composed of the polysaccharides cellulose, hemicellulose and pectin. Often, other polymers such as lignin, suberin or cutin are anchored to or embedded in plant cell walls. Algae possess walls made of glycoproteins and polysaccharides such as carrageenan and agar that are absent from land plants. In bacteria, the cell wall is composed of peptidoglycan. The cell walls of archaea have various compositions, and may be formed of glycoprotein S-layers, pseudopeptidoglycan, or polysaccharides. Fungi possess cell walls made of the glucosamine polymer chitin. Unusually, diatoms have a cell wall composed of biogenic silica.

History

A plant cell wall was first observed and named (simply as a "wall") by Robert Hooke in 1665. However, "the dead excrusion product of the living protoplast" was forgotten, for almost three centuries, being the subject of scientific interest mainly as a resource for industrial processing or in relation to animal or human health.

In 1804, Karl Rudolphi and J.H.F. Link proved that cells had independent cell walls. Before, it had been thought that cells shared walls and that fluid passed between them this way.

The mode of formation of the cell wall was controversial in the 19th century. Hugo von Mohl (1853, 1858) advocated the idea that the cell wall grows by apposition. Carl Nägeli (1858, 1862, 1863) believed that the growth of the wall in thickness and in area was due to a process termed intussusception. Each theory was improved in the following decades: the apposition (or lamination) theory by Eduard Strasburger (1882, 1889), and the intussusception theory by Julius Wiesner (1886).

In 1930, Ernst Münch coined the term *apoplast* in order to separate the "living" symplast from the "dead" plant region, the latter of which included the cell wall.

By the 1980s, some authors suggested replacing the term "cell wall", particularly as it was used for plants, with the more precise term "extracellular matrix", as used for animal cells, but others preferred the older term.

Properties

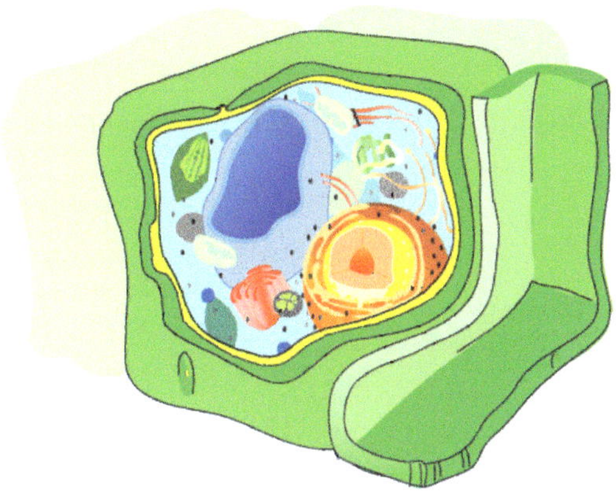

Diagram of the plant cell, with the cell wall in green.

Cell walls serve similar purposes in those organisms that possess them. They may give cells rigidity and strength, offering protection against mechanical stress. In multicellular organisms, they permit the organism to build and hold a definite shape (morphogenesis). Cell walls also limit the entry of large molecules that may be toxic to the cell. They further permit the creation of stable osmotic environments by preventing osmotic lysis and helping to retain water. Their composition, properties, and form may change during the cell cycle and depend on growth conditions.

Rigidity of Cell Walls

In most cells, the cell wall is flexible, meaning that it will bend rather than holding a fixed shape, but has considerable tensile strength. The apparent rigidity of primary plant tissues is enabled by cell walls, but is not due to the walls' stiffness. Hydraulic turgor pressure creates this rigidity, along with the wall structure. The flexibility of the cell walls is seen when plants wilt, so that the stems and leaves begin to droop, or in seaweeds that bend in water currents. As John Howland explains:

Think of the cell wall as a wicker basket in which a balloon has been inflated so that it exerts pressure from the inside. Such a basket is very rigid and resistant to mechanical damage. Thus does the prokaryote cell (and eukaryotic cell that possesses a cell wall) gain strength from a flexible plasma membrane pressing against a rigid cell wall.

The apparent rigidity of the cell wall thus results from inflation of the cell contained within. This inflation is a result of the passive uptake of water.

In plants, a secondary cell wall is a thicker additional layer of cellulose which increases wall rigidity. Additional layers may be formed by lignin in xylem cell walls, or suberin in cork cell walls. These compounds are rigid and waterproof, making the secondary wall stiff. Both wood and bark cells of trees have secondary walls. Other parts of plants such as the leaf stalk may acquire similar reinforcement to resist the strain of physical forces.

Permeability

The primary cell wall of most plant cells is freely permeable to small molecules including small proteins, with size exclusion estimated to be 30-60 kDa. The pH is an important factor governing the transport of molecules through cell walls.

Evolution

Cell walls evolved independently in many groups, even in the photosynthetic eukaryotes. In these lineages, the cell wall is closely related to the evolution of multicellularity, terrestrialization and vascularization.

Plant Cell Walls

The walls of plant cells must have sufficient tensile strength to withstand internal osmotic pressures of several times atmospheric pressure that result from the difference in solute concentration between the cell interior and external solutions. Plant cell walls vary from 0.1 to several μm in thickness.

Layers

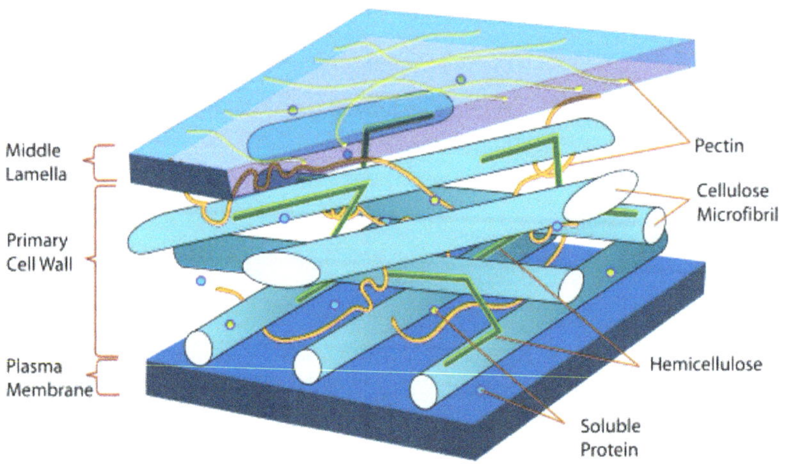

Molecular structure of the primary cell wall in plants.

Up to three strata or layers may be found in plant cell walls:

- The primary cell wall, generally a thin, flexible and extensible layer formed while the cell is growing.
- The secondary cell wall, a thick layer formed inside the primary cell wall after the cell is fully grown. It is not found in all cell types. Some cells, such as the conducting cells in xylem, possess a secondary wall containing lignin, which strengthens and waterproofs the wall.
- The middle lamella, a layer rich in pectins. This outermost layer forms the interface between adjacent plant cells and glues them together.

Composition

In the primary (growing) plant cell wall, the major carbohydrates are cellulose, hemicellulose and pectin. The cellulose microfibrils are linked via hemicellulosic tethers to form the cellulose-hemicellulose network, which is embedded in the pectin matrix. The most common hemicellulose in the primary cell wall is xyloglucan. In grass cell walls, xyloglucan and pectin are reduced in abundance and partially replaced by glucuronarabinoxylan, another type of hemicellulose. Primary cell walls characteristically extend (grow) by a mechanism called acid growth, which involves turgor-driven movement of the strong cellulose microfibrils within the weaker hemicellulose/pectin matrix, catalyzed by expansin proteins. The outer part of the primary cell wall of the plant epidermis is usually impregnated with cutin and wax, forming a permeability barrier known as the plant cuticle.

Secondary cell walls contain a wide range of additional compounds that modify their mechanical properties and permeability. The major polymers that make up wood (largely secondary cell walls) include:

- cellulose, 35-50%
- xylan, 20-35%, a type of hemicellulose
- lignin, 10-25%, a complex phenolic polymer that penetrates the spaces in the cell wall between cellulose, hemicellulose and pectin components, driving out water and strengthening the wall.

Additionally, structural proteins (1-5%) are found in most plant cell walls; they are classified as hydroxyproline-rich glycoproteins (HRGP), arabinogalactan proteins (AGP), glycine-rich proteins (GRPs), and proline-rich proteins (PRPs). Each class of glycoprotein is defined by a characteristic, highly repetitive protein sequence. Most are glycosylated, contain hydroxyproline (Hyp) and become cross-linked in the cell wall. These proteins are often concentrated in specialized cells and in cell corners. Cell walls of the epidermis may contain cutin. The Casparian strip in the endodermis roots and cork cells of plant bark contain suberin. Both cutin and suberin are polyesters that function as permeability barriers to the movement of water. The relative composition of carbohydrates, secondary compounds and proteins varies between plants and between the cell type and age. Plant cells walls also contain numerous enzymes, such as hydrolases, esterases, peroxidases, and transglycosylases, that cut, trim and cross-link wall polymers.

Secondary walls - especially in grasses - may also contain microscopic silica crystals, which may strengthen the wall and protect it from herbivores.

Cell walls in some plant tissues also function as storage deposits for carbohydrates that can be broken down and resorbed to supply the metabolic and growth needs of the plant. For example, endosperm cell walls in the seeds of cereal grasses, nasturtium, and other species, are rich in glucans and other polysaccharides that are readily digested by enzymes during seed germination to form simple sugars that nourish the growing embryo.

Formation

The middle lamella is laid down first, formed from the cell plate during cytokinesis, and the primary cell wall is then deposited inside the middle lamella. The actual structure of the cell wall is not clearly defined and several models exist - the covalently linked cross model, the tether model, the diffuse layer model and the stratified layer model. However, the primary cell wall, can be defined as composed of cellulose microfibrils aligned at all angles. Cellulose microfibrils are produced at the plasma membrane by the cellulose synthase complex, which is proposed to be made of a hexameric rosette that contains three cellulose synthase catalytic subunits for each of the six units. Microfibrils are held together by hydrogen bonds to provide a high tensile strength. The cells are held together and share the gelatinous membrane called the *middle lamella*, which contains magnesium and calcium pectates (salts of pectic acid). Cells interact though plasmodesmata, which are inter-connecting channels of cytoplasm that connect to the protoplasts of adjacent cells across the cell wall.

In some plants and cell types, after a maximum size or point in development has been reached, a *secondary wall* is constructed between the plasma membrane and primary wall. Unlike the primary wall, the cellulose microfibrils are aligned parallel in layers, the orientation changing slightly with each additional layer so that the structure becomes helicoidal. Cells with secondary cell walls can be rigid, as in the gritty sclereid cells in pear and quince fruit. Cell to cell communication is possible through pits in the secondary cell wall that allow plasmodesmata to connect cells through the secondary cell walls.

Fungal Cell Walls

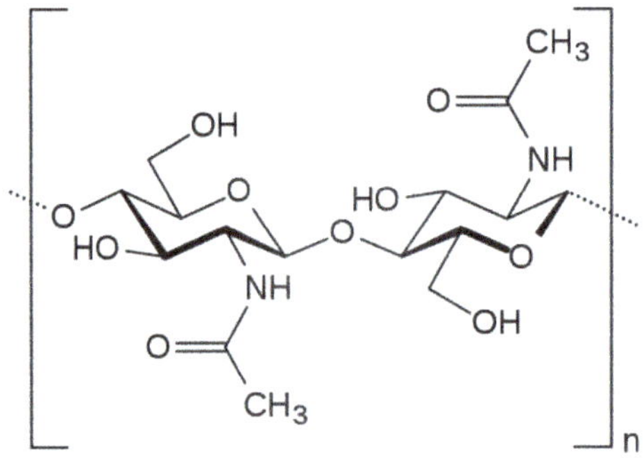

Chemical structure of a unit from a chitin polymer chain.

There are several groups of organisms that have been called "fungi". Some of these groups (Oomycete and Myxogastria) have been transferred out of the Kingdom Fungi, in part because of fundamental biochemical differences in the composition of the cell wall. Most true fungi have a cell wall consisting largely of chitin and other polysaccharides. True fungi do not have cellulose in their cell walls.

True Fungi

In fungi, the cell wall is the outer-most layer, external to the plasma membrane. The fungal cell wall is a matrix of three main components:

- chitin: polymers consisting mainly of unbranched chains of β-(1,4)-linked-N-Acetylglucosamine in the Ascomycota and Basidiomycota, or poly-β-(1,4)-linked-N-Acetylglucosamine (chitosan) in the Zygomycota. Both chitin and chitosan are synthesized and extruded at the plasma membrane.

- glucans: glucose polymers that function to cross-link chitin or chitosan polymers. β-glucans are glucose molecules linked via β-(1,3)- or β-(1,6)- bonds and provide rigidity to the cell wall while α-glucans are defined by α-(1,3)- and/or α-(1,4) bonds and function as part of the matrix.

- proteins: enzymes necessary for cell wall synthesis and lysis in addition to structural proteins are all present in the cell wall. Most of the structural proteins found in the cell wall are glycosylated and contain mannose, thus these proteins are called mannoproteins or mannans.

Other Eukaryotic Cell Walls

Algae

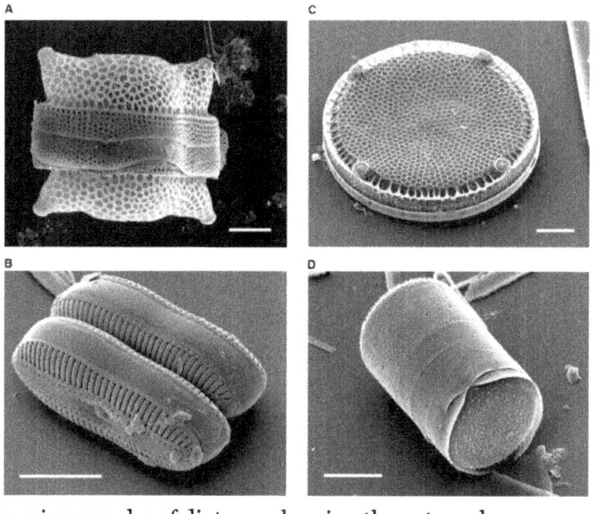

Scanning electron micrographs of diatoms showing the external appearance of the cell wall

Like plants, algae have cell walls. Algal cell walls contain either polysaccharides (such as cellulose (a glucan)) or a variety of glycoproteins (Volvocales) or both. The inclusion of additional polysaccharides in algal cells walls is used as a feature for algal taxonomy.

- Mannans: They form microfibrils in the cell walls of a number of marine green algae including those from the genera, *Codium*, *Dasycladus*, and *Acetabularia* as well as in the walls of some red algae, like *Porphyra* and *Bangia*.

- Xylans:

- Alginic acid: It is a common polysaccharide in the cell walls of brown algae.

- Sulfonated polysaccharides: They occur in the cell walls of most algae; those common in red algae include agarose, carrageenan, porphyran, furcelleran and funoran.

Other compounds that may accumulate in algal cell walls include sporopollenin and calcium ions.

The group of algae known as the diatoms synthesize their cell walls (also known as frustules or valves) from silicic acid (specifically orthosilicic acid, H_4SiO_4). The acid is polymerised intra-cellularly, then the wall is extruded to protect the cell. Significantly, relative to the organic cell walls produced by other groups, silica frustules require less energy to synthesize (approximately 8%), potentially a major saving on the overall cell energy budget and possibly an explanation for higher growth rates in diatoms.

In brown algae, phlorotannins may be a constituent of the cell walls.

Water Molds

The group Oomycetes, also known as water molds, are saprotrophic plant pathogens like fungi. Until recently they were widely believed to be fungi, but structural and molecular evidence has led to their reclassification as heterokonts, related to autotrophic brown algae and diatoms. Unlike fungi, oomycetes typically possess cell walls of cellulose and glucans rather than chitin, although some genera (such as *Achlya* and *Saprolegnia*) do have chitin in their walls. The fraction of cellulose in the walls is no more than 4 to 20%, far less than the fraction of glucans. Oomycete cell walls also contain the amino acid hydroxyproline, which is not found in fungal cell walls.

Slime Molds

The dictyostelids are another group formerly classified among the fungi. They are slime molds that feed as unicellular amoebae, but aggregate into a reproductive stalk and sporangium under certain conditions. Cells of the reproductive stalk, as well as the spores formed at the apex, possess a cellulose wall. The spore wall has been shown to possess three layers, the middle of which is composed primarily of cellulose, and the innermost is sensitive to cellulase and pronase.

Prokaryotic Cell Walls

Bacterial Cell Walls

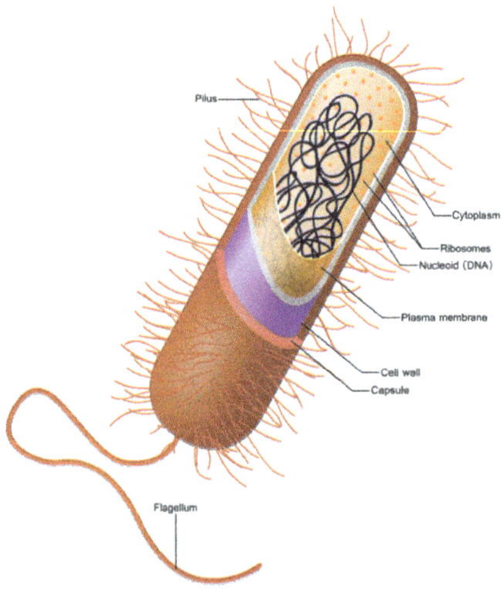

Diagram of a typical Gram-positive bacterium. The cell envelope comprises a plasma membrane, seen here in green, and a thick peptidoglycan-containing cell wall (the yellow layer). No outer lipid membrane is present, as would be the case in Gram-negative bacteria. The red layer, known as the capsule, is distinct from the cell envelope.

Around the outside of the cell membrane is the bacterial cell wall. Bacterial cell walls are made of peptidoglycan (also called murein), which is made from polysaccharide chains cross-linked by unusual peptides containing D-amino acids. Bacterial cell walls are different from the cell walls of plants and fungi which are made of cellulose and chitin, respectively. The cell wall of bacteria is also distinct from that of Archaea, which do not contain peptidoglycan. The cell wall is essential to the survival of many bacteria, although L-form bacteria can be produced in the laboratory that lack a cell wall. The antibiotic penicillin is able to kill bacteria by preventing the cross-linking of peptidoglycan and this causes the cell wall to weaken and lyse. The lysozyme enzyme can also damage bacterial cell walls.

There are broadly speaking two different types of cell wall in bacteria, called Gram-positive and Gram-negative. The names originate from the reaction of cells to the Gram stain, a test long-employed for the classification of bacterial species.

Gram-positive bacteria possess a thick cell wall containing many layers of peptidoglycan and teichoic acids. In contrast, Gram-negative bacteria have a relatively thin cell wall consisting of a few layers of peptidoglycan surrounded by a second lipid membrane containing lipopolysaccharides and lipoproteins. Most bacteria have the Gram-negative cell wall and only the Firmicutes and Actinobacteria (previously known as the low G+C and high G+C Gram-positive bacteria, respectively) have the alternative Gram-positive arrangement. These differences in structure can produce differences in antibiotic susceptibility, for instance vancomycin can kill only Gram-positive bacteria and is ineffective against Gram-negative pathogens, such as *Haemophilus influenzae* or *Pseudomonas aeruginosa*.

Archaeal Cell Walls

Although not truly unique, the cell walls of Archaea are unusual. Whereas peptidoglycan is a standard component of all bacterial cell walls, all archaeal cell walls lack peptidoglycan, with the exception of one group of methanogens. In that group, the peptidoglycan is a modified form very different from the kind found in bacteria. There are four types of cell wall currently known among the Archaea.

One type of archaeal cell wall is that composed of pseudopeptidoglycan (also called pseudomurein). This type of wall is found in some methanogens, such as *Methanobacterium* and *Methanothermus*. While the overall structure of archaeal *pseudo*peptidoglycan superficially resembles that of bacterial peptidoglycan, there are a number of significant chemical differences. Like the peptidoglycan found in bacterial cell walls, pseudopeptidoglycan consists of polymer chains of glycan cross-linked by short peptide connections. However, unlike peptidoglycan, the sugar N-acetylmuramic acid is replaced by N-acetyltalosaminuronic acid, and the two sugars are bonded with a β,1-3 glycosidic linkage instead of β,1-4. Additionally, the cross-linking peptides are L-amino acids rather than D-amino acids as they are in bacteria.

A second type of archaeal cell wall is found in *Methanosarcina* and *Halococcus*. This type of cell wall is composed entirely of a thick layer of polysaccharides, which may be sulfated in the case of *Halococcus*. Structure in this type of wall is complex and as yet is not fully investigated.

A third type of wall among the Archaea consists of glycoprotein, and occurs in the hyperthermophiles, *Halobacterium*, and some methanogens. In *Halobacterium*, the proteins in the wall have a high content of acidic amino acids, giving the wall an overall negative charge. The result is an unstable structure that is stabilized by the presence of large quantities of positive sodium ions that neutralize the charge. Consequently, *Halobacterium* thrives only under conditions with high salinity.

In other Archaea, such as *Methanomicrobium* and *Desulfurococcus*, the wall may be composed only of surface-layer proteins, known as an *S-layer*. S-layers are common in bacteria, where they serve as either the sole cell-wall component or an outer layer in conjunction with polysaccharides. Most Archaea are Gram-negative, though at least one Gram-positive member is known.

Other Cell Coverings

Many protists and bacteria produce other cell surface structures apart from cell walls, external (extracellular matrix) or internal. Many algae have a sheath or envelope of mucilage outside the cell made of exopolysaccharides. Diatoms build a frustule from silica extracted from the surrounding water; radiolarians, foraminiferans, testate amoebae and silicoflagellates also produce a skeleton from minerals, called test in some groups. Many green algae, such as *Halimeda* and the Dasycladales, and some red algae, the Corallinales, encase their cells in a secreted skeleton of calcium carbonate. In each case, the wall is rigid and essentially inorganic. It is the non-living component of cell. Some golden algae, ciliates and choanoflagellates produces a shell-like protective outer covering called lorica. Some dinoflagellates have a theca of cellulose plates, and coccolithophorids have coccoliths.

An extracellular matrix is also present in metazoans. Its composition varies between cells, but collagens are the most abundant protein in the ECM.

Vine

A vine (Latin *vīnea* "grapevine", "vineyard", from *vīnum* "wine") in the narrowest sense is the grapevine (*Vitis*), but more generally it can refer to any plant with a growth habit of trailing or scandent (that is, climbing) stems or runners. The word also can refer to such stems or runners themselves, for instance when used in wicker work.

A tendril

Vine twining around a steel fixed ladder

In the United Kingdom, the term "vine" applies almost exclusively to the grapevine. The term "climber" is used for all climbing plants.

Growth Forms

Climbing plant covering a chimney

Retaining wall covered by vines

Certain plants always grow as vines, while a few grow as vines only part of the time. For instance, poison ivy and bittersweet can grow as low shrubs when support is not available, but will become vines when support is available.

A vine displays a growth form based on long stems. This has two purposes. A vine may use rock exposures, other plants, or other supports for growth rather than investing energy in a lot of supportive tissue, enabling the plant to reach sunlight with a minimum investment of energy. This has been a highly successful growth form for plants such as kudzu and Japanese honeysuckle, both of which are invasive exotics in parts of North America. There are some tropical vines that develop skototropism, and grow away from the light, a type of negative phototropism. Growth away from light allows the vine to reach a tree trunk, which it can then climb to brighter regions.

The vine growth form may also enable plants to colonize large areas quickly, even without climbing high. This is the case with periwinkle and ground ivy. It is also an adaptation to life in areas where small patches of fertile soil are adjacent to exposed areas with more sunlight but little or no soil. A vine can root in the soil but have most of its leaves in the brighter, exposed area, getting the best of both worlds.

The evolution of a climbing habit has been implicated as a key innovation associated with the evolutionary success and diversification of a number of taxonomic groups of plants. It has evolved independently in several plant families, using many different climbing methods such as:

- twining their stems around a support (e.g., morning glories, *Ipomoea* species).
- by way of adventitious, clinging roots (e.g., ivy, *Hedera* species)
- with twining petioles (e.g., *Clematis* species)
- using tendrils, which can be specialized shoots (Vitaceae), leaves (Bignoniaceae), or even inflorescences (*Passiflora*)
- using tendrils which also produce adhesive pads at the end that attach themselves quite strongly to the support, (*Parthenocissus*)
- using thorns (e.g. climbing rose) or other hooked structures, such as hooked branches (e.g. *Artabotrys hexapetalus*)

The climbing fetterbush (*Pieris phillyreifolia*) is a woody shrub-vine which climbs without clinging roots, tendrils, or thorns. It directs its stem into a crevice in the bark of fibrous barked trees (such as bald cypress) where the stem adopts a flattened profile and grows up the tree underneath the host tree's outer bark. The fetterbush then sends out branches that emerge near the top of the tree.

Most vines are flowering plants. These may be divided into woody vines or lianas, such as wisteria, kiwifruit, and common ivy, and herbaceous (nonwoody) vines, such as morning glory.

One odd group of vining plants is the fern genus *Lygodium*, called climbing ferns. The stem does not climb, but rather the fronds (leaves) do. The fronds unroll from the tip, and theoretically never stop growing; they can form thickets as they unroll over other plants, rockfaces, and fences.

A large vine on the street in Sochi, Russia

Use as Garden Plants

Gardeners can use the tendency of climbing plants to grow quickly. If a plant display is wanted quickly, a climber can achieve this. Climbers can be trained over walls, pergolas, fences, etc. Climbers can be grown over other plants to provide additional attraction. Artificial support can also be provided. Some climbers climb by themselves; others need work, such as tying them in and training them.

Horticultural Climbing Plants

The term "Vine" also applies to cucurbitaceae like cucumbers where botanists refer to creeping vines; In commercial agriculture the natural tendency of coiling tendrils to attach themselves to pre-existing structures or espaliers is optimized by the installation of trellis netting.

Vines widely differ in size, form and evolutionary origin. Darwin classified climbing groups based on their climbing method. He classified five classes of vines including twining plants, leaf climbers, tendril bearers, root climbers and hook climbers.

Vines are unique in that they have multiple evolutionary origins and a wide range of phenotypic plasticity. They usually reside in tropical locations and have the unique ability to climb. Vines are able to grow in both deep shade and full sun due to their wide range of phenotypic plasticity. This climbing action prevents shading by neighbors and allows the vine to grow out of reach of herbivores The environment where a vine can grow successfully is determined by the climbing mechanism of a vine and how far it can spread across supports. There are many theories supporting the idea that photosynthetic responses are closely related to climbing mechanisms.

Temperate twining vines, which twist tightly around supports, are typically poorly adapted for climbing beneath closed canopies due to their smaller support diameter and shade intolerance. In contrast, tendril vines usually grow on the forest floor and onto trees until they reach the surface of the canopy, suggesting that they have greater physiological plasticity. It has also been suggested that twining vines revolving growth is mediated by changes in turgor pressure mediated by volume changes in the epidermal cells of the bending zone

Climbing vines possess many unique characteristics in response to changes in their environments.

Climbing vines can induce chemical defenses and modify their biomass allocation in response to herbivores. In particular, the twisting vine C. arvensis increases its twining in response to herbivore associated leaf damage, which may lead to reduced future herbivory. Additionally, the tendrils of perennial vine Cayratia japonica are more like to coil around nearby non-self plans than nearby self-plants in natural and experimental settings. This demonstrates the vine's ability to self-discriminate, which has only been previously documented in roots.

In tendrilled vines, the tendrils are highly sensitive to touch and the coiling action is mediated by the hormones octadecanoids, jasmonates and indole-3-acetic acid. The touch stimulus and hormones may interact via volatile compounds or internal oscillation patterns. Research has found the presence of ion translocating ATPases in the Bryonia dioica species of plants, which has implications for a possible ion mediation tendril curling mechanism. In response to a touch stimulus, vanadate sensitive K+, Mg2+ ATPase and a Ca2+ translocating ATPase both rapidly increase in their activity. This increases transmembrane ion fluxes that appear to be involved in the early stages of tendril coiling.

Examples

Spring growth of Virginia Creeper

- *Actinidia arguta*, the tara vine
- *Actinidia polygama*, the silver vine
- *Adlumia fungosa*, the Allegheny vine
- *Aeschynanthus radicans*, the lipstick vine
- *Akebia*, the chocolate vine
- *Ampelocissus acetosa*, known as wild grape or djabaru
- *Ampelopsis glandulosa* var. *brevipedunculata*, known as wild grape or porcelain berry
- *Anredera cordifolia*
- *Antigonon*, the coral vine

- *Antigonon leptopus*, the confederate vine
- *Berchemia scandens*, the rattan vine
- *Bignonia*, the cross vine
- *Bougainvillea*, a genus of thorny ornamental vines, bushes, and trees
- *Campsis*, the trumpet vine
    - o *Campsis grandiflora*, the Chinese trumpet vine
- *Cardiospermum halicacabum*, the balloon vine
- *Celastrus*, the staff vine
- *Ceropegia linearis*, the rosary vine or sweetheart vine
- *Cissus antarctica*, the kangaroo vine
- *Cissus hypoglauca*, the water vine
- *Cucumis sativus*, the cucumber
- *Cyphostemma juttae*, known as wild grape
- *Fallopia baldschuanica*, the Russian vine
- *Hedera helix*, known as common ivy, English ivy, European ivy, or ivy
- *Kennedia coccinea*, the common coral vine
- *Lagenaria siceraria*, known as the bottle gourd, calabash, opo squash, or long melon
- *Lathyrus odoratus*, the sweet pea
- *Lonicera japonica*, known as Suikazura or Japanese honeysuckle
- *Luffa*, a genus of tropical and subtropical vines classified in the cucumber (Cucurbitaceae) family
- *Lygodium*, a genus of about 40 species of ferns, known as climbing ferns
- *Momordica charantia*, the bitter gourd
- *Mikania scandens*, the hemp vine
- *Muehlenbeckia adpressa*, the macquarie vine
- *Nepenthes*, a genus of carnivorous plants known as tropical pitcher plants or monkey cups
- *Pandorea pandorana*, the wonga wonga vine
- *Parthenocissus quinquefolia*, known as the Virginia creeper, Victoria creeper, five-leaved ivy, or five-finger

- *Passiflora edulis*, the passionfruit
- *Periploca graeca*, the silk vine
- *Podranea ricasoliana*, the pink trumpet vine
- *Pueraria lobata*, the kudzu vine
- *Scindapsus pictus*, the silver vine
- *Sechium edule*, known as chayote, christophene, or several other names
- *Solandra*, a genus of flowering plants in the nightshade family
- *Solanum laxum*, the potato vine
- *Strongylodon macrobotrys*, the jade vine
- *Syngonium*, the goosefoot vine
    - *Syngonium podophyllum*, the arrowhead vine
- *Thunbergia grandiflora*, known as the Bengal clock vine or blue trumpet vine
- *Thunbergia erecta*, the bush clock vine
- *Toxicodendron radicans*, known as poison ivy
- *Vitis*, any of about sixty species of grape
- *Wisteria*, a genus of flowering plants in the pea family

Tree

Common ash (*Fraxinus excelsior*), a deciduous broad-leaved (angiosperm) tree

European larch (*Larix decidua*), a coniferous tree which is also deciduous

Lepidodendron, an extinct lycophyte tree

In botany, a tree is a perennial plant with an elongated stem, or trunk, supporting branches and leaves in most species. In some usages, the definition of a tree may be narrower, including only woody plants with secondary growth, plants that are usable as lumber or plants above a specified height. Trees are not a taxonomic group but include a variety of plant species that have independently evolved a woody trunk and branches as a way to tower above other plants to compete for sunlight. In looser senses, the taller palms, the tree ferns, bananas and bamboos are also trees. Trees tend to be long-lived, some reaching several thousand years old. The tallest known tree, a coast redwood named Hyperion, stands 115.6 m (379 ft) high. Trees have been in existence for 370 million years. It is estimated that there are just over 3 trillion mature trees in the world.

A tree typically has many secondary branches supported clear of the ground by the trunk. This trunk typically contains woody tissue for strength, and vascular tissue to carry materials from one part of the tree to another. For most trees it is surrounded by a layer of bark which serves as a protective barrier. Below the ground, the roots branch and spread out widely; they serve to anchor the tree and extract moisture and nutrients from the soil. Above ground, the branches divide into smaller branches and shoots. The shoots typically bear leaves, which capture light energy and convert it into sugars by photosynthesis, providing the food for the tree's growth and development. Flowers and fruit may also be present, but some trees, such as conifers, instead have pollen cones and seed cones; others, such as tree ferns, produce spores instead.

Trees play a significant role in reducing erosion and moderating the climate. They remove carbon dioxide from the atmosphere and store large quantities of carbon in their tissues. Trees and forests provide a habitat for many species of animals and plants. Tropical rainforests are one of the most biodiverse habitats in the world. Trees provide shade and shelter, timber for construction, fuel for cooking and heating, and fruit for food as well as having many other uses. In parts of the world, forests are shrinking as trees are cleared to increase the amount of land available for agriculture. Because of their longevity and usefulness, trees have always been revered, with sacred groves in various cultures, and they play a role in many of the world's mythologies.

Definition

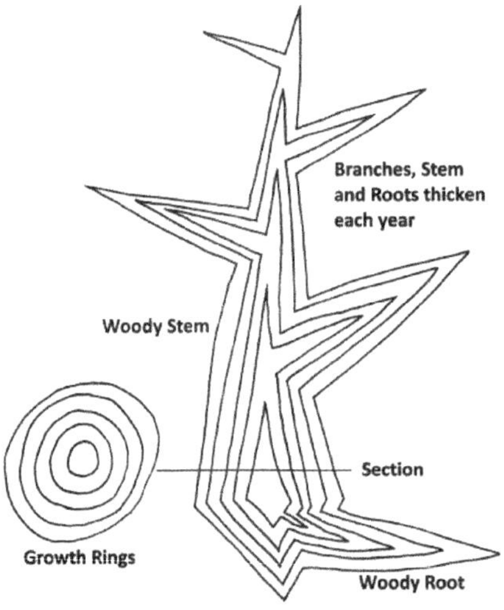

Diagram of secondary growth in a eudicot or coniferous tree showing idealised vertical and horizontal sections. A new layer of wood is added in each growing season, thickening the stem, existing branches and roots.

Although "tree" is a term of common parlance, there is no universally recognised precise definition of what a tree is, either botanically or in common language. In its broadest sense, a tree is any plant with the general form of an elongated stem, or trunk, which supports the photosynthetic leaves or branches at some distance above the ground. Trees are also typically defined by height, with smaller plants from 0.5 to 10 m (1.6 to 32.8 ft) being called shrubs, so the minimum height of a tree is only loosely defined. Large herbaceous plants such as papaya and bananas are trees in this broad sense.

A commonly applied narrower definition is that a tree has a woody trunk formed by secondary growth, meaning that the trunk thickens each year by growing outwards, in addition to the primary upwards growth from the growing tip. Under such a definition, herbaceous plants such as palms, bananas and papayas are not considered trees regardless of their height, growth form or stem girth. Certain monocots may be considered trees under a slightly looser definition; while the Joshua tree, bamboos and palms do not have secondary growth and never produce true wood with growth rings, they may produce "pseudo-wood" by lignifying cells formed by primary growth.

Aside from structural definitions, trees are commonly defined by use, for instance as those plants which yield lumber.

Overview

The tree growth habit is an evolutionary adaptation found in different groups of plants: by growing taller, trees are able to compete better for sunlight. Trees tend to be long-lived, some reaching several thousand years old, as well as tall. Trees have modified structures such as thicker stems composed of specialised cells that add structural strength and durability, allowing them to grow taller than non-woody plants and to spread out their foliage. They differ from shrubs, which are also woody plants, by usually growing larger and having a single main stem; but the distinction between a small tree and a large shrub is not always clear, made more confusing by the fact that trees may be reduced in size under harsher environmental conditions such as on mountains and subarctic areas. The tree form has evolved separately in unrelated classes of plants in response to similar environmental challenges, making it a classic example of parallel evolution. With an estimated 100,000 species, the number of trees worldwide might total twenty-five per cent of all living plant species. The greatest number of these grow in tropical regions and many of these areas have not yet been fully surveyed by botanists, making tree diversity and ranges poorly known.

Tall herbaceous monocotyledonous plants such as banana lack secondary growth, and are trees only in a loose sense.

Trees exist in two different groups of vascular or higher plants, the gymnosperms and the angiosperms. The gymnosperm trees include conifers, cycads, ginkgophytes and gnetales; they produce seeds which are not enclosed in fruits, but in open structures such as pine cones, and many have tough waxy leaves, such as pine needles. Most angiosperm trees are eudicots, the "true dicotyledons", so named because the seeds contain two cotyledons or seed leaves. There are also some trees among the old lineages of flowering plants called basal angiosperms or paleodicots; these include *Amborella*, *Magnolia*, nutmeg and avocado, while trees such as bamboo, palms and bananas are monocots.

Wood gives structural strength to the trunk of a tree; this supports the plant as it grows larger. The vascular system of trees allows water, nutrients and other chemicals to be distributed around the plant, and without it trees would not be able to grow as large as they do. Trees, as relatively tall plants, need to draw water up the stem through the xylem from the roots by the suction produced as water evaporates from the leaves. If insufficient water is available the leaves will die. The three main parts of trees include the root, stem, and leaves; they are integral parts of the vascular system which interconnects all the living cells. In trees and other plants that develop wood, the vascular cambium allows the expansion of vascular tissue that produces woody growth. Because this growth ruptures the epidermis of the stem, woody plants also have a cork cambium that develops among the phloem. The cork cambium gives rise to thickened cork cells to protect the surface of the plant and reduce water loss. Both the production of wood and the production of cork are forms of secondary growth.

Trees are either evergreen, having foliage that persists and remains green throughout the year, or deciduous, shedding their leaves at the end of the growing season and then having a dormant period without foliage. Most conifers are evergreens but larches (*Larix* and *Pseudolarix*) are deciduous, dropping their needles each autumn, and some species of cypress (*Glyptostrobus*, *Metasequoia* and *Taxodium*) shed small leafy shoots annually in a process known as cladoptosis. The crown is a name for the spreading top of a tree including the branches and leaves, while the uppermost layer in a forest, formed by the crowns of the trees, is known as the canopy. A sapling is a young tree.

Many tall palms are herbaceous monocots; these do not undergo secondary growth and never produce wood. In many tall palms, the terminal bud on the main stem is the only one to develop, so they have unbranched trunks with large spirally arranged leaves. Some of the tree ferns, order Cyatheales, have tall straight trunks, growing up to 20 metres (66 ft), but these are composed not of wood but of rhizomes which grow vertically and are covered by numerous adventitious roots.

Distribution

The Daintree rainforest

The number of trees in the world, according to a 2015 estimate, is 3.04 trillion, of which 1.39 trillion (46%) are in the tropics or sub-tropics, 0.61 trillion (20%) in the temperate zones, and 0.74 trillion (24%) in the coniferous boreal forests. The estimate is about eight times higher than previous estimates, and is based on tree densities measured on over 400,000 plots. It remains subject to a wide margin of error, not least because the samples are mainly from Europe and North America. The estimate suggests that about 15 billion trees are cut down annually and about 5 billion are planted. In the 12,000 years since the start of human agriculture, the number of trees worldwide has decreased by 46%.

In suitable environments, such as the Daintree Rainforest in Queensland, or the mixed podocarp and broadleaf forest of Ulva Island, New Zealand, forest is the more-or-less stable climatic climax community at the end of a plant succession, where open areas such as grassland are colonised by taller plants, which in turn give way to trees that eventually form a forest canopy.

In cool temperate regions, conifers often predominate; a widely distributed climax community in the far north of the northern hemisphere is moist taiga or northern coniferous forest (also called boreal forest). Taiga is the world's largest land biome, forming 29% of the world's forest cover. The long cold winter of the far north is unsuitable for plant growth and trees must grow rapidly in the short summer season when the temperature rises and the days are long. Light is very limited under their dense cover and there may be little plant life on the forest floor, although fungi may abound. Similar woodland is found on mountains where the altitude causes the average temperature to be lower thus reducing the length of the growing season.

Where rainfall is relatively evenly spread across the seasons in temperate regions, temperate broadleaf and mixed forest typified by species like oak, beech, birch and maple is found. Temperate forest is also found in the southern hemisphere, as for example in the Eastern Australia temperate forest, characterised by *Eucalyptus* forest and open acacia woodland.

In tropical regions with a monsoon or monsoon-like climate, where a drier part of the year alternates with a wet period as in the Amazon rainforest, different species of broad-leaved trees dominate the forest, some of them being deciduous. In tropical regions with a drier savanna climate and insufficient rainfall to support dense forests, the canopy is not closed, and plenty of sunshine reaches the ground which is covered with grass and scrub. *Acacia* and baobab are well adapted to living in such areas.

Parts and Function

A young red pine (*Pinus resinosa*) with spread of roots visible, as a result of soil erosion

Roots

The roots of a tree serve to anchor it to the ground and gather water and nutrients to transfer to all parts of the tree. They are also used for reproduction, defence, survival, energy storage and many other purposes. The radicle or embryonic root is the first part of a seedling to emerge from the seed during the process of germination. This develops into a taproot which goes straight downwards. Within a few weeks lateral roots branch out of the side of this and grow horizontally through the upper layers of the soil. In most trees, the taproot eventually withers away and the wide-spreading laterals remain. Near the tip of the finer roots are single cell root hairs. These are in immediate contact with the soil particles and can absorb water and nutrients such as potassium in solution. The roots require oxygen to respire and only a few species such as the mangrove and the pond cypress (*Taxodium ascendens*) can live in permanently waterlogged soil.

In the soil, the roots encounter the hyphae of fungi. Many of these are known as mycorrhiza and form a mutualistic relationship with the tree roots. Some are specific to a single tree species, which will not flourish in the absence of its mycorrhizal associate. Others are generalists and associate with many species. The tree acquires minerals such as phosphorus from the fungus while it obtains the carbohydrate products of photosynthesis from the tree. The hyphae of the fungus can link different trees and a network is formed, transferring nutrients from one place to another. The fungus promotes growth of the roots and helps protect the trees against predators and pathogens. It can also limit damage done to a tree by pollution as the fungus accumulate heavy metals within its tissues. Fossil evidence shows that roots have been associated with mycorrhizal fungi since the early Paleozoic, four hundred million years ago, when the first vascular plants colonised dry land.

Some trees such as the alders (*Alnus* species) have a symbiotic relationship with *Frankia* species, a filamentous bacterium that can fix nitrogen from the air, converting it into ammonia. They have actinorhizal root nodules on their roots in which the bacteria live. This process enables the tree to live in low nitrogen habitats where they would otherwise be unable to thrive. The plant hormones called cytokinins initiate root nodule formation, in a process closely related to mycorrhizal association.

Buttress roots of the kapok tree (*Ceiba pentandra*)

It has been demonstrated that some trees are interconnected through their root system, forming a colony. The interconnections are made by the inosculation process, a kind of natural grafting or welding of vegetal tissues. The tests to demonstrate this networking are performed by injecting chemicals, sometimes radioactive, into a tree, and then checking for its presence in neighbouring trees.

The roots are, generally, an underground part of the tree, but some tree species have evolved roots that are aerial. The common purposes for aerial roots may be of two kinds, to contribute to the mechanical stability of the tree, and to obtain oxygen from air. An instance of mechanical stability enhancement is the red mangrove that develops prop roots that loop out of the trunk and branches and descend vertically into the mud. A similar structure is developed by the Indian banyan. Many large trees have buttress roots which flare out from the lower part of the trunk. These brace the tree rather like angle brackets and provide stability, reducing sway in high winds. They are particularly prevalent in tropical rainforests where the soil is poor and the roots are close to the surface.

Some tree species have developed root extensions that pop out of soil, in order to get oxygen, when it is not available in the soil because of excess water. These root extensions are called pneumatophores, and are present, among others, in black mangrove and pond cypress.

Trunk

Northern beech (*Fagus sylvatica*) trunk in autumn

The main purpose of the trunk is to raise the leaves above the ground, enabling the tree to reach the light and survive: the tree can overtop other plants and shade them out. It also performs the task of transporting water and nutrients from the roots to the aerial parts of the tree and to distribute the food produced by the leaves to all other parts including the roots.

In the case of angiosperms and gymnosperms, the outermost layer of the trunk is the bark and is mostly composed of dead cells. It provides a thick, waterproof covering to the living inner tissue. It protects the trunk against the elements, disease, animal attack and fire. It is perforated by a large number of fine breathing pores called lenticels, through which oxygen diffuses. Bark is continually replaced by a living layer of cells called the cork cambium. The London plane (*Platanus × acerifolia*) periodically sheds its bark in large flakes. Similarly, the bark of the silver birch (*Betula pendula*) peels off in strips. As the tree's girth expands, newer layers of bark are larger in circumference, and the older layers develop fissures in many species. In some trees such as the pine (*Pinus* species) the bark exudes sticky resin which deters attackers whereas in rubber trees (*Hevea brasiliensis*) it is a milky latex that oozes out. The quinine bark tree (*Cinchona officinalis*) contains bitter substances to make the bark unpalatable. Large tree-like plants with lignified trunks in the

Pteridophyta, Arecales, Cycadophyta and Poales such as the tree ferns, palms, cycads and bamboos have no true bark, but they do have an outer protective covering of some form.

A section of yew (*Taxus baccata*) showing 27 annual growth rings, pale sapwood and dark heartwood

Although the bark functions as a protective barrier, it is itself attacked by boring insects such as beetles. These lay their eggs in crevices and the larvae chew their way through the cellulose tissues leaving a gallery of tunnels. This may allow fungal spores to gain admittance and attack the tree. Dutch elm disease is caused by a fungus (*Ophiostoma* species) carried from one elm tree to another by various beetles. The tree reacts to the growth of the fungus by blocking off the xylem tissue carrying sap upwards and the branch above, and eventually the whole tree, is deprived of nourishment and dies. In Britain in the 1990s, 25 million elm trees were killed by this disease.

The innermost layer of bark is known as the phloem and this is involved in the transport of the sap containing the sugars made by photosynthesis to other parts of the tree. It is a soft spongy layer of living cells, some of which are arranged end to end to form tubes. These are supported by parenchyma cells which provide padding and include fibres for strengthening the tissue. Inside the phloem is a layer of undifferentiated cells one cell thick called the vascular cambium layer. The cells are continually dividing, creating phloem cells on the outside and wood cells known as xylem on the inside.

The newly created xylem is the sapwood. It is composed of water-conducting cells and associated cells which are often living, and is usually pale in colour. It transports water and minerals from the roots to the upper parts of the tree. The oldest, inner part of the sapwood is progressively converted into heartwood as new sapwood is formed at the cambium. The conductive cells of the heartwood are blocked in some species, and the surrounding cells are more often dead. Heartwood is usually darker in colour than the sapwood. It is the dense central core of the trunk giving it rigidity. Three quarters of the dry mass of the xylem is cellulose, a polysaccharide, and most of the remainder is lignin, a complex polymer. A transverse section through a tree trunk or a horizontal core will show concentric circles or lighter or darker wood - tree rings. These rings are the annual growth rings There may also be rays running at right angles to growth rings. These are vascular rays which are thin sheets of living tissue permeating the wood. Many older trees may become hollow but may still stand upright for many years.

Buds and Growth

Dormant magnolia (*Magnolia sp.*) bud

Trees do not usually grow continuously throughout the year but mostly have spurts of active expansion followed by periods of rest. This pattern of growth is related to climatic conditions; growth normally ceases when conditions are either too cold or too dry. In readiness for the inactive period, trees form buds to protect the meristem, the zone of active growth. Before the period of dormancy, the last few leaves produced at the tip of a twig form scales. These are thick, small and closely wrapped and enclose the growing point in a waterproof sheath. Inside this bud there is a rudimentary stalk and neatly folded miniature leaves, ready to expand when the next growing season arrives. Buds also form in the axils of the leaves ready to produce new side shoots. A few trees, such as the eucalyptus, have "naked buds" with no protective scales and some conifers, such as the Lawson's cypress, have no buds but instead have little pockets of meristem concealed among the scale-like leaves.

When growing conditions improve, such as the arrival of warmer weather and the longer days associated with spring in temperate regions, growth starts again. The expanding shoot pushes its way out, shedding the scales in the process. These leave behind scars on the surface of the twig. The whole year's growth may take place in just a few weeks. The new stem is unlignified at first and may be green and downy. The Arecaceae (palms) have their leaves spirally arranged on an unbranched trunk. In some tree species in temperate climates, a second spurt of growth, a Lammas growth may occur which is believed to be a strategy to compensate for loss of early foliage to insect predators.

Primary growth is the elongation of the stems and roots. Secondary growth consists of a progressive thickening and strengthening of the tissues as the outer layer of the epidermis is converted into bark and the cambium layer creates new phloem and xylem cells. The bark is inelastic. Eventually the growth of a tree slows down and stops and it gets no taller. If damage occurs the tree may in time become hollow.

Leaves

Leaves are structures specialised for photosynthesis and are arranged on the tree in such a way as to maximise their exposure to light without shading each other. They are an important investment by the tree and may be thorny or contain phytoliths, lignins, tannins or poisons to discourage herbivory. Trees have evolved leaves in a wide range of shapes and sizes, in response to environmental pressures including climate and predation. They can be broad or needle-like, simple or compound, lobed or entire, smooth or hairy, delicate or tough, deciduous or evergreen. The needles of coniferous trees are compact but are structurally similar to those of broad-leaved trees. They are adapted for life in environments where resources are low or water is scarce. Frozen ground may limit water availability and conifers are often found in colder places at higher altitudes and higher latitudes than broad leaved trees. In conifers such as fir trees, the branches hang down at an angle to the trunk, enabling them to shed snow. In contrast, broad leaved trees in temperate regions deal with winter weather by shedding their leaves. When the days get shorter and the temperature begins to decrease, the leaves no longer make new chlorophyll and the red and yellow pigments already present in the blades become apparent. Synthesis in the leaf of a plant hormone called auxin also ceases. This causes the cells at the junction of the petiole and the twig to weaken until the joint breaks and the leaf floats to the ground. In tropical and subtropical regions, many trees keep their leaves all year round. Individual leaves may fall intermittently and be replaced by new growth but most leaves remain intact for some time. Other tropical species and those in arid regions may shed all their leaves annually, such as at the start of the dry season. Many deciduous trees flower before the new leaves emerge. A few trees do not have true leaves but instead have structures with similar external appearance such as Phylloclades – modified stem structures – as seen in the genus *Phyllocladus*.

Buds, leaves, flowers and fruit of oak (*Quercus robur*)

Buds, leaves and reproductive structures of white fir (*Abies alba*)

Reproduction

Form, leaves and reproductive structures of queen sago *Cycas circinalis*

Tree forms are found in a wide range of plants and their reproductive strategies are substantially the same as shrub or herbaceous plant forms. Many trees are wind pollinated which may be an evolutionary adaptation to take advantage of increased wind speeds high above the ground, particularly in the case of those that produce pollen before the leaves emerge. A vast quantity of pollen is produced because of the low likelihood of any particular grain landing on an appropriate female flower. Wind-pollinated flowers of broad-leaved trees are characterised by a lack of showy parts, no scent and a copious production of pollen, often with separate male and female flowers, or separate male and female trees. The male flowers may be high up in the tree, often in the form of dangling catkins. The female flowers may be lower down the tree. The pollen of pine trees contains air sacs which give it buoyancy and it has been known to travel as far as 800 kilometres (500 mi). Tree pollen can cause allergies and hay fever.

Seeds

Seeds are the primary way that trees reproduce and their seeds vary greatly in size and shape. Some of the largest seeds come from trees, but the largest tree, *Sequoiadendron giganteum*, produces one of the smallest tree seeds. The great diversity in tree fruits and seeds reflects the many different ways that tree species have evolved to disperse their offspring.

The single extant species of Ginkgophyta (*Ginkgo biloba*) has fleshy seeds produced at the ends of short branches on female trees, and *Gnetum*, a tropical and subtropical group of gymnosperms produce seeds at the tip of a shoot axis. The seeds of conifers, the largest group of gymnosperms, are enclosed in a cone and most species have seeds that are light and papery that can be blown considerable distances once free from the cone. Sometimes the seed remains in the cone for years waiting for a trigger event to liberate it. Fire stimulates release and germination of seeds of the jack pine, and also enriches the forest floor with wood ash and removes competing vegetation. Similarly, a number of angiosperms including *Acacia cyclops* and *Acacia mangium* have seeds that germinate better after exposure to high temperatures.

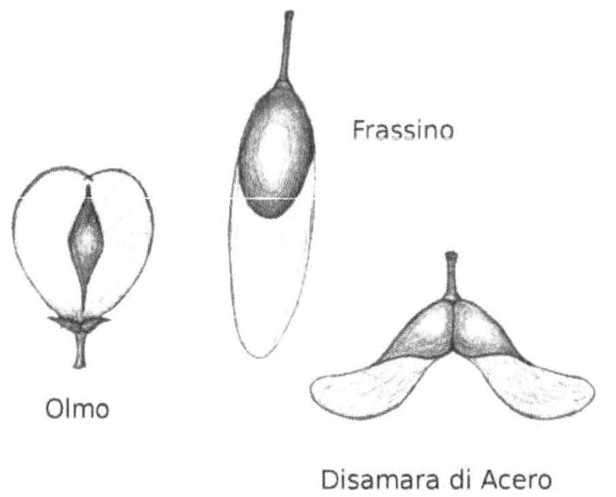

Wind dispersed seed of elm (*Ulmus*), ash (*Fraxinus*) and sycamore (*Acer pseudoplatanus*)

For a tree seedling to grow into an adult tree it needs light and space. If seeds only fell straight to the ground, competition among the concentrated saplings and the shade of the parent would likely prevent it from flourishing. Many seeds such as birch are small and have papery wings to aid dispersal by the wind. Ash trees and maples have larger seeds with blade shaped wings which spiral down to the ground when released. The kapok tree has cottony threads to catch the breeze. The flame tree does not rely on fire but shoots its seeds through the air when the two sides of its long pods crack apart explosively on drying. The miniature cone-like catkins of Alder trees produce seeds that contain small droplets of oil that help disperse the seeds on the surface of water. Mangroves often grow in water and some species have propagules, which are buoyant fruits with seeds that start germinating before becoming detached from the parent tree. These float on the water and may become lodged on emerging mudbanks and successfully take root. Other seeds, such as apple pips and plum stones, have fleshy receptacles and smaller fruits like hawthorns have seeds enclosed in edible tissue; animals including mammals and birds eat the fruits and either discard the seeds, or swallow them so they pass through the gut to be deposited in the animal's droppings well away from the parent tree. The germination of some seeds is improved when they are processed in this way. Nuts may be gathered by animals such as squirrels that cache any not immediately consumed. Many of these caches are never revisited, the nut-casing softens with rain and frost, and the seed germinates in the spring. Pine cones may similarly be hoarded by red squirrels, and grizzly bears may help to disperse the seed by raiding squirrel caches.

Evolutionary History

The earliest tree-like organisms were tree ferns, horsetails and lycophytes, which grew in forests in the Carboniferous period. The first tree may have been *Wattieza*, fossils of which have been found in New York State in 2007 dating back to the Middle Devonian (about 385 million years ago). Prior to this discovery, *Archaeopteris* was the earliest known tree. Both of these reproduced by spores rather than seeds and are considered to be links between ferns and the gymnosperms which evolved in the Triassic period. The gymnosperms include conifers, cycads, gnetales and ginkgos and these may have appeared as a result of a whole genome duplication event which took place about 319 million years ago. Ginkgophyta was once a widespread diverse group of which the only survivor is the maidenhair tree *Ginkgo biloba*. This is considered to be a living fossil because it is virtually unchanged from the fossilised specimens found in Triassic deposits.

Palms and cycads as they might have appeared in the middle Tertiary

During the Mesozoic (245 to 66 million years ago) the conifers flourished and became adapted to live in all the major terrestrial habitats. Subsequently, the tree forms of flowering plants evolved during the Cretaceous period. These began to dominate the conifers during the Tertiary era (66 to 2 million years ago) when forests covered the globe. When the climate cooled 1.5 million years ago and the first of four ice ages occurred, the forests retreated as the ice advanced. In the interglacials, trees recolonised the land that had been covered by ice, only to be driven back again in the next ice age.

Tree Ecology

Trees are an important part of the terrestrial ecosystem, providing essential habitats including many kinds of forest for communities of organisms. Epiphytic plants such as ferns, some mosses, liverworts, orchids and some species of parasitic plants (e.g., mistletoe) hang from branches; these along with arboreal lichens, algae, and fungi provide micro-habitats for themselves and for other organisms, including animals. Leaves, flowers and fruits are seasonally available. On the ground underneath trees there is shade, and often there is undergrowth, leaf litter, fallen branches and/ or decaying wood that provide other habitat. Trees stabilise the soil, prevent rapid run-off of rain water, help prevent desertification, have a role in climate control and help in the maintenance of biodiversity and ecosystem balance.

Many species of tree support their own specialised invertebrates. In their natural habitats, 284 different species of insect have been found on the English oak (*Quercus robur*) and 306 species of invertebrate on the Tasmanian oak (*Eucalyptus obliqua*). Non-native tree species provide a less biodiverse community, for example in the United Kingdom the sycamore (*Acer pseudoplatanus*), which originates from southern Europe, has few associated invertebrate species, though its bark supports a wide range of lichens, bryophytes and other epiphytes.

In ecosystems such as mangrove swamps, trees play a role in developing the habitat, since the roots of the mangrove trees reduce the speed of flow of tidal currents and trap water-borne sediment, reducing the water depth and creating suitable conditions for further mangrove colonisation. Thus mangrove swamps tend to extend seawards in suitable locations. Mangrove swamps also provide an effective buffer against the more damaging effects of cyclones and tsunamis.

Uses

Silviculture is the practice of controlling the establishment, growth, composition, health, and quality of forests, which are areas that have a high density of trees. Cultivated trees are planted and tended by humans, usually because they provide food (fruits or nuts), ornamental beauty, or some type of wood product that benefits people. A small wooded area, usually with no undergrowth, is called a grove and a small wood or thicket of trees and bushes is called a coppice or copse. A large area of land covered with trees and undergrowth is called woodland or forest. An area of woodland composed primarily of trees established by planting or artificial seeding is known as a plantation and an area of land planted with fruit or nut trees is an orchard.

Food

Forest honey

Trees are the source of many of the world's best known fleshy fruits. Apples, pears, plums, cherries and citrus are all grown commercially in temperate climates and a wide range of edible fruits are found in the tropics. Other commercially important fruit include dates, figs and olives. Palm oil is obtained from the fruits of the oil palm (*Elaeis guineensis*). The fruits of the cocoa tree (*Theobroma cacao*) are used to make cocoa and chocolate and the berries of coffee trees, *Coffea arabica* and *Coffea canephora*, are processed to extract the coffee beans. In many rural areas of the world, fruit is gathered from forest trees for consumption. Many trees bear edible nuts which can loosely be described as being large, oily kernels found inside a hard shell. These include coconuts (*Cocos nucifera*), Brazil nuts (*Bertholletia excelsa*), pecans (*Carya illinoinensis*), hazel nuts (*Corylus*), almonds (*Prunus dulcis*), walnuts (*Juglans regia*), pistachios (*Pistacia vera*) and many others. They are high in nutritive value and contain high-quality protein, vitamins and minerals as well as dietary fibre. Walnuts are particularly beneficial to health and contain a higher level of antioxidants than do other nuts. A variety of nut oils are extracted by pressing for culinary use; some such as walnut, pistachio and hazelnut oils are prized for their distinctive flavours, but they tend to spoil quickly.

Many trees have flowers rich in nectar which are attractive to bees. The production of forest honey is an important industry in rural areas of the developing world where it is undertaken by small-scale beekeepers using traditional methods. The flowers of the elder (*Sambucus*) are used to make elderflower cordial and petals of the plum (*Prunus spp.*) can be candied.

The leaves of trees are widely gathered as fodder for livestock and some can be eaten by humans but they tend to be high in tannins which makes them bitter. Leaves of the curry tree (*Murraya koenigii*) are eaten, those of kaffir lime *Citrus* × *hystrix* (e.g., Thai food) *Ailanthus* (e.g., in Korean dishes such as bugak) and those of the European bay tree (*Laurus nobilis*) and the California bay tree (*Umbellularia californica*) are used for flavouring food. *Camellia sinensis*, the source of tea, is a small tree but seldom reaches its full height, being heavily pruned to make picking the leaves easier.

Sugar maple (*Acer saccharum*) being tapped for the production of maple syrup

In temperate climates there is a sudden movement of sap at the end of the winter as trees prepare to burst into growth. In North America, the sap of the sugar maple (*Acer saccharum*) is most often used in the production of a sweet liquid, maple syrup. About 90% of the sap is water, the remaining 10% being a mixture of various sugars and certain minerals. The sap is harvested by drilling holes in the trunks of the trees and collecting the liquid that flows out of the inserted spigots. It is piped to a sugarhouse where it is heated to concentrate it and improve its flavour. One litre of maple syrup is obtained from every forty litres of sap and has a sugar content of exactly 66%. Similarly in northern Europe the spring rise in the sap of the silver birch (*Betula pendula*) is tapped and collected, either to be drunk fresh or fermented into an alcoholic drink. In Alaska, the sap of the sweet birch (*Betula lenta*) is made into a syrup with a sugar content of 67%. Sweet birch sap is more dilute than maple sap; a hundred litres are required to make one litre of birch syrup.

Various parts of trees are used as spices. These include cinnamon, made from the bark of the cinnamon tree (*Cinnamomum zeylanicum*) and allspice, the dried small fruits of the pimento tree (*Pimenta dioica*). Nutmeg is a seed found in the fleshy fruit of the nutmeg tree (*Myristica fragrans*) and cloves are the unopened flower buds of the clove tree (*Syzygium aromaticum*). Sassafras oil is an important flavouring obtained from distilling bark from the roots of the white sassafras tree (*Sassafras albidum*).

Fuel

Selling firewood at a market

Wood has traditionally been used for fuel, especially in rural areas. In less developed nations it may be the only fuel available and collecting firewood is often a time consuming task as it becomes necessary to travel further and further afield in the search for fuel. It is often burned inefficiently on an open fire. In more developed countries other fuels are available and burning wood is a choice rather than a necessity. Modern wood-burning stoves are very fuel efficient and new products such as wood pellets are available to burn.

Charcoal can be made by slow pyrolysis of wood by heating it in the absence of air in a kiln. The carefully stacked branches, often oak, are burned with a very limited amount of air. The process of converting them into charcoal takes about fifteen hours. Charcoal is used as a fuel in barbecues and by blacksmiths and has many industrial and other uses.

Wood smoke can be used to preserve food. In the hot smoking process the food is exposed to smoke and heat in a controlled environment. The food is ready to eat when the process is complete, having been tenderised and flavoured by the smoke it has absorbed. In the cold process, the temperature is not allowed to rise above 100 °F (38 °C). The flavour of the food is enhanced but raw food requires further cooking. If it is to be preserved, meat should be cured before cold smoking.

Timber

Joinery and roof trusses made from softwood

Timber, "trees that are grown in order to produce wood" is cut into lumber (sawn wood) for use in construction. Wood has been an important, easily available material for construction since humans started building shelters. Engineered wood products are available which bind the particles, fibres or veneers of wood together with adhesives to form composite materials. Plastics have taken over from wood for some traditional uses.

Wood is used in the construction of buildings, bridges, trackways, piles, poles for power lines, masts for boats, pit props, railway sleepers, fencing, hurdles, shuttering for concrete, pipes, scaffolding and pallets. In housebuilding it is used in joinery, for making joists, roof trusses, roofing shingles, thatching, staircases, doors, window frames, floor boards, parquet flooring, panelling and cladding.

Wood is used to construct carts, farm implements, boats, dugout canoes and in shipbuilding. It is used for making furniture, tool handles, boxes, ladders, musical instruments, bows, weapons, matches, clothes pegs, brooms, shoes, baskets, turnery, carving, toys, pencils, rollers, cogs, wooden screws, barrels, coffins, skittles, veneers, artificial limbs, oars, skis, wooden spoons, sports equipment and wooden balls.

Wood is pulped for paper and used in the manufacture of cardboard and made into engineered wood products for use in construction such as fibreboard, hardboard, chipboard and plywood. The wood of conifers is known as softwood while that of broad-leaved trees is hardwood.

Art

Informal upright style of bonsai on a juniper tree

Besides inspiring artists down the centuries, trees have been used to create art. Living trees have been used in bonsai and in tree shaping, and both living and dead specimens have been sculpted into sometimes fantastic shapes.

Bonsai

Bonsai (盆栽?, lit. The art of growing a miniature tree or trees in a low-sided pot or tray) is the practice of *hòn non bộ* originated in China and spread to Japan more than a thousand years ago, there are similar practices in other cultures like the living miniature landscapes of Vietnam *hòn non bộ*. The word *bonsai* is often used in English as an umbrella term for all miniature trees in containers or pots.

The purposes of bonsai are primarily contemplation (for the viewer) and the pleasant exercise of effort and ingenuity (for the grower). Bonsai practice focuses on long-term cultivation and shaping of one or more small trees growing in a container, beginning with a cutting, seedling, or small tree of a species suitable for bonsai development. Bonsai can be created from nearly any perennial woody-stemmed tree or shrub species that produces true branches and can be cultivated to remain small through pot confinement with crown and root pruning. Some species are popular as bonsai material because they have characteristics, such as small leaves or needles, that make them appropriate for the compact visual scope of bonsai and a miniature deciduous forest can even be created using such species as Japanese maple, Japanese zelkova or hornbeam.

Tree Shaping

People trees, by Pooktre

Tree shaping is the practice of changing living trees and other woody plants into man made shapes for art and useful structures. There are a few different methods of shaping a tree. There is a gradual method and there is an instant method, the gradual method slowly guides the growing tip along predetermined path ways over time where as the instant method bends and weaves saplings 2 to 3 m (6.6 to 9.8 ft) long into a shape that becomes more rigid as they thicken up. Most artists use grafting of living trunks, branches, and roots, for art or functional structures and there are plans to grow “living houses” with the branches of trees knitting together to give a solid, weatherproof exterior combined with an interior application of straw and clay to provide a stucco-like inner surface.

Tree shaping has been practised for at least several hundred years, the oldest known examples being the living root bridges built and maintained by the Khasi people of Meghalaya, India using the roots of the rubber tree (*Ficus elastica*).

Bark

Cork is produced from the thick bark of the cork oak (*Quercus suber*). It is harvested from the living trees about once every ten years in an environmentally sustainable industry. More than half the world’s cork comes from Portugal and is largely used to make stoppers for wine bottles. Other uses include floor tiles, bulletin boards, balls, footwear, cigarette tips, packaging, insulation and joints in woodwind instruments.

Recently stripped cork oak (*Quercus suber*)

The bark of other varieties of oak has traditionally been used in Europe for the tanning of hides though bark from other species of tree has been used elsewhere. The active ingredient, tannin, is extracted and after various preliminary treatments, the skins are immersed in a series of vats containing solutions in increasing concentrations. The tannin causes the hide to become supple, less affected by water and more resistant to bacterial attack.

At least 120 drugs come from plant sources, many of them from the bark of trees. Quinine originates from the cinchona tree (*Cinchona*) and was for a long time the remedy of choice for the treatment of malaria. Aspirin was synthesised to replace the sodium salicylate derived from the bark of willow trees (*Salix*) which had unpleasant side effects. The anti-cancer drug Paclitaxel is derived from taxol, a substance found in the bark of the Pacific yew (*Taxus brevifolia*). Other tree based drugs come from the paw-paw (*Carica papaya*), the cassia (*Cassia spp.*), the cocoa tree (*Theobroma cacao*), the tree of life (*Camptotheca acuminata*) and the downy birch (*Betula pubescens*).

The papery bark of the white birch tree (*Betula papyrifera*) was used extensively by Native Americans. Wigwams were covered by it and canoes were constructed from it. Other uses included food containers, hunting and fishing equipment, musical instruments, toys and sledges. Nowadays, bark chips, a by-product of the timber industry, are used as a mulch and as a growing medium for epiphytic plants that need a soil-free compost.

Alleé of London plane trees (*Platanus* × *acerifolia*) in garden

Ornamental Trees

Trees create a visual impact in the same way as do other landscape features and give a sense of maturity and permanence to park and garden. They are grown for the beauty of their forms, their foliage, flowers, fruit and bark and their siting is of major importance in creating a landscape. They can be grouped informally, often surrounded by plantings of bulbs, laid out in stately avenues or used as specimen trees. As living things, their appearance changes with the season and from year to year.

Trees are often planted in town environments where they are known as street trees or amenity trees. They can provide shade and cooling through evapotranspiration, absorb greenhouse gasses and pollutants, intercept rainfall and reduce the risk of flooding. It has been shown that they are beneficial to humans in creating a sense of well-being and reducing stress. Many towns have initiated tree-planting programmes. In London for example, there is an initiative to plant 20,000 new street trees and to have an increase in tree cover of 5% by 2025, equivalent to one tree for every resident.

Other Uses

Latex is a sticky defensive secretion that protects plants against herbivores. Many trees produce it when injured but the main source of the latex used to make natural rubber is the Pará rubber tree (*Hevea brasiliensis*). Originally used to create bouncy balls and for the waterproofing of cloth, natural rubber is now mainly used in tyres for which synthetic materials have proved less durable. The latex exuded by the balatá tree (*Manilkara bidentata*) is used to make golf balls and is similar to gutta-percha, made from the latex of the "getah perca" tree *Palaquium*. This is also used as an insulator, particularly of undersea cables, and in dentistry, walking sticks and gun butts. It has now largely been replaced by synthetic materials.

Resin is another plant exudate that may have a defensive purpose. It is a viscous liquid composed mainly of volatile terpenes and is produced mostly by coniferous trees. It is used in varnishes, for making small castings and in ten-pin bowling balls. When heated, the terpenes are driven off and the remaining product is called "rosin" and is used by stringed instrumentalists on their bows. Some resins contain essential oils and are used in incense and aromatherapy. Fossilised resin is known as amber and was mostly formed in the Cretaceous (145 to 66 million years ago) or more recently. The resin that oozed out of trees sometimes trapped insects or spiders and these are still visible in the interior of the amber.

The camphor tree (*Cinnamomum camphora*) produces an essential oil and the eucalyptus tree (*Eucalyptus globulus*) is the main source of eucalyptus oil which is used in medicine, as a fragrance and in industry.

Care

Dead trees pose a safety risk, especially during high winds and severe storms, and removing dead trees involves a financial burden, whereas the presence of healthy trees can clean the air, increase property values, and reduce the temperature of the built environment and thereby reduce building cooling costs. During times of drought, trees can fall into water stress, which may cause a tree to become more susceptible to disease and insect problems, and ultimately may lead to a tree's death. Irrigating trees during dry periods can reduce the risk of water stress and death. Irrigation can be accomplished by use of a garden hose, soaker hose, sprinkler, or modified five-gallon bucket.

Mythology

Yggdrasil, the World Ash of Norse mythology

Trees have been venerated since time immemorial. To the ancient Celts, certain trees, especially the oak, ash and thorn, held special significance as providing fuel, building materials, ornamental objects and weaponry. Other cultures have similarly revered trees, often linking the lives and fortunes of individuals to them or using them as oracles. In Greek mythology, dryads were believed to be shy nymphs who inhabited trees.

The Oubangui people of west Africa plant a tree when a child is born. As the tree flourishes, so does the child but if the tree fails to thrive, the health of the child is considered at risk. When it flowers it is time for marriage. Gifts are left at the tree periodically and when the individual dies, their spirit is believed to live on in the tree.

Trees have their roots in the ground and their trunk and branches extended towards the sky. This concept is found in many of the world's religions as a tree which links the underworld and the earth and holds up the heavens. In Norse mythology, Yggdrasil is a central cosmic tree whose roots and branches extend to various worlds. Various creatures live on it. In India, Kalpavriksha is a wish-fulfilling tree, one of the nine jewels that emerged from the primitive ocean. Icons are placed beneath it to be worshipped, tree nymphs inhabit the branches and it grants favours to the devout who tie threads round the trunk. Democracy started in North America when the Great Peacemaker formed the Iroquois Confederacy, inspiring the warriors of the original five American nations to bury their weapons under the Tree of Peace, an eastern white pine (*Pinus strobus*). In the creation story in the Bible, the tree of life and the knowledge of good and evil was planted by God in the Garden of Eden.

Sacred groves exist in China, India, Africa and elsewhere. They are places where the deities live and where all the living things are either sacred or are companions of the gods. Folklore lays down the supernatural penalties that will result if desecration takes place for example by the felling of trees. Because of their protected status, sacred groves may be the only relicts of ancient forest and have a biodiversity much greater than the surrounding area. Some Ancient Indian tree deities, such as Puliyidaivalaiyamman, the Tamil deity of the tamarind tree, or Kadambariyamman, associated with the kadamba tree were seen as manifestations of a goddess who offers her blessings by giving fruits in abundance.

Superlative Trees

The General Sherman Tree is thought to be the world's largest tree by volume.

Trees have a theoretical maximum height of 130 m (430 ft), but the tallest known specimen on earth is believed to be a coast redwood (*Sequoia sempervirens*) at Redwood National Park, California. It has been named Hyperion and is 115.6 metres (379 ft) tall. The tallest known broad-leaved tree is a mountain ash (*Eucalyptus regnans*) growing in Tasmania with a height of 99.8 m (327 ft).

The largest tree by volume is believed to be a giant sequoia (*Sequoiadendron giganteum*) known as the General Sherman Tree in the Sequoia National Park in Tulare County, California. Only the trunk is used in the calculation and the volume is estimated to be 1,487 m^3 (52,500 cu ft).

The oldest living tree with a verified age is also in California. It is a Great Basin bristlecone pine (*Pinus longaeva*) growing in the White Mountains. It has been dated by drilling a core sample and counting the annual rings. It is estimated to currently be 5,066 years old.[a]

A little further south, at Santa Maria del Tule, Oaxaca, Mexico, is the tree with the broadest trunk. It is a Montezuma cypress (*Taxodium mucronatum*) known as Árbol del Tule and its diameter at breast height is 11.62 m (38.1 ft) giving it a girth of 36.2 m (119 ft). The tree's trunk is far from round and the exact dimensions may be misleading as the circumference includes much empty space between the large buttress roots.

Forest

A forest is a large area dominated by trees. Hundreds of more precise definitions of forest are used throughout the world, incorporating factors such as tree density, tree height, land use, legal standing and ecological function. According to the widely used Food and Agriculture Organization definition, forests covered four billion hectares (15 million square miles) or approximately 30 percent of the world's land area in 2006.

A conifer forest in the Swiss Alps (National Park)

The Adirondack Mountains of Upstate New York form the southernmost part of the Eastern forest-boreal transition ecoregion.

Forest on Mount Dajt, Albania

Forests are the dominant terrestrial ecosystem of Earth, and are distributed across the globe. Forests account for 75% of the gross primary productivity of the Earth's biosphere, and contain 80% of the Earth's plant biomass.

Forests at different latitudes and elevations form distinctly different ecozones: boreal forests near the poles, tropical forests near the equator and temperate forests at mid-latitudes. Higher elevation areas tend to support forests similar to those at higher latitudes, and amount of precipitation also affects forest composition.

Human society and forests influence each other in both positive and negative ways. Forests provide ecosystem services to humans and serve as tourist attractions. Forests can also affect people's health. Human activities, including harvesting forest resources, can negatively affect forest ecosystems.

Definition

Forest in the Scottish Highlands

Although *forest* is a term of common parlance, there is no universally recognised precise definition, with more than 800 definitions of forest used around the world. Although a forest is usually defined by the presence of trees, under many definitions an area completely lacking trees may still be considered a forest if it grew trees in the past, will grow trees in the future, or was legally designated as a forest regardless of vegetation type.

There are three broad categories of forest definitions in use: administrative, land use, and land cover. Administrative definitions are based primarily upon the legal designations of land, and commonly bear little relationship to the vegetation growing on the land: land that is legally designated as a forest is defined as a forest even if no trees are growing on it. Land use definitions are based upon the primary purpose that the land serves. For example, a forest may defined as any land that is used primarily for production of timber. Under such a land use definition, cleared roads or infrastructure within an area used for forestry, or areas within the region that have been cleared by harvesting, disease or fire are still considered forests even if they contain no trees. Land cover definitions define forests based upon the type and density of vegetation growing on the land. Such definitions typically define a forest as an area growing trees above some threshold. These thresholds are typically the number of trees per area (density), the area of ground under the tree canopy (canopy cover) or the section of land that is occupied by the cross-section of tree trunks (basal area). Under such land cover definitions, and area of land only be defined as forest if it is growing trees. Areas that fail to meet the land cover definition may be still included under while immature trees are establishing if they are expected to meet the definition at maturity.

Under land use definitions, there is considerable variation on where the cutoff points are between a forest, woodland, and savanna. Under some definitions, forests require very high levels of tree canopy cover, from 60% to 100%, excluding savannas and woodlands in which trees have a lower canopy cover. Other definitions consider savannas to be a type of forest, and include all areas with tree canopies over 10%.

Conversely some areas covered in trees are legally defined as agricultural areas. For example, Norway spruce plantations in Austrian forest law when the trees are being grown as Christmas trees and below a certain height.

Etymology

Since the 13th century, the Niepołomice Forest in Poland has had special use and protection. In this view from space, different coloration can indicate different functions.

The word *forest* comes from Middle English, from Old French *forest* (also *forès*) "forest, vast expanse covered by trees"; first introduced in English as the word for wild land set aside for hunting without the necessity in definition for the existence of trees. Possibly a borrowing (probably via Frankish or Old High German) of the Medieval Latin word *foresta* "open wood", *foresta* was first used by Carolingian scribes in the Capitularies of Charlemagne to refer specifically to the king's royal hunting grounds. The term was not endemic to Romance languages (e.g. native words for "forest" in the Romance languages evolved out of the Latin word *silva* "forest, wood" (English *sylvan*); cf. Italian, Spanish, Portuguese *selva*; Romanian *silvă*; Old French *selve*); and cognates in Romance languages, such as Italian *foresta*, Spanish and Portuguese *floresta*, etc. are all ultimately borrowings of the French word.

The exact origin of Medieval Latin *foresta* is obscure. Some authorities claim the word derives from the Late Latin phrase *forestam silvam*, meaning "the outer wood"; others claim the term is a latinisation of the Frankish word **forhist* "forest, wooded country", assimilated to *forestam silvam* (a common practice among Frankish scribes). Frankish **forhist* is attested by Old High German *forst* "forest", Middle Low German *vorst* "forest", Old English *fyrhþ* "forest, woodland, game preserve, hunting ground" (English *frith*), and Old Norse *fýri* "coniferous forest", all of which derive from Proto-Germanic **furhísa-*, **furhíþija-* "a fir-wood, coniferous forest", from Proto-Indo-European **perkʷu-* "a coniferous or mountain forest, wooded height".

Uses of the word "forest" in English to denote any uninhabited area of non-enclosure are now considered archaic. The word was introduced by the Norman rulers of England as a legal term (appearing in Latin texts like the Magna Carta) denoting an uncultivated area legally set aside for hunting by feudal nobility (see Royal Forest).

These hunting forests were not necessarily wooded much, if at all. However, as hunting forests did often include considerable areas of woodland, the word "forest" eventually came to mean wooded land more generally. By the start of the 14th century, the word appeared in English texts, indicating all three senses: the most common one, the legal term and the archaic usage. Other terms used to mean "an area with a high density of trees" are *wood, woodland, wold, weald, holt, frith* and *firth*. Unlike *forest*, these are all derived from Old English and were not borrowed from another language. Some classifications now reserve the term *woodland* for an area with more open space between trees and distinguish among woodlands, *open forests*, and *closed forests* based on crown cover.

Tywi forest, in Wales

Evolution

The first known forests on Earth arose in the Late Devonian (approximately 380 million years ago), with the evolution of *Archaeopteris*. *Archaeopteris* was a plant that was both tree-like and fern-like, growing to 10 metres (33 ft) in height. *Archaeopteris* quickly spread throughout the world, from the equator to subpolar latitudes. *Archaeopteris* formed the first forest by being the first known species to cast shade due to its fronds and forming soil from its roots. *Archaeopteris* was deciduous, dropping its fronds onto the forest floor. The shade, soil, and forest duff from the dropped fronds created the first forest. The shed organic matter altered the freshwater environment, slowing it down and providing food. This promoted freshwater fish.

Ecology

Temperate rainforest in Tasmania's Hellyer Gorge

Forests account for 75% of the gross primary productivity of the Earth's biosphere, and contain

80% of the Earth's plant biomass. Forest ecosystems can be found in all regions capable of sustaining tree growth, at altitudes up to the tree line, except where natural fire frequency or other disturbance is too high, or where the environment has been altered by human activity.

The latitudes 10° north and south of the equator are mostly covered in tropical rainforest, and the latitudes between 53°N and 67°N have boreal forest. As a general rule, forests dominated by angiosperms (*broadleaf forests*) are more species-rich than those dominated by gymnosperms (*conifer*, *montane*, or *needleleaf forests*), although exceptions exist.

Forests sometimes contain many tree species within a small area (as in tropical rain and temperate deciduous forests), or relatively few species over large areas (e.g., taiga and arid montane coniferous forests). Forests are often home to many animal and plant species, and biomass per unit area is high compared to other vegetation communities. Much of this biomass occurs below ground in the root systems and as partially decomposed plant detritus. The woody component of a forest contains lignin, which is relatively slow to decompose compared with other organic materials such as cellulose or carbohydrate.

Components

Even, dense old-growth stand of beech trees (*Fagus sylvatica*) prepared to be regenerated by their saplings in the understory, in the Brussels part of the Sonian Forest.

A forest consists of many components that can be broadly divided into two categories that are biotic (living) and abiotic (non-living) components. The living parts include trees, shrubs, vines, grasses and other herbaceous (non-woody) plants, mosses, algae, fungi, insects, mammals, birds, reptiles, amphibians, and microorganisms living on the plants and animals and in the soil.

Layers

Biogradska forest in Montenegro

Spiny forest at Ifaty, Madagascar, featuring various *Adansonia* (baobab) species, *Alluaudia procera* (Madagascar ocotillo) and other vegetation

A forest is made up of many layers. Starting from the ground level and moving up, the main layers of all forest types are the forest floor, the understory and the canopy. The emergent layer exists in tropical rainforests. Each layer has a different set of plants and animals depending upon the availability of sunlight, moisture and food.

- Forest floor contains decomposing leaves, animal droppings, and dead trees. All of decay on the forest floor and form the new soil and provides nutrients to the plants. The forest floor supports ferns, grasses, mushroom and trees seedlings.
- Understory is made up of bushes, shrubs, and young trees that are adapted to living in the shades of the canopy.
- Canopy is formed by the mass of intertwined branches, twigs and leaves of the mature trees. The crowns of the dominant trees receive most of the sunlight. This is the most productive part of the trees where maximum food is produced. The canopy forms a shady, protective "umbrella" over the rest of the forest.
- Emergent layer exists in the tropical rain forest and is composed of a few scattered trees that tower over the canopy.

Types

Forests can be classified in different ways and to different degrees of specificity. One such way is in terms of the biome in which they exist, combined with leaf longevity of the dominant species (whether they are evergreen or deciduous). Another distinction is whether the forests are composed predominantly of broadleaf trees, coniferous (needle-leaved) trees, or mixed.

- Boreal forests occupy the subarctic zone and are generally evergreen and coniferous.
- Temperate zones support both broadleaf deciduous forests (*e.g.*, temperate deciduous forest) and evergreen coniferous forests (*e.g.*, temperate coniferous forests and temperate rainforests). Warm temperate zones support broadleaf evergreen forests, including laurel forests.

- Tropical and subtropical forests include tropical and subtropical moist forests, tropical and subtropical dry forests, and tropical and subtropical coniferous forests.
- Physiognomy classifies forests based on their overall physical structure or developmental stage (e.g. old growth vs. second growth).
- Forests can also be classified more specifically based on the climate and the dominant tree species present, resulting in numerous different forest types (e.g., Ponderosa pine/Douglas-fir forest).

The number of trees in the world, according to a 2015 estimate, is 3 trillion, of which 1.4 trillion are in the tropics or sub-tropics, 0.6 trillion in the temperate zones, and 0.7 trillion in the coniferous boreal forests. The estimate is about eight times higher than previous estimates, and is based on tree densities measured on over 400,000 plots. It remains subject to a wide margin of error, not least because the samples are mainly from Europe and North America.

Forests can also be classified according to the amount of human alteration. Old-growth forest contains mainly natural patterns of biodiversity in established seral patterns, and they contain mainly species native to the region and habitat. In contrast, secondary forest is regrowing forest following timber harvest and may contain species originally from other regions or habitats.

Different global forest classification systems have been proposed, but none has gained universal acceptance. UNEP-WCMC's forest category classification system is a simplification of other more complex systems (e.g. UNESCO's forest and woodland 'subformations'). This system divides the world's forests into 26 major types, which reflect climatic zones as well as the principal types of trees. These 26 major types can be reclassified into 6 broader categories: temperate needleleaf; temperate broadleaf and mixed; tropical moist; tropical dry; sparse trees and parkland; and forest plantations. Each category is described as a separate section below.

Boreal

Boreal forests occur in the highest latitudes, and are characterized by coniferous trees consisting mostly of pines, spruces and larches. The boreal forest is the world's largest biome apart from the oceans, and makes up 29% of the world's forest cover; the largest areas are located in Russia and Canada. The boreal forest is the terrestrial biome with the lowest annual average temperatures after the tundra and permanent ice caps.

Temperate Needleleaf

Temperate needleleaf forests mostly occupy the higher latitude regions of the northern hemisphere, as well as high altitude zones and some warm temperate areas, especially on nutrient-poor or otherwise unfavourable soils. These forests are composed entirely, or nearly so, of coniferous species (Coniferophyta). In the Northern Hemisphere pines *Pinus*, spruces *Picea*, larches *Larix*, firs *Abies*, Douglas firs *Pseudotsuga* and hemlocks *Tsuga*, make up the canopy, but other taxa are also important. In the Southern Hemisphere, most coniferous trees (members of the Araucariaceae and Podocarpaceae) occur in mixtures with broadleaf species, and are classed as broadleaf and mixed forests.

Temperate Broadleaf and Mixed

Broadleaf forest in Bhutan

Temperate broadleaf and mixed forests include a substantial component of trees in the Anthophyta. They are generally characteristic of the warmer temperate latitudes, but extend to cool temperate ones, particularly in the southern hemisphere. They include such forest types as the mixed deciduous forests of the United States and their counterparts in China and Japan, the broadleaf evergreen rainforests of Japan, Chile and Tasmania, the sclerophyllous forests of Australia, central Chile, the Mediterranean and California, and the southern beech Nothofagus forests of Chile and New Zealand.

Tropical Moist

There are many different types of tropical moist forests, with lowland evergreen broad leaf tropical rainforests, for example várzea and igapó forests and the terra firma forests of the Amazon Basin; the peat swamp forests, dipterocarp forests of Southeast Asia; and the high forests of the Congo Basin. Seasonal tropical forests, perhaps the best description for the colloquial term "jungle", typically range from the rainforest zone 10 degrees north or south of the equator, to the Tropic of Cancer and Tropic of Capricorn. Forests located on mountains are also included in this category, divided largely into upper and lower montane formations on the basis of the variation of physiognomy corresponding to changes in altitude.

Tropical Dry

Tropical dry forests are characteristic of areas in the tropics affected by seasonal drought. The seasonality of rainfall is usually reflected in the deciduousness of the forest canopy, with most trees being leafless for several months of the year. However, under some conditions, e.g. less fertile soils or less predictable drought regimes, the proportion of evergreen species increases and the forests are characterised as "sclerophyllous". Thorn forest, a dense forest of low stature with a high frequency of thorny or spiny species, is found where drought is prolonged, and especially where grazing animals are plentiful. On very poor soils, and especially where fire or herbivory are a recurrent phenomenon, savannas develop (see 'sparse trees and savanna').

Sparse Trees and Savanna

Taiga forest near Saranpaul in the northeast Ural Mountains, Khanty–Mansia, Russia. Trees include *Picea obovata* (dominant on right bank), *Larix sibirica*, *Pinus sibirica*, and *Betula pendula.*

Sparse trees and savanna are forests with lower canopy cover of trees. They occur principally in areas of transition from forested to non-forested landscapes. The two major zones in which these ecosystems occur are in the boreal region and in the seasonally dry tropics. At high latitudes, north of the main zone of boreal forest, growing conditions are not adequate to maintain a continuous closed forest cover, so tree cover is both sparse and discontinuous. This vegetation is variously called open taiga, open lichen woodland, and forest tundra. A savanna is a mixed woodland grassland ecosystem characterized by the trees being sufficiently widely spaced so that the canopy does not close. The open canopy allows sufficient light to reach the ground to support an unbroken herbaceous layer consisting primarily of grasses. Savannas maintain an open canopy despite a high tree density.

Forest Plantations

Forest plantations are generally intended for the production of timber and pulpwood. Commonly mono-specific and/or composed of introduced tree species, these ecosystems are not generally important as habitat for native biodiversity. However, they can be managed in ways that enhance their biodiversity protection functions and they can provide ecosystem services such as maintaining nutrient capital, protecting watersheds and soil structure, and storing carbon.

Societal Significance

Coastal Douglas fir woodland in northwest Oregon

Redwood tree in northern California redwood forest, where many redwood trees are managed for preservation and longevity, rather than being harvested for wood production

A forest near Vinitsa, Republic of Macedonia

Forests provide a diversity of ecosystem services including converting carbon dioxide into oxygen and biomass, acting as a carbon sink, aiding in regulating climate, purifying water, mitigating natural hazards such as floods, and serving as a genetic reserve. Forests also serve as a source of lumber and as recreational areas.

Some researchers state that forests do not only provide benefits, but can in certain cases also incur costs to humans. Forests may impose an economic burden, diminish the enjoyment of natural areas, reduce the food producing capacity of grazing land and cultivated land, reduce biodiversity reduce available water for humans and wildlife, harbour dangerous or destructive wildlife, and act as reservoirs of human and livestock disease.

The management of forests is often referred to as forestry. Forest management has changed considerably over the last few centuries, with rapid changes from the 1980s onwards culminating in a practice now referred to as sustainable forest management. Forest ecologists concentrate on forest patterns and processes, usually with the aim of elucidating cause-and-effect relationships. Foresters who practice sustainable forest management focus on the integration of ecological, social, and economic values, often in consultation with local communities and other stakeholders.

Humans have generally decreased the amount of forest worldwide. Anthropogenic factors that can affect forests include logging, urban sprawl, human-caused forest fires, acid rain, invasive species, and the slash and burn practices of swidden agriculture or shifting cultivation. The loss and re-growth of forest leads to a distinction between two broad types of forest, primary or old-growth forest and secondary forest. There are also many natural factors that can cause changes in forests over time including forest fires, insects, diseases, weather, competition between species, etc. In 1997, the World Resources Institute recorded that only 20% of the world's original forests remained in large intact tracts of undisturbed forest. More than 75% of these intact forests lie in three countries—the boreal forests of Russia and Canada and the rainforest of Brazil.

In 2010, the Food and Agriculture Organization of the United Nations reported that world deforestation, mainly the conversion of tropical forests to agricultural land, had decreased over the past ten years but still continues at a high rate in many countries. Globally, around 13 million hectares of forests were converted to other uses or lost through natural causes each year between 2000 and 2010 as compared to around 16 million hectares per year during the 1990s. The study covered 233 countries and areas. Brazil and Indonesia, which had the highest loss of forests in the 1990s, have significantly reduced their deforestation rates. China instituted a ban on logging, beginning in 1998, due to the erosion and flooding that it caused. In addition, ambitious tree planting programmes in countries such as China, India, the United States and Vietnam - combined with natural expansion of forests in some regions - have added more than seven million hectares of new forests annually. As a result, the net loss of forest area was reduced to 5.2 million hectares per year between 2000 and 2010, down from 8.3 million hectares annually in the 1990s. In 2015, a study for *Nature Climate Change* showed that the trend has recently been reversed, leading to an "overall gain" in global biomass and forests. This gain is due especially to reforestation in China and Russia. However new forests are not completely equivalent to old growth forests in terms of species diversity, resilience and carbon capture. On September 7, 2015, the Food and Agriculture Organization of the United Nations released a new study stating that, over the last 25 years, the global deforestation rate has decreased by 50% due to improved management of forests and greater government protection.

Smaller areas of woodland in cities may be managed as urban forestry, sometimes within public parks. These are often created for human benefits; Attention Restoration Theory argues that spending time in nature reduces stress and improves health, while forest schools and kindergartens help young people to develop social as well as scientific skills in forests. These typically need to be close to where the children live, for practical logistics.

Canada

Garibaldi Provincial Park, British Columbia

Canada has about 4,020,000 square kilometres (1,550,000 sq mi) of forest land. More than 90% of forest land is publicly owned and about 50% of the total forest area is allocated for harvesting. These allocated areas are managed using the principles of sustainable forest management, which includes extensive consultation with local stakeholders. About eight percent of Canada's forest is legally protected from resource development. Much more forest land—about 40 percent of the total forest land base—is subject to varying degrees of protection through processes such as integrated land use planning or defined management areas such as certified forests.

By December 2006, over 1,237,000 square kilometers of forest land in Canada (about half the global total) had been certified as being sustainably managed. Clearcutting, first used in the latter half of the 20th century, is less expensive, but devastating to the environment, and companies are required by law to ensure that harvested areas are adequately regenerated. Most Canadian provinces have regulations limiting the size of clearcuts, although some older clearcuts can range upwards of 110 square kilometres (27,000 acres) in size which were cut over several years.

United States

Trees on a mountain in northern Utah during early autumn.

In the United States, most forests have historically been affected by humans to some degree, though in recent years improved forestry practices has helped regulate or moderate large scale or severe impacts. However, the United States Forest Service estimates a net loss of about 2 million hectares (4,942,000 acres) between 1997 and 2020; this estimate includes conversion of forest land to other uses, including urban and suburban development, as well as afforestation and natural reversion of abandoned crop and pasture land to forest. However, in many areas of the United States, the area of forest is stable or increasing, particularly in many northern states. The opposite problem from flooding has plagued national forests, with loggers complaining that a lack of thinning and proper forest management has resulted in large forest fires.

Land Area

Forest land area										
	2008		2009		2010		2011		2012	
	('000 km²)	('000 mi²)	('000 km²)	('000 mi²)	('000 km²)	('000 mi²)	('000 km²)	('000 mi²)	('000 km²)	('000 mi²)
Australia	1,511	583	1,502	580	1,493	576	1,473	569	1,467	566
Brazil	5,239	2,023	5,217	2,014	5,195	2,006	5,183	2,001	5,173	1,997

Canada	3,101	1,197	3,101	1,197	3,101	1,197	3,101	1,197	3,101	1,197
China	2,013	777	2,041	788	2,069	799	2,159	834	2,168	837
European Union	1,559	602	1,564	604	1,569	606	1,573	607	1,578	609
Germany	111	43	111	43	111	43	113	44	113	44
India	681	263	683	264	684	264	693	268	693	268
Indonesia	958	370	951	367	944	364	937	362	931	359
Japan	250	97	250	97	250	97	251	97	251	97
Russia	8,090	3,120	8,090	3,120	8,091	3,124	8,092	3,124	8,093	3,125
United States	3,033	1,171	3,036	1,172	3,040	1,170	3,049	1,177	3,051	1,178
World total	40,318	15,567	40,261	15,545	40,204	15,523	40,184	15,515	39,519	15,258

References

- Rost, Thomas L.; Weier, T. Elliot; Weier, Thomas Elliot (1979). Botany: a brief introduction to plant biology. New York: Wiley. p. 319. ISBN 0-471-02114-8.
- "Carol C. Baskin, Jerry M. Baskin. Seeds: Ecology, Biogeography, and Evolution of Dormancy and Germination. Elsevier, 2001". google.ca. p. 27. ISBN 0-12-080263-5.
- Morhardt, Sia; Morhardt, Emil; Emil Morhardt, J. (2004). California desert flowers: an introduction to families, genera, and species. Berkeley: University of California Press. p. 24. ISBN 0-520-24003-0.
- Black, Michael H.; Halmer, Peter (2006). The encyclopedia of seeds: science, technology and uses. Wallingford, UK: CABI. p. 224. ISBN 978-0-85199-723-0.
- Hartmann, Hudson Thomas, and Dale E. Kester. 1983. Plant propagation principles and practices. Englewood Cliffs, N.J.: Prentice-Hall. ISBN 0-13-681007-1.
- Taylor EL; Taylor TMC (1993). The biology and evolution of fossil plants. Englewood Cliffs, N.J: Prentice Hall. p. 466. ISBN 0-13-651589-4.
- Schlegel, Rolf H J (January 1, 2003). Encyclopedic Dictionary of Plant Breeding and Related Subjects. Haworth Press. p. 177. ISBN 1-56022-950-0.
- Mauseth, James D. (April 1, 2003). Botany: An Introduction to Plant Biology. Jones and Bartlett. pp. 271–272. ISBN 0-7637-2134-4.
- Spiegel-Roy, P.; E. E. Goldschmidt (August 28, 1996). The Biology of Citrus. Cambridge University Press. pp. 87–88. ISBN 0-521-33321-0.
- Heiser, Charles B. (April 1, 2003). Weeds in My Garden: Observations on Some Misunderstood Plants. Timber Press. pp. 93–95. ISBN 0-88192-562-4.
- Adams, Denise Wiles (February 1, 2004). Restoring American Gardens: An Encyclopedia of Heirloom Ornamental Plants, 1640–1940. Timber Press. ISBN 0-88192-619-1.
- Cothran, James R. (November 1, 2003). Gardens and Historic Plants of the Antebellum South. University of South Carolina Press. p. 221. ISBN 1-57003-501-6.
- Adrosko, Rita J. (June 1, 1971). Natural Dyes and Home Dyeing: A Practical Guide with over 150 Recipes. Courier Dover Publications. ISBN 0-486-22688-3.
- Wake, Warren (March 13, 2000). Design Paradigms: A Sourcebook for Creative Visualization. John Wiley and Sons. pp. 162–163. ISBN 978-0-471-29976-9.

4

Principles and Processes of Botany

The technology of using plants for food and fuel is agronomy. Agronomy combines within it subjects like biology, chemistry, economics and Earth science. Agronomists deal with a number of issues such as providing food, managing environmental issues and the extraction of energy from plants. The following unit focuses on the principles and the processes of botany.

Agronomy

Agronomy is the science and technology of producing and using plants for food, fuel, fiber, and land reclamation. Agronomy has come to encompass work in the areas of plant genetics, plant physiology, meteorology, and soil science. It is the application of a combination of sciences like biology, chemistry, economics, ecology, earth science, and genetics. Agronomists of today are involved with many issues, including producing food, creating healthier food, managing the environmental impact of agriculture, and extracting energy from plants. Agronomists often specialise in areas such as crop rotation, irrigation and drainage, plant breeding, plant physiology, soil classification, soil fertility, weed control, and insect and pest control.

Plant Breeding

An agronomist field sampling a trial plot of flax.

This area of agronomy involves selective breeding of plants to produce the best crops under various conditions. Plant breeding has increased crop yields and has improved the nutritional value of numerous crops, including corn, soybeans, and wheat. It has also led to the development of new types of plants. For example, a hybrid grain called triticale was produced by crossbreeding rye and wheat. Triticale contains more usable protein than does either rye or wheat. Agronomy has also been instrumental in fruit and vegetable production research.

Biotechnology

Purdue University agronomy professor George Van Scoyoc explains the difference between forest and prairie soils to soldiers of the Indiana National Guard's Agribusiness Development Team at the Beck Agricultural Center in West Lafayette, Indiana

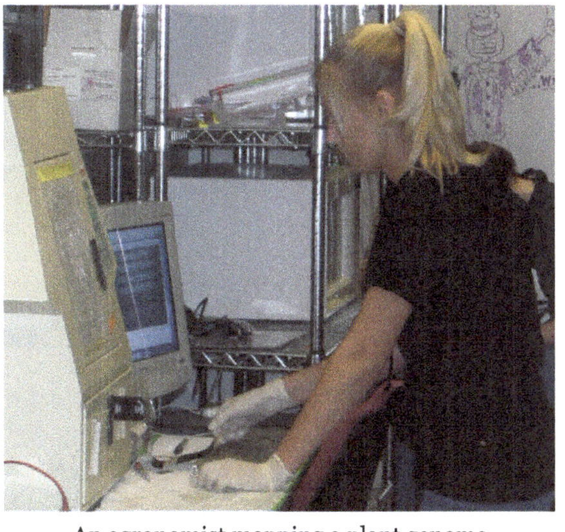

An agronomist mapping a plant genome

Agronomists use biotechnology to extend and expedite the development of desired characteristic. Biotechnology is often a lab activity requiring field testing of the new crop varieties that are developed.

In addition to increasing crop yields agronomic biotechnology is increasingly being applied for novel uses other than food. For example, oilseed is at present used mainly for margarine and other food oils, but it can be modified to produce fatty acids for detergents, substitute fuels and petrochemicals.

Soil Science

Agronomists study sustainable ways to make soils more productive and profitable. They classify soils and analyze them to determine whether they contain nutrients vital to plant growth. Common macronutrients analyzed include compounds of nitrogen, phosphorus, potassium, calcium, magnesium, and sulfur. Soil is also assessed for several micronutrients, like zinc and boron. The percentage of organic matter, soil pH, and nutrient holding capacity (cation exchange capacity) are tested in a regional laboratory. Agronomists will interpret these lab reports and make recommendations to balance soil nutrients for optimal plant growth.

Soil Conservation

In addition, agronomists develop methods to preserve the soil and to decrease the effects of erosion by wind and water. For example, a technique called contour plowing may be used to prevent soil erosion and conserve rainfall. Researchers in agronomy also seek ways to use the soil more effectively in solving other problems. Such problems include the disposal of human and animal manure, water pollution, and pesticide build-up in the soil. Techniques include no-tilling crops, planting of soil-binding grasses along contours on steep slopes, and contour drains of depths up to 1 metre.

Agroecology

Agroecology is the management of agricultural systems with an emphasis on ecological and environmental perspectives. This area is closely associated with work in the areas of sustainable agriculture, organic farming, alternative food systems and the development of alternative cropping systems.

Theoretical Modeling

Agronomy Schools

Agronomy programs, offered at colleges, universities, and specialized agricultural schools, often involve classes across a range of departments, including agriculture, biology, chemistry, economics, and physiology. Completing the coursework usually takes from four to twelve years. Many companies will pay agronomists-in-training's college expenses if they agree to work for them when they graduate.

Photosynthesis

Photosynthesis is a process used by plants and other organisms to convert light energy into chemical energy that can later be released to fuel the organisms' activities (energy transformation). This chemical energy is stored in carbohydrate molecules, such as sugars, which are synthesized from carbon dioxide and water. In most cases, oxygen is also released as a waste product. Most plants, most algae, and cyanobacteria perform photosynthesis; such organisms are called photoautotrophs. Photosynthesis is largely responsible for producing and maintaining the oxygen content of the Earth's atmosphere, and supplies all of the organic compounds and most of the energy necessary for life on Earth.

Schematic of photosynthesis in plants. The carbohydrates produced are stored in or used by the plant.

$$\underset{\text{Carbon dioxide}}{6CO_2} + \underset{\text{Water}}{6H_2O} \xrightarrow{\text{Light}} \underset{\text{Sugar}}{C_6H_{12}O_6} + \underset{\text{Oxygen}}{6O_2}$$

Overall equation for the type of photosynthesis that occurs in plants

Although photosynthesis is performed differently by different species, the process always begins when energy from light is absorbed by proteins called reaction centres that contain green chlorophyll pigments. In plants, these proteins are held inside organelles called chloroplasts, which are most abundant in leaf cells, while in bacteria they are embedded in the plasma membrane. In these light-dependent reactions, some energy is used to strip electrons from suitable substances, such as water, producing oxygen gas. The hydrogen freed by the splitting of water is used in the creation of two further compounds that act as an immediate energy storage means: reduced nicotinamide adenine dinucleotide phosphate (NADPH) and adenosine triphosphate (ATP), the "energy currency" of cells.

In plants, algae and cyanobacteria, long-term energy storage in the form of sugars is produced by a subsequent sequence of light-independent reactions called the Calvin cycle; some bacteria use different mechanisms, such as the reverse Krebs cycle, to achieve the same end. In the Calvin cycle, atmospheric carbon dioxide is incorporated into already existing organic carbon compounds, such as ribulose bisphosphate (RuBP). Using the ATP and NADPH produced by the light-dependent reactions, the resulting compounds are then reduced and removed to form further carbohydrates, such as glucose.

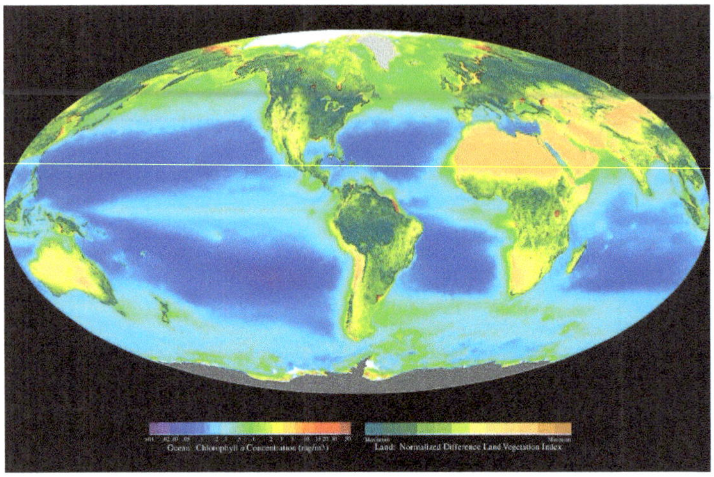

Composite image showing the global distribution of photosynthesis, including both oceanic phytoplankton and terrestrial vegetation. Dark red and blue-green indicate regions of high photosynthetic activity in the ocean and on land, respectively.

The first photosynthetic organisms probably evolved early in the evolutionary history of life and most likely used reducing agents such as hydrogen or hydrogen sulfide, rather than water, as sources of electrons. Cyanobacteria appeared later; the excess oxygen they produced contributed directly to the oxygenation of the Earth, which rendered the evolution of complex life possible. Today, the average rate of energy capture by photosynthesis globally is approximately 130 terawatts, which is about three times the current power consumption of human civilization. Photosynthetic organisms also convert around 100–115 thousand million metric tonnes of carbon into biomass per year.

Overview

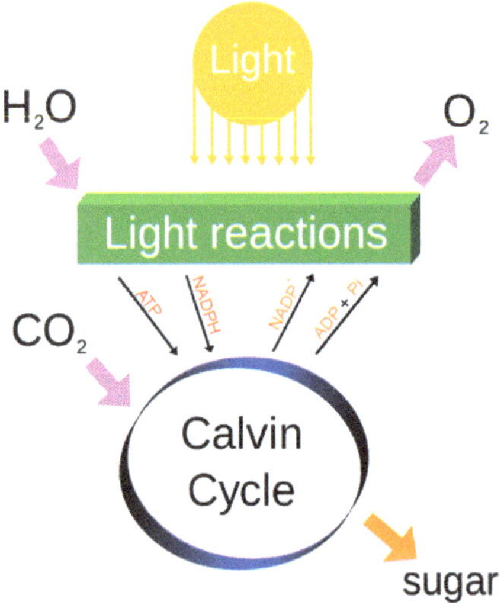

Photosynthesis changes sunlight into chemical energy, splits water to liberate O_2, and fixes CO_2 into sugar.

Photosynthetic organisms are photoautotrophs, which means that they are able to synthesize food directly from carbon dioxide and water using energy from light. However, not all organisms that use light as a source of energy carry out photosynthesis; photoheterotrophs use organic compounds, rather than carbon dioxide, as a source of carbon. In plants, algae and cyanobacteria, photosynthesis releases oxygen. This is called *oxygenic photosynthesis* and is by far the most common type of photosynthesis used by living organisms. Although there are some differences between oxygenic photosynthesis in plants, algae, and cyanobacteria, the overall process is quite similar in these organisms. There are also many varieties of anoxygenic photosynthesis, used mostly by certain types of bacteria, which consume carbon dioxide but do not release oxygen.

Carbon dioxide is converted into sugars in a process called carbon fixation. Carbon fixation is an endothermic redox reaction, so photosynthesis needs to supply both a source of energy to drive this process, and the electrons needed to convert carbon dioxide into a carbohydrate via a reduction reaction. The addition of electrons to a chemical species is called reduction. In general outline and in effect, photosynthesis is the opposite of cellular respiration, in which glucose and other compounds are oxidized to produce carbon dioxide and water, and to release chemical energy (an exothermic reaction) to drive the organism's metabolism. The two processes, reduction of carbon dioxide to carbohydrate and then later oxidation of the carbohydrate, are distinct: photosynthesis and cellular respiration take place through a different sequence of chemical reactions and in different cellular compartments.

The general equation for photosynthesis as first proposed by Cornelius van Niel is therefore:

$$CO_2 + 2H_2A + \text{photons} \rightarrow [CH_2O] + 2A + H_2O$$

carbon dioxide + electron donor + light energy → carbohydrate + oxidized electron donor + water

Since water is used as the electron donor in oxygenic photosynthesis, the equation for this process is:

$$CO_2 + 2H_2O + \text{photons} \rightarrow [CH_2O] + O_2 + H_2O$$

carbon dioxide + water + light energy → carbohydrate + oxygen + water

This equation emphasizes that water is both a reactant in the light-dependent reaction and a product of the light-independent reaction, but canceling *n* water molecules from each side gives the net equation:

$$CO_2 + H_2O + \text{photons} \rightarrow [CH_2O] + O_2$$

carbon dioxide + water + light energy → carbohydrate + oxygen

Other processes substitute other compounds (such as arsenite) for water in the electron-supply role; for example some microbes use sunlight to oxidize arsenite to arsenate: The equation for this reaction is:

$$CO_2 + (AsO3{-}3) + \text{photons} \rightarrow (AsO3{-}4) + CO$$

carbon dioxide + arsenite + light energy → arsenate + carbon monoxide (used to build other compounds in subsequent reactions)

Photosynthesis occurs in two stages. In the first stage, *light-dependent reactions* or *light reactions* capture the energy of light and use it to make the energy-storage molecules ATP and NADPH. During the second stage, the *light-independent reactions* use these products to capture and reduce carbon dioxide.

Most organisms that utilize oxygenic photosynthesis use visible light for the light-dependent reactions, although at least three use shortwave infrared or, more specifically, far-red radiation.

Some organisms employ even more radical variants of photosynthesis. Some archea use a simpler method that employs a pigment similar to those used for vision in animals. The bacteriorhodopsin changes its configuration in response to sunlight, acting as a proton pump. This produces a proton gradient more directly, which is then converted to chemical energy. The process does not involve carbon dioxide fixation and does not release oxygen, and seems to have evolved separately from the more common types of photosynthesis.

Photosynthetic Membranes and Organelles

Chloroplast ultrastructure:
1. outer membrane
2. intermembrane space
3. inner membrane (1+2+3: envelope)
4. stroma (aqueous fluid)
5. thylakoid lumen (inside of thylakoid)
6. thylakoid membrane
7. granum (stack of thylakoids)
8. thylakoid (lamella)
9. starch
10. ribosome
11. plastidial DNA
12. plastoglobule (drop of lipids)

In photosynthetic bacteria, the proteins that gather light for photosynthesis are embedded in cell membranes. In its simplest form, this involves the membrane surrounding the cell itself. However, the membrane may be tightly folded into cylindrical sheets called thylakoids, or bunched up into round vesicles called *intracytoplasmic membranes*. These structures can fill most of the interior of a cell, giving the membrane a very large surface area and therefore increasing the amount of light that the bacteria can absorb.

In plants and algae, photosynthesis takes place in organelles called chloroplasts. A typical plant cell contains about 10 to 100 chloroplasts. The chloroplast is enclosed by a membrane. This membrane is composed of a phospholipid inner membrane, a phospholipid outer membrane, and an intermembrane space. Enclosed by the membrane is an aqueous fluid called the stroma. Embedded within the stroma are stacks of thylakoids (grana), which are the site of photosynthesis. The thylakoids appear as flattened disks. The thylakoid itself is enclosed by the thylakoid membrane, and within the enclosed volume is a lumen or thylakoid space. Embedded in the thylakoid membrane are integral and peripheral membrane protein complexes of the photosynthetic system.

Plants absorb light primarily using the pigment chlorophyll. The green part of the light spectrum is not absorbed but is reflected which is the reason that most plants have a green color. Besides chlorophyll, plants also use pigments such as carotenes and xanthophylls. Algae also use chlorophyll, but various other pigments are present, such as phycocyanin, carotenes, and xanthophylls in green algae, phycoerythrin in red algae (rhodophytes) and fucoxanthin in brown algae and diatoms resulting in a wide variety of colors.

These pigments are embedded in plants and algae in complexes called antenna proteins. In such proteins, the pigments are arranged to work together. Such a combination of proteins is also called a light-harvesting complex.

Although all cells in the green parts of a plant have chloroplasts, the majority of those are found in specially adapted structures called leaves. Certain species adapted to conditions of strong sunlight and aridity, such as many Euphorbia and cactus species, have their main photosynthetic organs in their stems. The cells in the interior tissues of a leaf, called the mesophyll, can contain between 450,000 and 800,000 chloroplasts for every square millimeter of leaf. The surface of the leaf is coated with a water-resistant waxy cuticle that protects the leaf from excessive evaporation of water and decreases the absorption of ultraviolet or blue light to reduce heating. The transparent epidermis layer allows light to pass through to the palisade mesophyll cells where most of the photosynthesis takes place.

Light-dependent Reactions

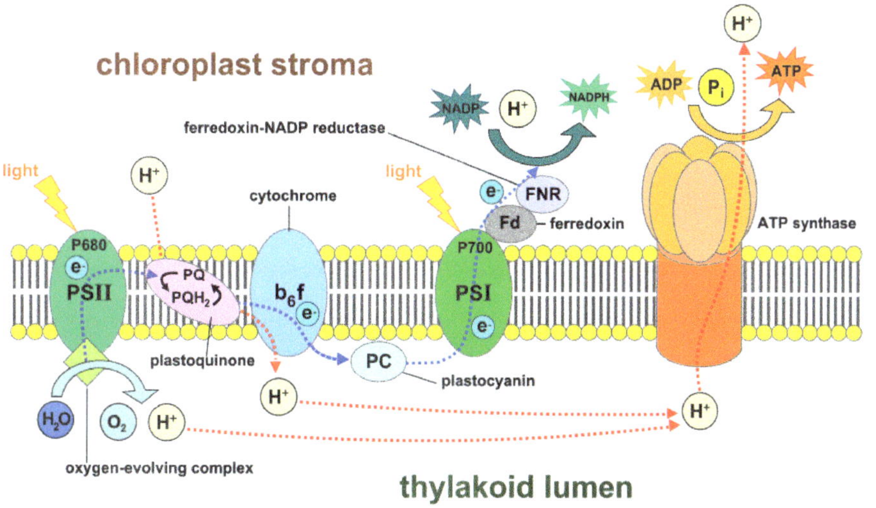

Light-dependent reactions of photosynthesis at the thylakoid membrane

In the light-dependent reactions, one molecule of the pigment chlorophyll absorbs one photon and loses one electron. This electron is passed to a modified form of chlorophyll called pheophytin, which passes the electron to a quinone molecule, starting the flow of electrons down an electron transport chain that leads to the ultimate reduction of NADP to NADPH. In addition, this creates a proton gradient (energy gradient) across the chloroplast membrane, which is used by ATP synthase in the synthesis of ATP. The chlorophyll molecule ultimately regains the electron it lost when a water molecule is split in a process called photolysis, which releases a dioxygen (O_2) molecule as a waste product.

The overall equation for the light-dependent reactions under the conditions of non-cyclic electron flow in green plants is:

$$2\ H_2O + 2\ NADP^+ + 3\ ADP + 3\ P_i + \text{light} \rightarrow 2\ NADPH + 2\ H^+ + 3\ ATP + O_2$$

Not all wavelengths of light can support photosynthesis. The photosynthetic action spectrum depends on the type of accessory pigments present. For example, in green plants, the action spectrum resembles the absorption spectrum for chlorophylls and carotenoids with peaks for violet-blue and red light. In red algae, the action spectrum is blue-green light, which allows these algae to use the blue end of the spectrum to grow in the deeper waters that filter out the longer wavelengths (red light) used by above ground green plants. The non-absorbed part of the light spectrum is what gives photosynthetic organisms their color (e.g., green plants, red algae, purple bacteria) and is the least effective for photosynthesis in the respective organisms.

Z scheme

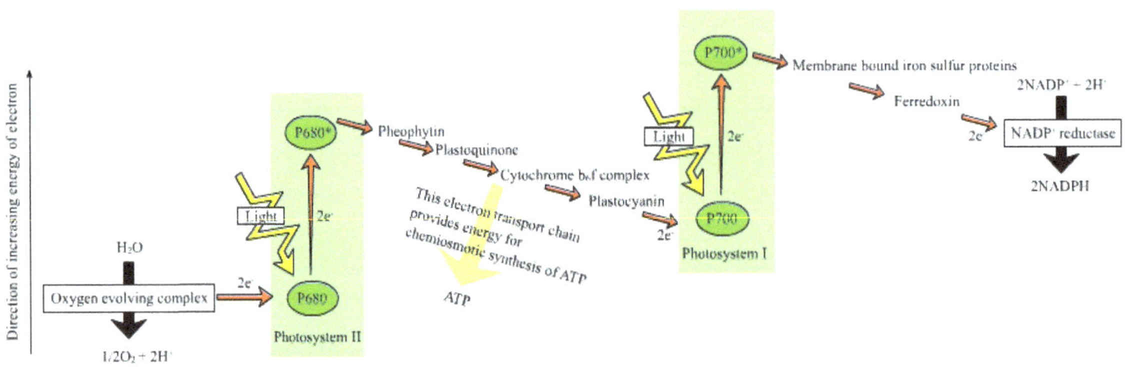

The "Z scheme"

In plants, light-dependent reactions occur in the thylakoid membranes of the chloroplasts where they drive the synthesis of ATP and NADPH. The light-dependent reactions are of two forms: cyclic and non-cyclic.

In the non-cyclic reaction, the photons are captured in the light-harvesting antenna complexes of photosystem II by chlorophyll and other accessory pigments (see diagram at right). The absorption of a photon by the antenna complex frees an electron by a process called photoinduced charge separation. The antenna system is at the core of the chlorophyll molecule of the photosystem II reaction center. That freed electron is transferred to the primary electron-acceptor molecule, pheophytin. As the electrons are shuttled through an electron transport chain (the so-called *Z-scheme* shown in the diagram), it initially functions to generate a chemiosmotic potential by pumping proton cations (H^+)

across the membrane and into the thylakoid space. An ATP synthase enzyme uses that chemiosmotic potential to make ATP during photophosphorylation, whereas NADPH is a product of the terminal redox reaction in the *Z-scheme*. The electron enters a chlorophyll molecule in Photosystem I. There it is further excited by the light absorbed by that photosystem. The electron is then passed along a chain of electron acceptors to which it transfers some of its energy. The energy delivered to the electron acceptors is used to move hydrogen ions across the thylakoid membrane into the lumen. The electron is eventually used to reduce the co-enzyme NADP with a H^+ to NADPH (which has functions in the light-independent reaction); at that point, the path of that electron ends.

The cyclic reaction is similar to that of the non-cyclic, but differs in that it generates only ATP, and no reduced NADP (NADPH) is created. The cyclic reaction takes place only at photosystem I. Once the electron is displaced from the photosystem, the electron is passed down the electron acceptor molecules and returns to photosystem I, from where it was emitted, hence the name *cyclic reaction*.

Water Photolysis

The NADPH is the main reducing agent produced by chloroplasts, which then goes on to provide a source of energetic electrons in other cellular reactions. Its production leaves chlorophyll in photosystem I with a deficit of electrons (chlorophyll has been oxidized), which must be balanced by some other reducing agent that will supply the missing electron. The excited electrons lost from chlorophyll from photosystem I are supplied from the electron transport chain by plastocyanin. However, since photosystem II is the first step of the *Z-scheme*, an external source of electrons is required to reduce its oxidized chlorophyll *a* molecules. The source of electrons in green-plant and cyanobacterial photosynthesis is water. Two water molecules are oxidized by four successive charge-separation reactions by photosystem II to yield a molecule of diatomic oxygen and four hydrogen ions; the electrons yielded are transferred to a redox-active tyrosine residue that then reduces the oxidized chlorophyll *a* (called P680) that serves as the primary light-driven electron donor in the photosystem II reaction center. That photo receptor is in effect reset and is then able to repeat the absorption of another photon and the release of another photo-dissociated electron. The oxidation of water is catalyzed in photosystem II by a redox-active structure that contains four manganese ions and a calcium ion; this oxygen-evolving complex binds two water molecules and contains the four oxidizing equivalents that are used to drive the water-oxidizing reaction. Photosystem II is the only known biological enzyme that carries out this oxidation of water. The hydrogen ions released contribute to the transmembrane chemiosmotic potential that leads to ATP synthesis. Oxygen is a waste product of light-dependent reactions, but the majority of organisms on Earth use oxygen for cellular respiration, including photosynthetic organisms.

Light-independent Reactions

Calvin Cycle

In the light-independent (or "dark") reactions, the enzyme RuBisCO captures CO_2 from the atmosphere and, in a process called the Calvin-Benson cycle, it uses the newly formed NADPH and releases three-carbon sugars, which are later combined to form sucrose and starch. The overall equation for the light-independent reactions in green plants is

$$3\ CO_2 + 9\ ATP + 6\ NADPH + 6\ H^+ \rightarrow C_3H_6O_3\text{-phosphate} + 9\ ADP + 8\ P_i + 6\ NADP^+ + 3\ H_2O$$

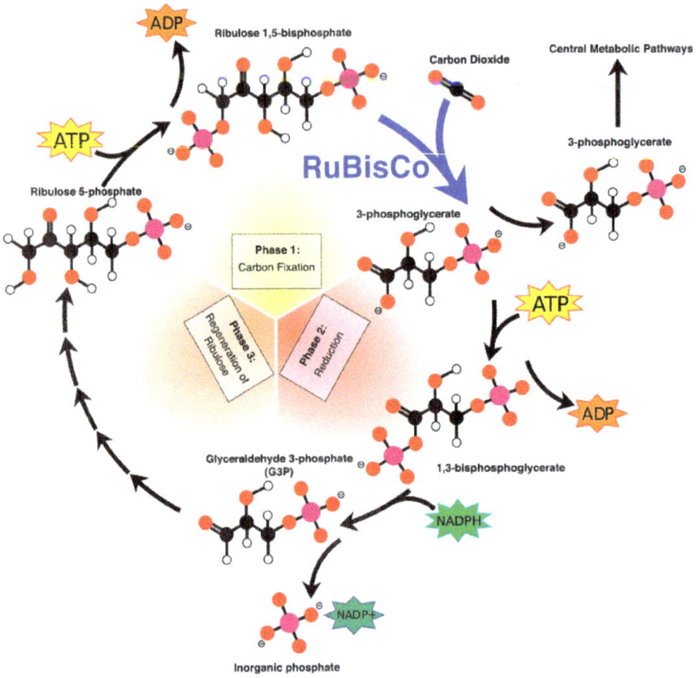

Overview of the Calvin cycle and carbon fixation

Carbon fixation produces the intermediate three-carbon sugar product, which is then converted to the final carbohydrate products. The simple carbon sugars produced by photosynthesis are then used in the forming of other organic compounds, such as the building material cellulose, the precursors for lipid and amino acid biosynthesis, or as a fuel in cellular respiration. The latter occurs not only in plants but also in animals when the energy from plants is passed through a food chain.

The fixation or reduction of carbon dioxide is a process in which carbon dioxide combines with a five-carbon sugar, ribulose 1,5-bisphosphate, to yield two molecules of a three-carbon compound, glycerate 3-phosphate, also known as 3-phosphoglycerate. Glycerate 3-phosphate, in the presence of ATP and NADPH produced during the light-dependent stages, is reduced to glyceraldehyde 3-phosphate. This product is also referred to as 3-phosphoglyceraldehyde (PGAL) or, more generically, as triose phosphate. Most (5 out of 6 molecules) of the glyceraldehyde 3-phosphate produced is used to regenerate ribulose 1,5-bisphosphate so the process can continue. The triose phosphates not thus "recycled" often condense to form hexose phosphates, which ultimately yield sucrose, starch and cellulose. The sugars produced during carbon metabolism yield carbon skeletons that can be used for other metabolic reactions like the production of amino acids and lipids.

Carbon Concentrating Mechanisms

On Land

In hot and dry conditions, plants close their stomata to prevent water loss. Under these conditions, CO_2 will decrease and oxygen gas, produced by the light reactions of photosynthesis, will increase, causing an increase of photorespiration by the oxygenase activity of ribulose-1,5-bisphosphate carboxylase/oxygenase and decrease in carbon fixation. Some plants have evolved mechanisms to increase the CO_2 concentration in the leaves under these conditions.

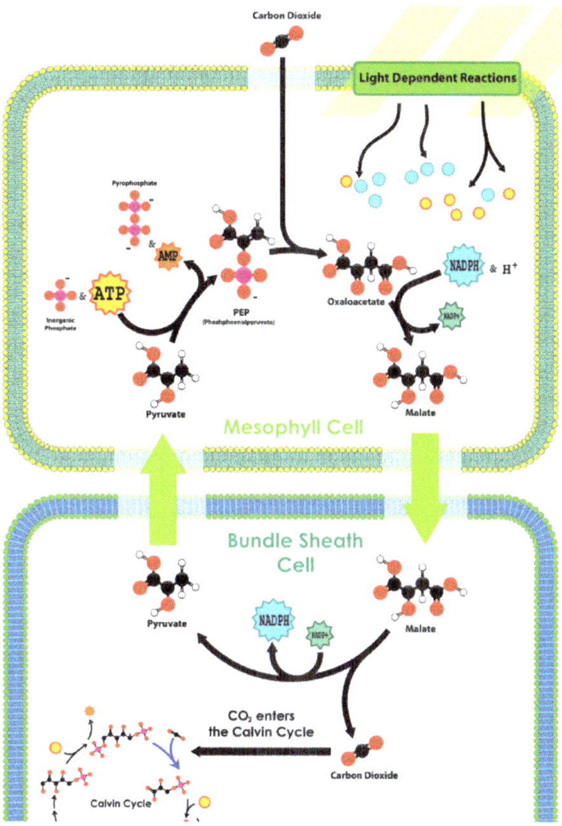

Overview of C4 carbon fixation

Plants that use the C_4 carbon fixation process chemically fix carbon dioxide in the cells of the mesophyll by adding it to the three-carbon molecule phosphoenolpyruvate (PEP), a reaction catalyzed by an enzyme called PEP carboxylase, creating the four-carbon organic acid oxaloacetic acid. Oxaloacetic acid or malate synthesized by this process is then translocated to specialized bundle sheath cells where the enzyme RuBisCO and other Calvin cycle enzymes are located, and where CO_2 released by decarboxylation of the four-carbon acids is then fixed by RuBisCO activity to the three-carbon 3-phosphoglyceric acids. The physical separation of RuBisCO from the oxygen-generating light reactions reduces photorespiration and increases CO_2 fixation and, thus, the photosynthetic capacity of the leaf. C_4 plants can produce more sugar than C_3 plants in conditions of high light and temperature. Many important crop plants are C_4 plants, including maize, sorghum, sugarcane, and millet. Plants that do not use PEP-carboxylase in carbon fixation are called C_3 plants because the primary carboxylation reaction, catalyzed by RuBisCO, produces the three-carbon 3-phosphoglyceric acids directly in the Calvin-Benson cycle. Over 90% of plants use C_3 carbon fixation, compared to 3% that use C_4 carbon fixation; however, the evolution of C_4 in over 60 plant lineages makes it a striking example of convergent evolution.

Xerophytes, such as cacti and most succulents, also use PEP carboxylase to capture carbon dioxide in a process called Crassulacean acid metabolism (CAM). In contrast to C_4 metabolism, which *physically* separates the CO_2 fixation to PEP from the Calvin cycle, CAM *temporally* separates these two processes. CAM plants have a different leaf anatomy from C_3 plants, and fix the CO_2 at night, when their stomata are open. CAM plants store the CO_2 mostly in the form of malic acid via carboxylation of phosphoenolpyruvate to oxaloacetate, which is then reduced to malate. Decarboxylation of malate during the day releases CO_2 inside the leaves, thus allowing carbon fixation to 3-phosphoglycerate by RuBisCO. Sixteen thousand species of plants use CAM.

In Water

Cyanobacteria possess carboxysomes, which increase the concentration of CO_2 around RuBisCO to increase the rate of photosynthesis. An enzyme, carbonic anhydrase, located within the carboxysome releases CO_2 from the dissolved hydrocarbonate ions (HCO−3). Before the CO_2 diffuses out it is quickly sponged up by RuBisCO, which is concentrated within the carboxysomes. HCO−3 ions are made from CO_2 outside the cell by another carbonic anhydrase and are actively pumped into the cell by a membrane protein. They cannot cross the membrane as they are charged, and within the cytosol they turn back into CO_2 very slowly without the help of carbonic anhydrase. This causes the HCO−3 ions to accumulate within the cell from where they diffuse into the carboxysomes. Pyrenoids in algae and hornworts also act to concentrate CO_2 around rubisco.

Order and Kinetics

The overall process of photosynthesis takes place in four stages:

Stage	Description	Time scale
1	Energy transfer in antenna chlorophyll (thylakoid membranes)	femtosecond to picosecond
2	Transfer of electrons in photochemical reactions (thylakoid membranes)	picosecond to nanosecond
3	Electron transport chain and ATP synthesis (thylakoid membranes)	microsecond to millisecond
4	Carbon fixation and export of stable products	millisecond to second

Efficiency

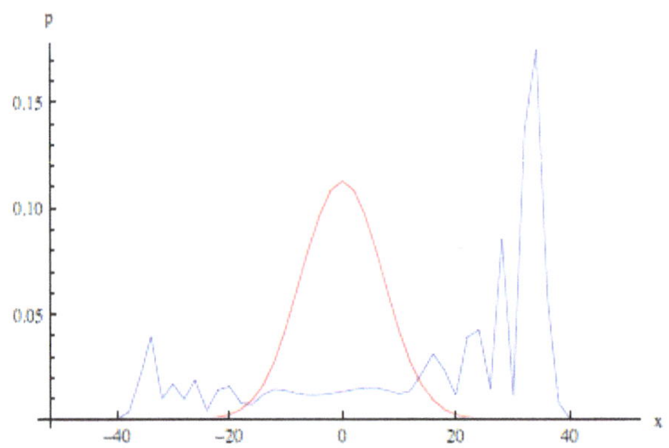

Probability distribution resulting from one-dimensional discrete time random walks. The quantum walk created using the Hadamard coin is plotted (blue) vs a classical walk (red) after 50 time steps.

Plants usually convert light into chemical energy with a photosynthetic efficiency of 3–6%. Absorbed light that is unconverted is dissipated primarily as heat, with a small fraction (1–2%) re-emitted as chlorophyll fluorescence at longer (redder) wavelengths. A fact that allows measurement of the light reaction of photosynthesis by using chlorophyll fluorometers.

Actual plants' photosynthetic efficiency varies with the frequency of the light being converted, light intensity, temperature and proportion of carbon dioxide in the atmosphere, and can vary from 0.1% to 8%. By comparison, solar panels convert light into electric energy at an efficiency of approximately 6–20% for mass-produced panels, and above 40% in laboratory devices.

The efficiency of both light and dark reactions can be measured but the relationship between the two can be complex. For example, the ATP and NADPH energy molecules, created by the light reaction, can be used for carbon fixation or for photorespiration in C_3 plants. Electrons may also flow to other electron sinks. For this reason, it is not uncommon for authors to differentiate between work done under non-photorespiratory conditions and under photorespiratory conditions.

Chlorophyll fluorescence of photosystem II can measure the light reaction, and Infrared gas analyzers can measure the dark reaction. It is also possible to investigate both at the same time using an integrated chlorophyll fluorometer and gas exchange system, or by using two separate systems together. Infrared gas analyzers and some moisture sensors are sensitive enough to measure the photosynthetic assimilation of CO_2, and of ΔH_2O using reliable methods CO_2 is commonly measured in $\mu mols/m^2/s^{-1}$, parts per million or volume per million and H_2o is commonly measured in $mmol/m^2/s^{-1}$ or in mbars. By measuring CO_2 assimilation, ΔH_2O, leaf temperature, barometric pressure, leaf area, and photosynthetically active radiation or PAR, it becomes possible to estimate, "A" or carbon assimilation, "E" or transpiration, "gs" or stomatal conductance, and Ci or intracellular CO_2. However, it is more common to used chlorophyll fluorescence for plant stress measurement, where appropriate, because the most commonly used measuring parameters FV/FM and Y(II) or F/FM' can be made in a few seconds, allowing the measurement of larger plant populations.

Gas exchange systems that offer control of CO_2 levels, above and below ambient, allow the common practice of measurement of A/Ci curves, at different CO_2 levels, to characterize a plant's photosynthetic response.

Integrated chlorophyll fluorometer – gas exchange systems allow a more precise measure of photosynthetic response and mechanisms. While standard gas exchange photosynthesis systems can measure Ci, or substomatal CO_2 levels, the addition of integrated chlorophyll fluorescence measurements allows a more precise measurement of C_C to replace Ci. The estimation of CO_2 at the site of carboxylation in the chloroplast, or C_C, becomes possible with the measurement of mesophyll conductance or g_m using an integrated system.

Photosynthesis measurement systems are not designed to directly measure the amount of light absorbed by the leaf. But analysis of chlorophyll-fluorescence, P700- and P515-absorbance and gas exchange measurements reveal detailed information about e.g. the photosystems, quantum efficiency and the CO_2 assimilation rates. With some instruments even wavelength-dependency of the photosynthetic efficiency can be analyzed.

A phenomenon known as quantum walk increases the efficiency of the energy transport of light significantly. In the photosynthetic cell of an algae, bacterium, or plant, there are light-sensitive molecules called chromophores arranged in an antenna-shaped structure named a photocomplex. When a photon is absorbed by a chromophore, it is converted into a quasiparticle referred to as an exciton, which jumps from chromophore to chromophore towards the reaction center of the photocomplex, a collection of molecules that traps its energy in a chemical form that makes it accessible for the cell's metabolism. The exciton's wave properties enable it to cover a wider area and try out several possible paths simultaneously, allowing it to instantaneously "choose" the most efficient route, where it will have the highest probability of arriving at its destination in the minimum possible time. Because that quantum walking takes place at temperatures far higher than quantum phenomena usually occur, it is only possible over very short distances, due to obstacles in the form of destructive interference that come into play. These obstacles cause the particle to lose its wave properties for an instant before it regains them once again after it is freed from its locked position through a classic "hop". The movement of the electron towards the photo center is therefore covered in a series of conventional hops and quantum walks.

Evolution

Early photosynthetic systems, such as those in green and purple sulfur and green and purple nonsulfur bacteria, are thought to have been anoxygenic, and used various other molecules as electron donors rather than water. Green and purple sulfur bacteria are thought to have used hydrogen and sulfur as electron donors. Green nonsulfur bacteria used various amino and other organic acids as an electron donor. Purple nonsulfur bacteria used a variety of nonspecific organic molecules. The use of these molecules is consistent with the geological evidence that Earth's early atmosphere was highly reducing at that time.

Fossils of what are thought to be filamentous photosynthetic organisms have been dated at 3.4 billion years old.

The main source of oxygen in the Earth's atmosphere derives from oxygenic photosynthesis, and its first appearance is sometimes referred to as the oxygen catastrophe. Geological evidence suggests that oxygenic photosynthesis, such as that in cyanobacteria, became important during the Paleoproterozoic era around 2 billion years ago. Modern photosynthesis in plants and most photosynthetic prokaryotes is oxygenic. Oxygenic photosynthesis uses water as an electron donor, which is oxidized to molecular oxygen (O2) in the photosynthetic reaction center.

Symbiosis and the Origin of Chloroplasts

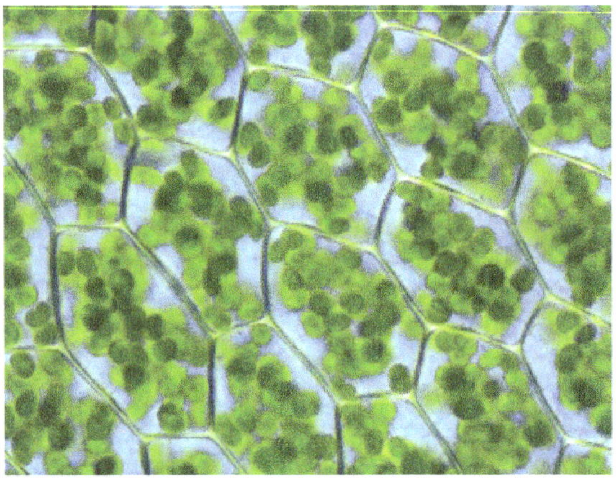

Plant cells with visible chloroplasts (from a moss, *Plagiomnium affine*)

Several groups of animals have formed symbiotic relationships with photosynthetic algae. These are most common in corals, sponges and sea anemones. It is presumed that this is due to the particularly simple body plans and large surface areas of these animals compared to their volumes.

In addition, a few marine mollusks *Elysia viridis* and *Elysia chlorotica* also maintain a symbiotic relationship with chloroplasts they capture from the algae in their diet and then store in their bodies. This allows the mollusks to survive solely by photosynthesis for several months at a time. Some of the genes from the plant cell nucleus have even been transferred to the slugs, so that the chloroplasts can be supplied with proteins that they need to survive.

An even closer form of symbiosis may explain the origin of chloroplasts. Chloroplasts have many similarities with photosynthetic bacteria, including a circular chromosome, prokaryotic-type ribosome, and similar proteins in the photosynthetic reaction center. The endosymbiotic theory suggests that photosynthetic bacteria were acquired (by endocytosis) by early eukaryotic cells to form the first plant cells. Therefore, chloroplasts may be photosynthetic bacteria that adapted to life inside plant cells. Like mitochondria, chloroplasts possess their own DNA, separate from the nuclear DNA of their plant host cells and the genes in this chloroplast DNA resemble those found in cyanobacteria. DNA in chloroplasts codes for redox proteins such as those found in the photosynthetic reaction centers. The CoRR Hypothesis proposes that this Co-location is required for Redox Regulation.

Cyanobacteria and the Evolution of Photosynthesis

The biochemical capacity to use water as the source for electrons in photosynthesis evolved once, in a common ancestor of extant cyanobacteria. The geological record indicates that this transforming event took place early in Earth's history, at least 2450–2320 million years ago (Ma), and, it is speculated, much earlier. Because the Earth's atmosphere contained almost no oxygen during the estimated development of photosynthesis, it is believed that the first photosynthetic cyanobacteria did not generate oxygen. Available evidence from geobiological studies of Archean (>2500 Ma) sedimentary rocks indicates that life existed 3500 Ma, but the question of when oxygenic photosynthesis evolved is still unanswered. A clear paleontological window on cyanobacterial evolution opened about 2000 Ma, revealing an already-diverse biota of blue-green algae. Cyanobacteria remained the principal primary producers of oxygen throughout the Proterozoic Eon (2500–543 Ma), in part because the redox structure of the oceans favored photoautotrophs capable of nitrogen fixation. Green algae joined blue-green algae as the major primary producers of oxygen on continental shelves near the end of the Proterozoic, but it was only with the Mesozoic (251–65 Ma) radiations of dinoflagellates, coccolithophorids, and diatoms did the primary production of oxygen in marine shelf waters take modern form. Cyanobacteria remain critical to marine ecosystems as primary producers of oxygen in oceanic gyres, as agents of biological nitrogen fixation, and, in modified form, as the plastids of marine algae.

The Oriental hornet (*Vespa orientalis*) converts sunlight into electric power using a pigment called xanthopterin. This is the first evidence of a member of the animal kingdom engaging in photosynthesis.

Discovery

Although some of the steps in photosynthesis are still not completely understood, the overall photosynthetic equation has been known since the 19th century.

Jan van Helmont began the research of the process in the mid-17th century when he carefully measured the mass of the soil used by a plant and the mass of the plant as it grew. After noticing that the soil mass changed very little, he hypothesized that the mass of the growing plant must come from the water, the only substance he added to the potted plant. His hypothesis was partially accurate — much of the gained mass also comes from carbon dioxide as well as water. However, this was a signaling point to the idea that the bulk of a plant's biomass comes from the inputs of photosynthesis, not the soil itself.

Joseph Priestley, a chemist and minister, discovered that, when he isolated a volume of air under an inverted jar, and burned a candle in it, the candle would burn out very quickly, much before it ran out of wax. He further discovered that a mouse could similarly "injure" air. He then showed that the air that had been "injured" by the candle and the mouse could be restored by a plant.

In 1778, Jan Ingenhousz, repeated Priestley's experiments. He discovered that it was the influence of sunlight on the plant that could cause it to revive a mouse in a matter of hours.

In 1796, Jean Senebier, a Swiss pastor, botanist, and naturalist, demonstrated that green plants consume carbon dioxide and release oxygen under the influence of light. Soon afterward, Nicolas-Théodore de Saussure showed that the increase in mass of the plant as it grows could not be due only to uptake of CO_2 but also to the incorporation of water. Thus, the basic reaction by which photosynthesis is used to produce food (such as glucose) was outlined.

Cornelis Van Niel made key discoveries explaining the chemistry of photosynthesis. By studying purple sulfur bacteria and green bacteria he was the first to demonstrate that photosynthesis is a light-dependent redox reaction, in which hydrogen reduces carbon dioxide.

Melvin Calvin works in his photosynthesis laboratory.

Robert Emerson discovered two light reactions by testing plant productivity using different wavelengths of light. With the red alone, the light reactions were suppressed. When blue and red were combined, the output was much more substantial. Thus, there were two photosystems, one absorbing up to 600 nm wavelengths, the other up to 700 nm. The former is known as PSII, the latter is PSI. PSI contains only chlorophyll "a", PSII contains primarily chlorophyll "a" with most of the available chlorophyll "b", among other pigment. These include phycobilins, which are the red and blue pigments of red and blue algae respectively, and fucoxanthol for brown algae and diatoms.

The process is most productive when the absorption of quanta are equal in both the PSII and PSI, assuring that input energy from the antenna complex is divided between the PSI and PSII system, which in turn powers the photochemistry.

Robert Hill thought that a complex of reactions consisting of an intermediate to cytochrome b_6 (now a plastoquinone), another is from cytochrome f to a step in the carbohydrate-generating mechanisms. These are linked by plastoquinone, which does require energy to reduce cytochrome f for it is a sufficient reductant. Further experiments to prove that the oxygen developed during the photosynthesis of green plants came from water, were performed by Hill in 1937 and 1939. He showed that isolated chloroplasts give off oxygen in the presence of unnatural reducing agents like iron oxalate, ferricyanide or benzoquinone after exposure to light. The Hill reaction is as follows:

$$2\ H_2O + 2\ A + (\text{light, chloroplasts}) \rightarrow 2\ AH_2 + O_2$$

where A is the electron acceptor. Therefore, in light, the electron acceptor is reduced and oxygen is evolved.

Samuel Ruben and Martin Kamen used radioactive isotopes to determine that the oxygen liberated in photosynthesis came from the water.

Melvin Calvin and Andrew Benson, along with James Bassham, elucidated the path of carbon assimilation (the photosynthetic carbon reduction cycle) in plants. The carbon reduction cycle is known as the Calvin cycle, which ignores the contribution of Bassham and Benson. Many scientists refer to the cycle as the Calvin-Benson Cycle, Benson-Calvin, and some even call it the Calvin-Benson-Bassham (or CBB) Cycle.

Nobel Prize-winning scientist Rudolph A. Marcus was able to discover the function and significance of the electron transport chain.

Otto Heinrich Warburg and Dean Burk discovered the I-quantum photosynthesis reaction that splits the CO_2, activated by the respiration.

Louis N.M. Duysens and Jan Amesz discovered that chlorophyll a will absorb one light, oxidize cytochrome f, chlorophyll a (and other pigments) will absorb another light, but will reduce this same oxidized cytochrome, stating the two light reactions are in series.

Development of the concept

In 1893, Charles Reid Barnes proposed two terms, *photosyntax* and *photosynthesis*, for the biological process of *synthesis of complex carbon compounds out of carbonic acid, in the presence of chlorophyll, under the influence of light*. Over time, the term *photosynthesis* came into common usage as the term of choice. Later discovery of anoxygenic photosynthetic bacteria and photophosphorylation necessitated redefinition of the term.

Factors

There are three main factors affecting photosynthesis and several corollary factors. The three main are:

- Light irradiance and wavelength
- Carbon dioxide concentration
- Temperature.

The leaf is the primary site of photosynthesis in plants.

Light Intensity (Irradiance), Wavelength and Temperature

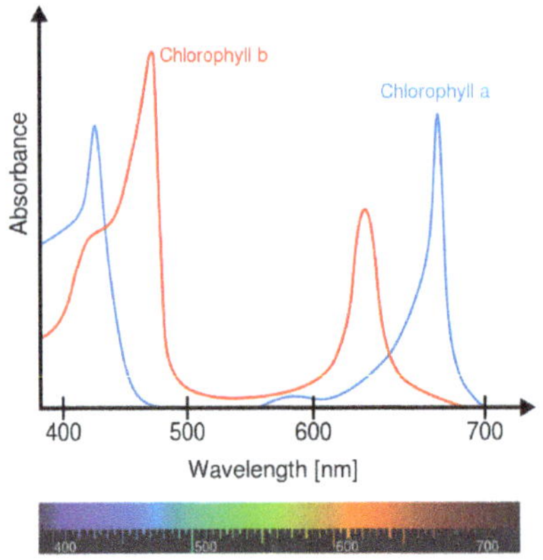

Absorbance spectra of free chlorophyll *a* (green) and *b* (red) in a solvent. The action spectra of chlorophyll molecules are slightly modified *in vivo* depending on specific pigment-protein interactions.

The process of photosynthesis provides the main input of free energy into the biosphere, and is one of four main ways in which radiation is important for plant life.

The radiation climate within plant communities is extremely variable, with both time and space.

In the early 20th century, Frederick Blackman and Gabrielle Matthaei investigated the effects of light intensity (irradiance) and temperature on the rate of carbon assimilation.

- At constant temperature, the rate of carbon assimilation varies with irradiance, increasing as the irradiance increases, but reaching a plateau at higher irradiance.

- At low irradiance, increasing the temperature has little influence on the rate of carbon assimilation. At constant high irradiance, the rate of carbon assimilation increases as the temperature is increased.

These two experiments illustrate several important points: First, it is known that, in general, photochemical reactions are not affected by temperature. However, these experiments clearly show that temperature affects the rate of carbon assimilation, so there must be two sets of reactions in the full process of carbon assimilation. These are, of course, the light-dependent 'photochemical' temperature-independent stage, and the light-independent, temperature-dependent stage. Second, Blackman's experiments illustrate the concept of limiting factors. Another limiting factor is the wavelength of light. Cyanobacteria, which reside several meters underwater, cannot receive the correct wavelengths required to cause photoinduced charge separation in conventional photosynthetic pigments. To combat this problem, a series of proteins with different pigments surround the reaction center. This unit is called a phycobilisome.

Carbon Dioxide Levels and Photorespiration

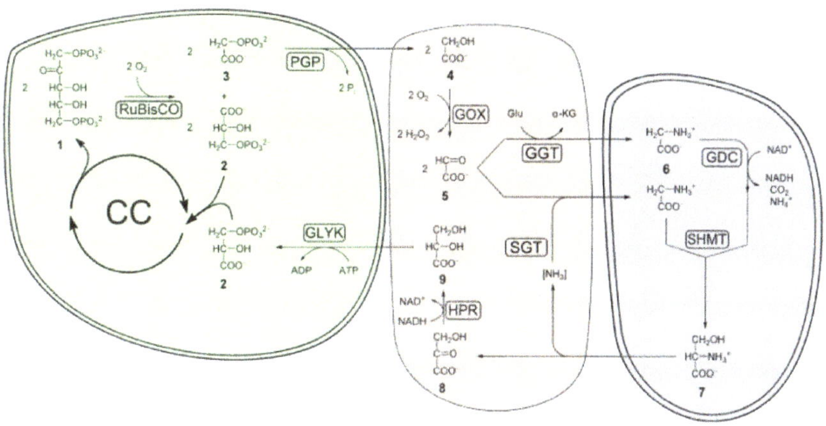

Photorespiration

As carbon dioxide concentrations rise, the rate at which sugars are made by the light-independent reactions increases until limited by other factors. RuBisCO, the enzyme that captures carbon dioxide in the light-independent reactions, has a binding affinity for both carbon dioxide and oxygen. When the concentration of carbon dioxide is high, RuBisCO will fix carbon dioxide. However, if the carbon dioxide concentration is low, RuBisCO will bind oxygen instead of carbon dioxide. This process, called photorespiration, uses energy, but does not produce sugars.

RuBisCO oxygenase activity is disadvantageous to plants for several reasons:

1. One product of oxygenase activity is phosphoglycolate (2 carbon) instead of 3-phosphoglycerate (3 carbon). Phosphoglycolate cannot be metabolized by the Calvin-Benson cycle and represents carbon lost from the cycle. A high oxygenase activity, therefore, drains the sugars that are required to recycle ribulose 5-bisphosphate and for the continuation of the Calvin-Benson cycle.

2. Phosphoglycolate is quickly metabolized to glycolate that is toxic to a plant at a high concentration; it inhibits photosynthesis.

3. Salvaging glycolate is an energetically expensive process that uses the glycolate pathway, and only 75% of the carbon is returned to the Calvin-Benson cycle as 3-phosphoglycerate. The reactions also produce ammonia (NH_3), which is able to diffuse out of the plant, leading to a loss of nitrogen.

   A highly simplified summary is:

   2 glycolate + ATP → 3-phosphoglycerate + carbon dioxide + ADP + NH_3

The salvaging pathway for the products of RuBisCO oxygenase activity is more commonly known as photorespiration, since it is characterized by light-dependent oxygen consumption and the release of carbon dioxide.

Arboriculture

Arboriculture /ˈɑːrbərɨkʌltʃər/ is the cultivation, management, and study of individual trees, shrubs, vines, and other perennial woody plants. The science of arboriculture studies how these plants grow and respond to cultural practices and to their environment. The practice of arboriculture includes cultural techniques such as selection, planting, training, fertilization, pest and pathogen control, pruning, shaping, and removal.

A person who practices or studies arboriculture can be termed an 'arborist' or an 'arboriculturist'. A 'tree surgeon' is more typically someone who is trained in the physical maintenance and manipulation of trees and therefore more a part of the arboriculture process rather than an arborist. Risk management, legal issues, and aesthetic considerations have come to play prominent roles in the practice of arboriculture. Businesses often need to hire arboriculturists to complete "tree hazard surveys" and generally manage the trees on-site to fulfill occupational safety and health obligations.

Arboriculture is primarily focused on individual woody plants and trees maintained for permanent landscape and amenity purposes, usually in gardens, parks or other populated settings, by arborists, for the enjoyment, protection, and benefit of human beings. It falls under the general umbrella of horticulture.

Micropropagation

Micropropagation is the practice of rapidly multiplying stock plant material to produce a large number of progeny plants, using modern plant tissue culture methods.

Micropropagation is used to multiply noble plants such as those that have been genetically modified or bred through conventional plant breeding methods. It is also used to provide a sufficient number of plantlets for planting from a stock plant which does not produce seeds, or does not respond well to vegetative reproduction.

Cornell University botanist Frederick Campion Steward discovered and pioneered micropropagation and plant tissue culture in the late 1950s and early 1960s.

Method (Stages of Micropropagation)

Establishment

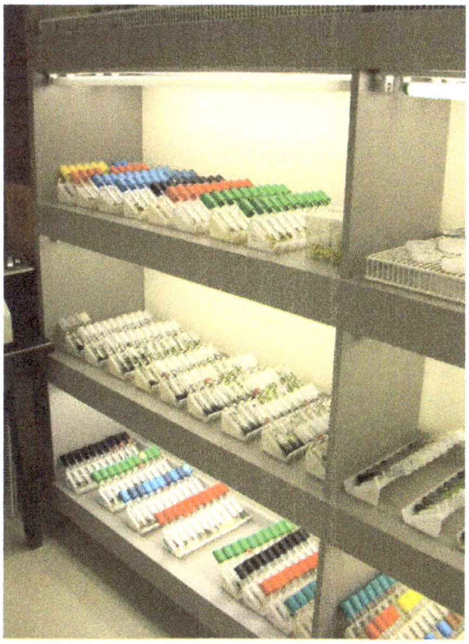

In vitro culture of plants in a controlled, sterile environment

Micropropagation begins with the selection of plant material to be propagated. The plant tissues are removed from an intact plant in a sterile condition. Clean stock materials that are free of viruses and fungi are important in the production of the healthiest plants. Once the plant material is chosen for culture, the collection of explant(s) begins and is dependent on the type of tissue to be used; including stem tips, anthers, petals, pollen and others plant tissues. The explant material is then surface sterilized, usually in multiple courses of bleach and alcohol washes, and finally rinsed in sterilized water. This small portion of plant tissue, sometimes only a single cell, is placed on a growth medium, typically containing sucrose as an energy source and one or more plant growth regulators (plant hormones). Usually the medium is thickened with agar to create a gel which supports the explant during growth. Some plants are easily grown on simple media, but others require more complicated media for successful growth; the plant tissue grows and differentiates into new tissues depending on the medium. For example, media containing cytokinin are used to create branched shoots from plant buds.

Multiplication

Multiplication is the taking of tissue samples produced during the first stage and increasing their number. Following the successful introduction and growth of plant tissue, the establishment stage is followed by multiplication. Through repeated cycles of this process, a single explant sample may be increased from one to hundreds and thousands of plants. Depending on the type of tissue grown, multiplication can involve different methods and media. If the plant material grown is callus tissue, it can be placed in a blender and cut into smaller pieces and recultured on the same type of culture medium to grow more callus tissue. If the tissue is grown as small plants called plantlets, hormones are often added that cause the plantlets to produce many small offshoots. After the formation of multiple shoots, these shoots are transferred to rooting medium with a high auxin\ cytokinin ratio. After the development of roots, plantlets can be used for hardening.

Pretransplant

Banana plantlets transferred to soil (with vermicompost) from plant media. This process is done for acclimatization of plantlets to the soil as they were previously grown in plant media. After growing for some days the plantlets are transferred to the field.

This stage involves treating the plantlets/shoots produced to encourage root growth and "hardening." It is performed *in vitro*, or in a sterile "test tube" environment.

"Hardening" refers to the preparation of the plants for a natural growth environment. Until this stage, the plantlets have been grown in "ideal" conditions, designed to encourage rapid growth. Due to the controlled nature of their maturation, the plantlets often do not have fully functional dermal coverings. This causes them to be highly susceptible to disease and inefficient in their use of water and energy. In vitro conditions are high in humidity, and plants grown under these conditions often do not form a working cuticle and stomata that keep the plant from drying out. When taken out of culture, the plantlets need time to adjust to more natural environmental conditions. Hardening typically involves slowly weaning the plantlets from a high-humidity, low light, warm environment to what would be considered a normal growth environment for the species in question.

Transfer from Culture

Plant tissue cultures being grown at a USDA seed bank, the National Center for Genetic Resources Preservation.

In the final stage of plant micropropagation, the plantlets are removed from the plant media and transferred to soil or (more commonly) potting compost for continued growth by conventional methods.

This stage is often combined with the "pretransplant" stage.

Methods of Micro-propagation

Meristem Culture

In Meristem culture the Meristem and a few subtending leaf primordial are placed into a suitable growing media. An elongated rooted platelet is produced after some weeks, and is transferred to the soil when it has attained a considerable height. A disease free plant can be produced by this method. Experimental result also suggest that this technique can be successfully utilized for rapid multiplication of various herbaceous Plants.

Callus Culture

A callus is mass of undifferentiated parenchymatous cells. When a living plant tissue is placed in an artificial growing medium with other conditions favorable, callus is formed. The growth of callus varies with the homogenous levels of auxin and Cyotkininn and can be manipulated by endogenous supply of these growth regulators in the culture medium. The callus growth and its organogenesis or embryogenesis can be referred into three different stages.

- Stage I: Rapid production of callus after placing the explants in culture medium
- Stage II: The callus is transferred to other medium containing growth regulators for the induction of adventitious organs.
- Stage III: The new plantlet is then exposed gradually to the environmental condition.

Suspension Culture

A cell suspension culture refers to cells and or groups of cells dispersed and growing in an aerated liquid culture medium (Street, 1997, Thorpe1981) is placed in a liquid medium and shaken vigorously and balanced dose of hormones. Suezawa et al. (1988) reported Cyotkininn induced adventitious buds in kiwi fruit in a suspension culture sub- culture for about a week.

Embryo Culture

In embryo culture, the embryo is excised and placed into a culture medium with proper nutrient in aseptic condition. To obtain a quick and optimum growth into plantlets, it is transferred to soil. It is particularly important for the production of interspecific and intergeneric hybrids and to overcome the embryo abortion.

Protoplast Culture

In protoplast culture, the plant cell can be isolated with the help of wall degrading enzymes and growth in a suitable culture medium in a controlled condition for regeneration of plantlets. Under suitable conditions the protoplast develops a cell wall followed by an increase in cell division and differentiation and grows into a new plant. The protoplast are fist culture in liquid medium at 25 to 28 C with a light intensity of 100 to 500 lux or in dark and after undergoing substantial cell division, they are transferred into solid medium congenial or morphogenesis in many horticultural crops response well to protoplast culture.

Advantages

Micropropagation has a number of advantages over traditional plant propagation techniques:

- The main advantage of micropropagation is the production of many plants that are clones of each other.
- Micropropagation can be used to produce disease-free plants.
- It can have an extraordinarily high fecundity rate, producing thousands of propagules while conventional techniques might only produce a fraction of this number.
- It is the only viable method of regenerating genetically modified cells or cells after protoplast fusion.
- It is useful in multiplying plants which produce seeds in uneconomical amounts, or when plants are sterile and do not produce viable seeds or when seed cannot be stored (see recalcitrant seeds).
- Micropropagation often produces more robust plants, leading to accelerated growth compared to similar plants produced by conventional methods - like seeds or cuttings.
- Some plants with very small seeds, including most orchids, are most reliably grown from seed in sterile culture.
- A greater number of plants can be produced per square meter and the propagules can be stored longer and in a smaller area.

Disadvantages

Micropropagation is not always the perfect means of multiplying plants. Conditions that limits its use include:

- It is *very* expensive, and can have a labour cost of more than 70%.
- A monoculture is produced after micropropagation, leading to a lack of overall disease resilience, as all progeny plants may be vulnerable to the same infections.
- An infected plant sample can produce infected progeny. This is uncommon as the stock plants are carefully screened and vetted to prevent culturing plants infected with virus or fungus.
- Not all plants can be successfully tissue cultured, often because the proper medium for growth is not known or the plants produce secondary metabolic chemicals that stunt or kill the explant.

- Sometimes plants or cultivars do not come true to type after being tissue cultured. This is often dependent on the type of explant material utilized during the initiation phase or the result of the age of the cell or propagule line.
- Some plants are very difficult to disinfect of fungal organisms.

The major limitation in the use of micropropagation for many plants is the cost of production; for many plants the use of seeds, which are normally disease free and produced in good numbers, readily produce plants (see orthodox seed) in good numbers at a lower cost. For this reason, many plant breeders do not utilize micropropagation because the cost is prohibitive. Other breeders use it to produce stock plants that are then used for seed multiplication.

Mechanisation of the process could reduce labour costs, but has proven difficult to achieve, despite active attempts to develop technological solutions.

Plant Breeding

Plant breeding is the art and science of changing the traits of plants in order to produce desired characteristics. Plant breeding can be accomplished through many different techniques ranging from simply selecting plants with desirable characteristics for propagation, to more complex molecular techniques (see cultigen and cultivar).

Plant breeding has been practiced for thousands of years, since near the beginning of human civilization. It is practiced worldwide by individuals such as gardeners and farmers, or by professional plant breeders employed by organizations such as government institutions, universities, crop-specific industry associations or research centers.

International development nation agencies believe that breeding new crops is important for ensuring food security by developing new varieties that are higher-yielding, disease resistant, drought-resistant or regionally adapted to different environments and growing conditions.

History

Plant breeding started with sedentary agriculture and particularly the domestication of the first agricultural plants, a practice which is estimated to date back 9,000 to 11,000 years. Initially early farmers simply selected food plants with particular desirable characteristics, and employed these as progenitors for subsequent generations, resulting in an accumulation of valuable traits over time.

Gregor Mendel's experiments with plant hybridization led to his establishing laws of inheritance. Once this work became well known, it formed the basis of the new science of genetics, which stimulated research by many plant scientists dedicated to improving crop production through plant breeding.

Modern plant breeding is applied genetics, but its scientific basis is broader, covering molecular biology, cytology, systematics, physiology, pathology, entomology, chemistry, and statistics (biometrics). It has also developed its own technology.

Classical Plant Breeding

One major technique of plant breeding is selection, the process of selectively propagating plants with desirable characteristics and eliminating or "culling" those with less desirable characteristics.

Another technique is the deliberate interbreeding (crossing) of closely or distantly related individuals to produce new crop varieties or lines with desirable properties. Plants are crossbred to introduce traits/genes from one variety or line into a new genetic background. For example, a mildew-resistant pea may be crossed with a high-yielding but susceptible pea, the goal of the cross being to introduce mildew resistance without losing the high-yield characteristics. Progeny from the cross would then be crossed with the high-yielding parent to ensure that the progeny were most like the high-yielding parent, (backcrossing). The progeny from that cross would then be tested for yield (selection, as described above) and mildew resistance and high-yielding resistant plants would be further developed. Plants may also be crossed with themselves to produce inbred varieties for breeding. Pollinators may be excluded through the use of pollination bags.

Classical breeding relies largely on homologous recombination between chromosomes to generate genetic diversity. The classical plant breeder may also make use of a number of *in vitro* techniques such as protoplast fusion, embryo rescue or mutagenesis (see below) to generate diversity and produce hybrid plants that would not exist in nature.

Traits that breeders have tried to incorporate into crop plants include:

1. Improved quality, such as increased nutrition, improved flavor, or greater beauty
2. Increased yield of the crop
3. Increased tolerance of environmental pressures (salinity, extreme temperature, drought)
4. Resistance to viruses, fungi and bacteria
5. Increased tolerance to insect pests
6. Increased tolerance of herbicides
7. Longer storage period for the harvested crop

Before World War II

Successful commercial plant breeding concerns were founded from the late 19th century. Gartons Agricultural Plant Breeders in England was established in the 1890s by John Garton, who was one of the first to commercialize new varieties of agricultural crops created through cross-pollination. The firm's first introduction was Abundance Oat, one of the first agricultural grain varieties bred from a *controlled* cross, introduced to commerce in 1892.

In the early 20th century, plant breeders realized that Mendel's findings on the non-random nature of inheritance could be applied to seedling populations produced through deliberate pollinations to predict the frequencies of different types. Wheat hybrids were bred to increase the crop production of Italy during the so-called "Battle for Grain" (1925–1940). Heterosis was explained by George Harrison Shull. It describes the tendency of the progeny of a specific cross to outperform both parents. The detection of the usefulness of heterosis for plant breeding has led to the development of inbred lines that reveal a heterotic yield advantage when they are crossed. Maize was the first species where heterosis was widely used to produce hybrids.

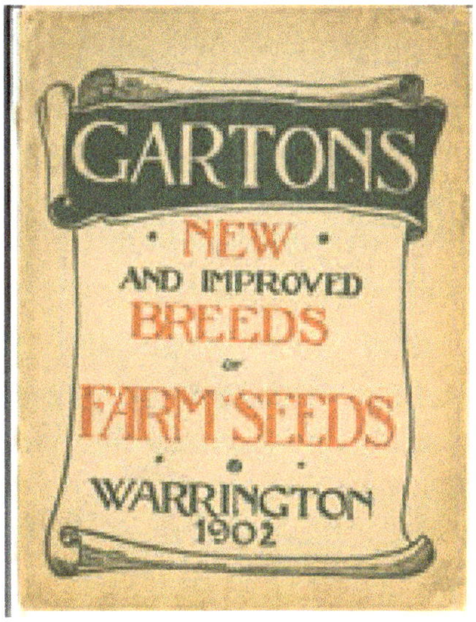

Garton's catalogue from 1902

Statistical methods were also developed to analyze gene action and distinguish heritable variation from variation caused by environment. In 1933 another important breeding technique, cytoplasmic male sterility (CMS), developed in maize, was described by Marcus Morton Rhoades. CMS is a maternally inherited trait that makes the plant produce sterile pollen. This enables the production of hybrids without the need for labor-intensive detasseling.

These early breeding techniques resulted in large yield increase in the United States in the early 20th century. Similar yield increases were not produced elsewhere until after World War II, the Green Revolution increased crop production in the developing world in the 1960s.

After World War II

In vitro-culture of Vitis (grapevine), Geisenheim Grape Breeding Institute

Following World War II a number of techniques were developed that allowed plant breeders to hybridize distantly related species, and artificially induce genetic diversity.

When distantly related species are crossed, plant breeders make use of a number of plant tissue culture techniques to produce progeny from otherwise fruitless mating. Interspecific and intergeneric hybrids are produced from a cross of related species or genera that do not normally sexually reproduce with each other. These crosses are referred to as *Wide crosses*. For example, the cereal triticale is a wheat and rye hybrid. The cells in the plants derived from the first generation created from the cross contained an uneven number of chromosomes and as result was sterile. The cell division inhibitor colchicine was used to double the number of chromosomes in the cell and thus allow the production of a fertile line.

Failure to produce a hybrid may be due to pre- or post-fertilization incompatibility. If fertilization is possible between two species or genera, the hybrid embryo may abort before maturation. If this does occur the embryo resulting from an interspecific or intergeneric cross can sometimes be rescued and cultured to produce a whole plant. Such a method is referred to as Embryo Rescue. This technique has been used to produce new rice for Africa, an interspecific cross of Asian rice *(Oryza sativa)* and African rice *(Oryza glaberrima)*.

Hybrids may also be produced by a technique called protoplast fusion. In this case protoplasts are fused, usually in an electric field. Viable recombinants can be regenerated in culture.

Chemical mutagens like EMS and DMS, radiation and transposons are used to generate mutants with desirable traits to be bred with other cultivars - a process known as *Mutation Breeding*. Classical plant breeders also generate genetic diversity within a species by exploiting a process called somaclonal variation, which occurs in plants produced from tissue culture, particularly plants derived from callus. Induced polyploidy, and the addition or removal of chromosomes using a technique called chromosome engineering may also be used.

When a desirable trait has been bred into a species, a number of crosses to the favored parent are made to make the new plant as similar to the favored parent as possible. Returning to the example of the mildew resistant pea being crossed with a high-yielding but susceptible pea, to make the mildew resistant progeny of the cross most like the high-yielding parent, the progeny will be crossed back to that parent for several generations (See backcrossing). This process removes most of the genetic contribution of the mildew resistant parent. Classical breeding is therefore a cyclical process.

With classical breeding techniques, the breeder does not know exactly what genes have been introduced to the new cultivars. Some scientists therefore argue that plants produced by classical breeding methods should undergo the same safety testing regime as genetically modified plants. There have been instances where plants bred using classical techniques have been unsuitable for human consumption, for example the poison solanine was unintentionally increased to unacceptable levels in certain varieties of potato through plant breeding. New potato varieties are often screened for solanine levels before reaching the marketplace.

Modern Plant Breeding

Modern plant breeding may use techniques of molecular biology to select, or in the case of genetic modification, to insert, desirable traits into plants. Application of biotechnology or molecular biology is also known as molecular breeding (see: Molecular breeding).

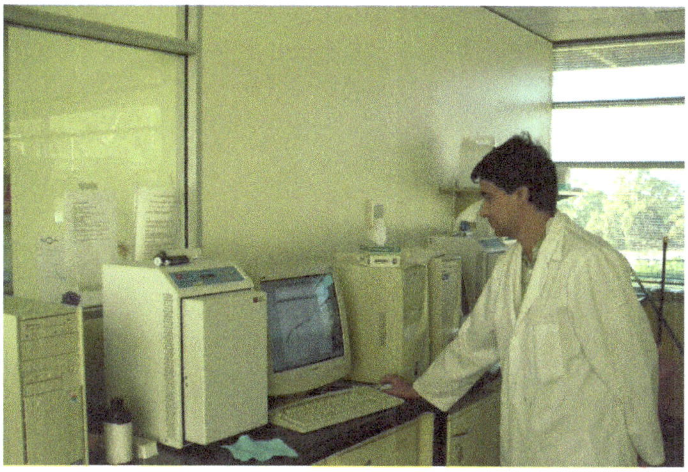

Modern facilities in molecular biology have converted classical plant breeding to molecular plant breeding

Steps of Plant Breeding

The following are the major activities of plant breeding:

1. Collection of variation
2. Selection
3. Evaluation
4. Release
5. Multiplication
6. Distribution of the new variety

Marker Assisted Selection

Sometimes many different genes can influence a desirable trait in plant breeding. The use of tools such as molecular markers or DNA fingerprinting can map thousands of genes. This allows plant breeders to screen large populations of plants for those that possess the trait of interest. The screening is based on the presence or absence of a certain gene as determined by laboratory procedures, rather than on the visual identification of the expressed trait in the plant.

Reverse Breeding and Doubled Haploids (DH)

A method for efficiently producing homozygous plants from a heterozygous starting plant, which has all desirable traits. This starting plant is induced to produce doubled haploid from haploid cells, and later on creating homozygous/doubled haploid plants from those cells. While in natural offspring genetic recombination occurs and traits can be unlinked from each other, in doubled haploid cells and in the resulting DH plants recombination is no longer an issue. There, a recombination between two corresponding chromosomes does not lead to un-linkage of alleles or traits, since it just leads to recombination with its identical copy. Thus, traits on one chromosome stay linked. Selecting those offspring having the desired set of chromosomes and crossing them will result in a final F1 hybrid plant, having exactly the same set of chromosomes, genes and traits as the starting hybrid plant. The homozygous parental lines can reconstitute the original heterozygous plant by crossing, if desired even in a large quantity. An individual heterozygous plant can be converted into a heterozygous variety (F1 hybrid) without the necessity of vegetative propagation but as the result of the cross of two homozygous/doubled haploid lines derived from the originally selected plant. patent

Genetic Modification

Genetic modification of plants is achieved by adding a specific gene or genes to a plant, or by knocking down a gene with RNAi, to produce a desirable phenotype. The plants resulting from adding a gene are often referred to as transgenic plants. If for genetic modification genes of the species or of a crossable plant are used under control of their native promoter, then they are called cisgenic plants. Sometimes genetic modification can produce a plant with the desired trait or traits faster than classical breeding because the majority of the plant's genome is not altered.

To genetically modify a plant, a genetic construct must be designed so that the gene to be added or removed will be expressed by the plant. To do this, a promoter to drive transcription and a termination sequence to stop transcription of the new gene, and the gene or genes of interest must be introduced to the plant. A marker for the selection of transformed plants is also included. In the laboratory, antibiotic resistance is a commonly used marker: Plants that have been successfully transformed will grow on media containing antibiotics; plants that have not been transformed will die. In some instances markers for selection are removed by backcrossing with the parent plant prior to commercial release.

The construct can be inserted in the plant genome by genetic recombination using the bacteria *Agrobacterium tumefaciens* or *A. rhizogenes*, or by direct methods like the gene gun or microinjection. Using plant viruses to insert genetic constructs into plants is also a possibility, but the technique is limited by the host range of the virus. For example, Cauliflower mosaic virus (CaMV) only infects cauliflower and related species. Another limitation of viral vectors is that the virus is not usually passed on the progeny, so every plant has to be inoculated.

The majority of commercially released transgenic plants are currently limited to plants that have introduced resistance to insect pests and herbicides. Insect resistance is achieved through incorporation of a gene from *Bacillus thuringiensis* (Bt) that encodes a protein that is toxic to some insects. For example, the cotton bollworm, a common cotton pest, feeds on Bt cotton it will ingest the toxin and die. Herbicides usually work by binding to certain plant enzymes and inhibiting their action. The enzymes that the herbicide inhibits are known as the herbicides *target site*. Herbicide resistance can be engineered into crops by expressing a version of *target site* protein that is not inhibited by the herbicide. This is the method used to produce glyphosate resistant crop plants (See Glyphosate)

Genetic modification of plants that can produce pharmaceuticals (and industrial chemicals), sometimes called *pharming*, is a rather radical new area of plant breeding.

Issues and Concerns

Modern plant breeding, whether classical or through genetic engineering, comes with issues of concern, particularly with regard to food crops. The question of whether breeding can have a negative effect on nutritional value is central in this respect. Although relatively little direct research in this area has been done, there are scientific indications that, by favoring certain aspects of a plant's development, other aspects may be retarded. A study published in the *Journal of the American College of Nutrition* in 2004, entitled *Changes in USDA Food Composition Data for 43 Garden Crops, 1950 to 1999*, compared nutritional analysis of vegetables done in 1950 and in 1999, and found substantial decreases in six of 13 nutrients measured, including 6% of protein and 38% of riboflavin. Reductions in calcium, phosphorus, iron and ascorbic acid were also found. The study, conducted at the Biochemical Institute, University of Texas at Austin, concluded in summary: *"We suggest that any real declines are generally most easily explained by changes in cultivated varieties between 1950 and 1999, in which there may be trade-offs between yield and nutrient content."*

The debate surrounding genetically modified food during the 1990s peaked in 1999 in terms of media coverage and risk perception, and continues today - for example, "*Germany has thrown its weight behind a growing European mutiny over genetically modified crops by banning the planting of a widely grown pest-resistant corn variety.*" The debate encompasses the ecological impact of genetically modified plants, the safety of genetically modified food and concepts used for safety evaluation like substantial equivalence. Such concerns are not new to plant breeding. Most countries have regulatory processes in place to help ensure that new crop varieties entering the marketplace are both safe and meet farmers' needs. Examples include variety registration, seed schemes, regulatory authorizations for GM plants, etc.

Plant breeders' rights is also a major and controversial issue. Today, production of new varieties is dominated by commercial plant breeders, who seek to protect their work and collect royalties through national and international agreements based in intellectual property rights. The range of related issues is complex. In the simplest terms, critics of the increasingly restrictive regulations argue that, through a combination of technical and economic pressures, commercial breeders are reducing biodiversity and significantly constraining individuals (such as farmers) from developing and trading seed on a regional level. Efforts to strengthen breeders' rights, for example, by lengthening periods of variety protection, are ongoing.

When new plant breeds or cultivars are bred, they must be maintained and propagated. Some plants are propagated by asexual means while others are propagated by seeds. Seed propagated cultivars require specific control over seed source and production procedures to maintain the integrity of the plant breeds results. Isolation is necessary to prevent cross contamination with related plants or the mixing of seeds after harvesting. Isolation is normally accomplished by planting distance but in certain crops, plants are enclosed in greenhouses or cages (most commonly used when producing F1 hybrids.)

Role of Plant Breeding in Organic Agriculture

Critics of organic agriculture claim it is too low-yielding to be a viable alternative to conventional agriculture. However, part of that poor performance may be the result of growing poorly adapted varieties. It is estimated that over 95% of organic agriculture is based on conventionally adapted varieties, even though the production environments found in organic vs. conventional farming systems are vastly different due to their distinctive management practices. Most notably, organic farmers have fewer inputs available than conventional growers to control their production environments. Breeding varieties specifically adapted to the unique conditions of organic agriculture is critical for this sector to realize its full potential. This requires selection for traits such as:

- Water use efficiency
- Nutrient use efficiency (particularly nitrogen and phosphorus)
- Weed competitiveness
- Tolerance of mechanical weed control
- Pest/disease resistance
- Early maturity (as a mechanism for avoidance of particular stresses)
- Abiotic stress tolerance (i.e. drought, salinity, etc...)

Currently, few breeding programs are directed at organic agriculture and until recently those that did address this sector have generally relied on indirect selection (i.e. selection in conventional environments for traits considered important for organic agriculture). However, because the difference between organic and conventional environments is large, a given genotype may perform very differently in each environment due to an interaction between genes and the environment (see gene-environment interaction). If this interaction is severe enough, an important trait required for the organic environment may not be revealed in the conventional environment, which can result in the selection of poorly adapted individuals. To ensure the most adapted varieties are identified, advocates of organic breeding now promote the use of direct selection (i.e. selection in the target environment) for many agronomic traits.

There are many classical and modern breeding techniques that can be utilized for crop improvement in organic agriculture despite the ban on genetically modified organisms. For instance, controlled crosses between individuals allow desirable genetic variation to be recombined and transferred to seed progeny via natural processes. Marker assisted selection can also be employed as a diagnostics tool to facilitate selection of progeny who possess the desired trait(s), greatly speeding up the breeding process. This technique has proven particularly useful for the introgression of resistance genes into new backgrounds, as well as the efficient selection of many resistance genes pyramided into a single individual. Unfortunately, molecular markers are not currently available for many important traits, especially complex ones controlled by many genes.

Addressing Global Food Security Through Plant Breeding

For future agriculture to thrive there are necessary changes which must be made in accordance to arising global issues. These issues are arable land, harsh cropping conditions and food security which involves, being able to provide the world population with food containing sufficient nutrients. These crops need to be able to mature in several environments allowing for worldwide access, this is involves issues such as drought tolerance. These global issues are achievable through the process of plant breeding, as it offers the ability to select specific genes allowing the crop to perform at a level which yields the desired results.

Minimal Land Degradation

Land degradation is a major issue, as it can negatively impact the capability of the land to be productive. Poor agricultural management has a huge impact on the degradation of soil worldwide and it is Africa and Asia that are most affected. Through education and development of modified plants, these statistics can be reduced and agricultural land can become more productive. Plant breeding allows for an increase in yield with out the extra strain on the land. The genetically modified, Bt white maize, was introduced to South Africa and was surveyed in 33 large commercial farms and 368 small landholders properties and in both cases a higher yield was recorded.

Increased Yield Without Expansion

With an increasing population, the production of food needs to increase with it. It is estimated that a 70% increase in food production is needed by 2050 in order to meet the Declaration of the World Summit on Food Security. But with the natural degradation of agricultural land, simply planting more crops is no longer a viable option. Therefore, new varieties of plants need to be developed through plant breeding that generates an increase of yield without relying on an increase in land area. An example of this can be seen in Asia, where food production per capita has increased two-fold. This has been achieved through not only the use of fertilisers, but through the use of better crops that have been specifically designed for the area.

Breeding for Increased Nutritional Value

Plant breeding can contribute to global food security as it is a cost-effective tool for increasing nutritional value of forage and crops. Improvements in nutritional value for forage crops from the use of analytical chemistry and rumen fermentation technology have been recorded since 1960; this science and technology gave breeders the ability to screen thousands of samples within a small amount of time, meaning breeders could identify a high performing hybrid quicker. The main area genetic increases were made was in vitro dry matter digestibility (IVDMD) resulting in 0.7-2.5% increase, at just 1% increase in IVDMD a single Bos Taurus also known as beef cattle reported 3.2% increase in daily gains. This improvement indicates plant breeding is an essential tool in gearing future agriculture to perform at a more advanced level.

Breeding for Tolerance

Plant breeding of hybrid crops has become extremely popular worldwide in an effort to combat the harsh environment. With long periods of drought and lack of water or nitrogen stress tolerance has become a significant part of agriculture. Plant breeders have focused on identifying crops which will ensure crops perform under these conditions; a way to achieve this is finding strains of the crop that is resistance to drought conditions with low nitrogen. It is evident from this that plant breeding is vital for future agriculture to survive as it enables farmers to produce stress resistant crops hence improving food security.

Participatory Plant Breeding

The development of agricultural science, with phenomenon like the Green Revolution arising, have left millions of farmers in developing countries, most of whom operate small farms under unstable and difficult growing conditions, in a precarious situation. The adoption of new plant varieties by this group has been hampered by the constraints of poverty and the international policies promoting an industrialized model of agriculture. Their response has been the creation of a novel and promising set of research methods collectively known as participatory plant breeding. Participatory means that farmers are more involved in the breeding process and breeding goals are defined by farmers instead of international seed companies with their large-scale breeding programs. Farmers' groups and NGOs, for example, may wish to affirm local people's rights over genetic resources, produce seeds themselves, build farmers' technical expertise, or develop new products for niche markets, like organically grown food.

List of Notable Plant Breeders

- Gartons Agricultural Plant Breeders
- Gregor Mendel
- Keith Downey
- Luther Burbank
- Nazareno Strampelli
- Niels Ebbesen Hansen
- Edger McFadden
- Norman Borlaug

Plant Propagation

Plant propagation is the process of creating new plants from a variety of sources: seeds, cuttings, bulbs and other plant parts. Plant propagation can also refer to the artificial or natural dispersal of plants.

Sexual Propagation (Seed)

Seeds and spores can be used for reproduction (through e.g. sowing). Seeds are typically produced from sexual reproduction within a species, because genetic recombination has occurred. A plant grown from seeds may have different characteristics from its parents. Some species produce seeds that require special conditions to germinate, such as cold treatment. The seeds of many Australian plants and plants from southern Africa and the American west require smoke or fire to germinate. Some plant species, including many trees do not produce seeds until they reach maturity, which may take many years. Seeds can be difficult to acquire and some plants do not produce seed at all.

Some plants (like certain F1/F2 hybrids and GMO plants) may produce seed, but not fertile seed. In certain cases (like with GMO's), this is done to prevent the accidental spreading of these plants (which are generally non-native crops), for example by birds and other animals.

One way to propagate an avocado seed

Asexual Propagation

Plants have a number of mechanisms for asexual or vegetative reproduction. Some of these have been taken advantage of by horticulturists and gardeners to multiply or clone plants rapidly. People also use methods that plants do not use, such as tissue culture and grafting. Plants are produced using material from a single parent and as such there is no exchange of genetic material, therefore vegetative propagation methods almost always produce plants that are identical to the parent. Vegetative reproduction uses plants parts such as roots, stems and leaves. In some plants seeds can be produced without fertilization and the seeds contain only the genetic material of the parent plant. Therefore, propagation via asexual seeds or apomixis is asexual reproduction but not vegetative propagation.

Softwood stemcuttings rooting in a controlled environment

Techniques for vegetative propagation include:

- Air or ground layering

- Division
- Grafting and bud grafting, widely used in fruit tree propagation
- Micropropagation
- Stolons or runners
- Storage organs such as bulbs, corms, tubers and rhizomes
- Striking or cuttings
- Twin-scaling
- Offsets

Heated Propagator

A heated propagator is a horticultural device to maintain a warm and damp environment for seeds and cuttings to grow in.

This can be in the form of a clear enclosed bin sitting over a hotpad, or even a portable heater pointed at the bin. The key is to keep the moisture in the clear bin, while keeping lighting over the top of it, usually.

Seed Propagation Mat

An electric seed-propagation mat is a heated rubber mat covered by a metal cage which is used in gardening. The mats are made so that planters containing seedlings can be placed on top of the metal cage without the risk of starting a fire. In extreme cold, gardeners place a loose plastic cover over the planters/mats which creates a sort of miniature greenhouse. The constant and predictable heat allows people to garden in the winter months when the weather is generally too cold for seedlings to survive naturally. When combined with a lighting system, many plants can be grown indoors using these mats.

Plant Reproduction

Bryophyllum, a plant that reproduces asexually via new shoots from the leaves

Plant reproduction is the production of new individuals or offspring in plants, which can be accomplished by sexual or asexual reproduction. Sexual reproduction produces offspring by the fusion of gametes, resulting in offspring genetically different from the parent or parents. Asexual reproduction produces new individuals without the fusion of gametes, genetically identical to the parent plants and each other, except when mutations occur. In seed plants, the offspring can be packaged in a protective seed, which is used as an agent of dispersal.

Asexual Reproduction

Plants have two main types of asexual reproduction in which new plants are produced that are genetically identical clones of the parent individual. Vegetative reproduction involves a vegetative piece of the original plant (budding, tillering, etc.) and is distinguished from *apomixis*, which is a replacement for sexual reproduction, and in some cases involves seeds. Apomixis occurs in many plant species and also in some non-plant organisms. For apomixis and similar processes in non-plant organisms, see parthenogenesis.

Natural vegetative reproduction is mostly a process found in herbaceous and woody perennial plants, and typically involves structural modifications of the stem or roots and in a few species leaves. Most plant species that employ vegetative reproduction do so as a means to perennialize the plants, allowing them to survive from one season to the next and often facilitating their expansion in size. A plant that persists in a location through vegetative reproduction of individuals constitutes a clonal colony; a single ramet, or apparent individual, of a clonal colony is genetically identical to all others in the same colony. The distance that a plant can move during vegetative reproduction is limited, though some plants can produce ramets from branching rhizomes or stolons that cover a wide area, often in only a few growing seasons. In a sense, this process is not one of reproduction but one of survival and expansion of biomass of the individual. When an individual organism increases in size via cell multiplication and remains intact, the process is called vegetative growth. However, in vegetative reproduction, the new plants that result are new individuals in almost every respect except genetic. A major disadvantage to vegetative reproduction, is the transmission of pathogens from parent to offspring; it is uncommon for pathogens to be transmitted from the plant to its seeds (in sexual reproduction or in apomixis), though there are occasions when it occurs.

Seeds generated by apomixis are a means of asexual reproduction, involving the formation and dispersal of seeds that do not originate from the fertilization of the embryos. Hawkweed (*Hieracium*), dandelion (*Taraxacum*), some Citrus (*Citrus*) and Kentucky blue grass (*Poa pratensis*) all use this form of asexual reproduction. Pseudogamy occurs in some plants that have apomictic seeds, where pollination is often needed to initiate embryo growth, though the pollen contributes no genetic material to the developing offspring. Other forms of apomixis occur in plants also, including the generation of a plantlet in replacement of a seed or the generation of bulbils instead of flowers, where new cloned individuals are produced.

Structures

A rhizome is a modified underground stem serving as an organ of vegetative reproduction; the growing tips of the rhizome can separate as new plants, e.g., Polypody, Iris, Couch Grass and Nettles.

Prostrate aerial stems, called runners or stolons are important vegetative reproduction organs in some species, such as the strawberry, numerous grasses, and some ferns.

Adventitious buds form on roots near the ground surface, on damaged stems (as on the stumps of cut trees), or on old roots. These develop into above-ground stems and leaves. A form of budding called suckering is the reproduction or regeneration of a plant by shoots that arise from an existing root system. Species that characteristically produce suckers include Elm (*Ulmus*), Dandelion (*Taraxacum*), and many members of the Rose family such as *Rosa* and *Rubus*.

Plants like onion (*Allium cepa*), hyacinth (*Hyacinth*), narcissus (*Narcissus*) and tulips (*Tulipa*) reproduce by dividing their underground bulbs into more bulbs. Other plants like potatoes (*Solanum tuberosum*) and dahlia (*Dahlia*) reproduce by a similar method involving underground tubers. Gladioli and crocuses (*Crocus*) reproduce in a similar way with corms.

Usage

The most common form of plant reproduction utilized by people is seeds, but a number of asexual methods are utilized which are usually enhancements of natural processes, including: cutting, grafting, budding, layering, division, sectioning of rhizomes, roots, tubers, bulbs, stolons, tillers (suckers), etc., and artificial propagation by laboratory tissue cloning. Asexual methods are most often used to propagate cultivars with individual desirable characteristics that do not come true from seed. Fruit tree propagation is frequently performed by budding or grafting desirable cultivars (clones), onto rootstocks that are also clones, propagated by stooling.

In horticulture, a "cutting" is a branch that has been cut off from a mother plant below an internode and then rooted, often with the help of a rooting liquid or powder containing hormones. When a full root has formed and leaves begin to sprout anew, the clone is a self-sufficient plant, genetically identical to the mother plant. Examples include cuttings from the stems of blackberries (*Rubus occidentalis*), African violets (*Saintpaulia*), verbenas (*Verbena*) to produce new plants. A related use of cuttings is grafting, where a stem or bud is joined onto a different stem. Nurseries offer for sale trees with grafted stems that can produce four or more varieties of related fruits, including apples. The most common usage of grafting is the propagation of cultivars onto already rooted plants, sometimes the rootstock is used to dwarf the plants or protect them from root damaging pathogens.

Since vegetatively propagated plants are clones, they are important tools in plant research. When a clone is grown in various conditions, differences in growth can be ascribed to environmental effects instead of genetic differences.

Sexual Reproduction

Sexual reproduction involves two fundamental processes: meiosis, which rearranges the genes and reduces the number of chromosomes, and fertilization, which restores the chromosome to a complete diploid number. In between these two processes, different types of plants and algae vary, but many of them, including all land plants, undergo alternation of generations, with two different multicellular structures (phases), a gametophyte and a sporophyte. The evolutionary origin and adaptive significance of sexual reproduction are discussed in the pages "Evolution of sexual reproduction" and "Origin and function of meiosis."

The gametophyte is the multicellular structure (plant) that is haploid, containing a single set of chromosomes in each cell. The gametophyte produces male or female gametes (or both), by a process of cell division called mitosis. In vascular plants with separate gametophytes, female gametophytes are known as mega gametophytes (mega=large, they produce the large egg cells) and the male gametophytes are called micro gametophytes (micro=small, they produce the small sperm cells).

The fusion of male and female gametes (fertilization) produces a diploid zygote, which develops by mitotic cell divisions into a multicellular sporophyte.

The mature sporophyte produces spores by meiosis, sometimes referred to as "reduction division" because the chromosome pairs are separated once again to form single sets.

In mosses and liverworts the gametophyte is relatively large, and the sporophyte is a much smaller structure that is never separated from the gametophyte. In ferns, gymnosperms, and flowering plants (angiosperms), the gametophytes are relatively small and the sporophyte is much larger. In gymnosperms and flowering plants the mega gametophyte is contained within the ovule (that may develop into a seed) and the micro gametophyte is contained within a pollen grain.

History of Sexual Reproduction

Unlike animals, plants are immobile, and cannot seek out sexual partners for reproduction. In the evolution of early plants, abiotic means, including water and wind, transported sperm for reproduction. The first plants were aquatic, as described in the page "Evolutionary history of plants", and released sperm freely into the water to be carried with the currents. Primitive land plants like liverworts and mosses had motile sperm that swam in a thin film of water or were splashed in water droplets from the male reproduction organs onto the female organs. As taller and more complex plants evolved, modifications in the alternation of generations evolved; in the Paleozoic era progymnosperms reproduced by using spores dispersed on the wind. The seed plants including seed ferns, conifers and cordaites, which were all gymnosperms, evolved 350 million years ago; they had pollen grains that contained the male gametes for protection of the sperm during the process of transfer from the male to female parts. It is believed that insects fed on the pollen, and plants thus evolved to use insects to actively carry pollen from one plant to the next. Seed producing plants, which include the angiosperms and the gymnosperms, have heteromorphic alternation of generations with large sporophytes containing much reduced gametophytes. Angiosperms have distinctive reproductive organs called flowers, with carpels, and the female gametophyte is greatly reduced to a female embryo sac, with as few as eight cells. The male gametophyte consists of the pollen grains. The sperm of seed plants are non-motile, except for two older groups of plants, the Cycadophyta and the Ginkgophyta, which have flagellated sperm.

Flowering Plants

Flowering plants are the dominant plant form on land and they reproduce by sexual and asexual means. Often their most distinguishing feature is their reproductive organs, commonly called flowers. Sexual reproduction in flowering plants involves the production of male and female gametes, the transfer of the male gametes to the female ovules in a process called pollination. After pollination occurs, fertilization happens and the ovules grow into seeds within a fruit. After the seeds are ready for dispersal, the fruit ripens and by various means the seeds are freed from the fruit and after varying amounts of time and under specific conditions the seeds germinate and grow into the next generation.

The anther produces male gametophytes which are pollen grains, which attach to the stigma on top of a carpel, in which the female gametophytes (inside ovules) are located. After the pollen tube grows through the carpel's style, the sperm from the pollen grain migrate into the ovule to fertilize the egg cell and central cell within the female gametophyte in a process termed double fertilization. The resulting zygote develops into an embryo, while the triploid endosperm (one sperm cell plus a binucleate female cell) and female tissues of the ovule give rise to the surrounding tissues in the developing seed. The ovary, which produced the female gametophyte(s), then grows into a fruit, which surrounds the seed(s). Plants may either self-pollinate or cross-pollinate.

Pollination

An orchid flower

In plants that use insects or other animals to move pollen from one flower to the next, plants have developed greatly modified flower parts to attract pollinators and to facilitate the movement of pollen from one flower to the insect and from the insect back to the next flower. Flowers of wind pollinated plants tend to lack petals and or sepals; typically large amounts of pollen are produced and pollination often occurs early in the growing season before leaves can interfere with the dispersal of the pollen. Many trees and all grasses and sedges are wind pollinated, as such they have no need for large fancy flowers.

Plants have a number of different means to attract pollinators including colour, scent, heat, nectar glands, edible pollen and flower shape. Along with modifications involving the above structures two other conditions play a very important role in the sexual reproduction of flowering plants, the first is timing of flowering and the other is the size or number of flowers produced. Often plant species have a few large, very showy flowers while others produce many small flowers, often flowers are collected together into large inflorescences to maximize their visual effect, becoming more noticeable to passing pollinators. Flowers are attraction strategies and sexual expressions are functional strategies used to produce the next generation of plants, with pollinators and plants having co-evolved, often to some extraordinary degrees, very often rendering mutual benefit.

Flower heads showing disk and ray florets.

The largest family of flowering plants is the orchids (Orchidaceae), estimated by some specialists to include up to 35,000 species, which often have highly specialized flowers that attract particular insects for pollination. The stamens are modified to produce pollen in clusters called pollinia, which become attached to insects that crawl into the flower. The flower shapes may force insects to pass by the pollen, which is "glued" to the insect. Some orchids are even more highly specialized, with flower shapes that mimic the shape of insects to attract them to 'mate' with the flowers, a few even have scents that mimic insect pheromones.

Another large group of flowering plants is the Asteraceae or sunflower family with close to 22,000 species, which also have highly modified inflorescences that are flowers collected together in heads composed of a composite of individual flowers called florets. Heads with florets of one sex, when the flowers are pistillate or functionally staminate, or made up of all bisexual florets, are called homogamous and can include discoid and liguliflorous type heads. Some radiate heads may be homogamous too. Plants with heads that have florets of two or more sexual forms are called heterogamous and include radiate and disciform head forms, though some radiate heads may be heterogamous too.

Ferns

Ferns typically produce large diploid sporophytes with rhizomes, roots and leaves; and on fertile leaves called sporangium, spores are produced. The spores are released and germinate to produce short, thin gametophytes that are typically heart shaped, small and green in color. The gametophytes or thallus, produce both motile sperm in the antheridia and egg cells in separate archegonia. After rains or when dew deposits a film of water, the motile sperm are splashed away from the antheridia, which are normally produced on the top side of the thallus, and swim in the film of water to the antheridia where they fertilize the egg. To promote out crossing or cross fertilization the sperm are released before the eggs are receptive of the sperm, making it more likely that the sperm will fertilize the eggs of different thallus. A zygote is formed after fertilization, which grows into a new sporophytic plant. The condition of having separate sporophyte and gametophyte plants is call alternation of generations. Other plants with similar reproductive means include the *Psilotum*, *Lycopodium*, *Selaginella* and *Equisetum*.

Bryophytes

The bryophytes, which include liverworts, hornworts and mosses, reproduce both sexually and vegetatively. The gametophyte is the most commonly known phase of the plant. All are small plants found growing in moist locations and like ferns, have motile sperm with flagella and need water to facilitate sexual reproduction. These plants start as a haploid spore that grows into the dominate form, which is a multicellular haploid body with leaf-like structures that photosynthesize. Haploid gametes are produced in antherida and archegonia by mitosis. The sperm released from the antheridia respond to chemicals released by ripe archegonia and swim to them in a film of water and fertilize the egg cells, thus producing zygotes that are diploid. The zygote divides by mitotic division and grows into a sporophyte that is diploid. The multicellular diploid sporophyte produces structures called spore capsules. The spore capsules produce spores by meiosis, and when ripe, the capsules burst open and the spores are released. Bryophytes show considerable variation in their breeding structures and the above is a basic outline. In some species each gametophyte is one sex while other species produce both antheridia and archegonia on the same gametophyte which is thus hermaphrodite.

Sexual Morphology

Many plants have evolved complex sexual reproductive systems, which is expressed in different combinations of their reproductive organs. Some species have separate male and female plants, and some have separate male and female flowers on the same plant, but the majority of plants have both male and female parts in the same flower. Some plants change their morphological expression depending on a number of factors like age, time of day, or because of environmental conditions. Plant sexual morphology also varies within different populations of some species.

References

- "Photosynthesis". McGraw-Hill Encyclopedia of Science & Technology. 13. New York: McGraw-Hill. 2007. ISBN 0-07-144143-3.
- Whitmarsh J, Govindjee (1999). "Chapter 2: The Basic Photosynthetic Process". In Singhal GS, Renger G, Sopory SK, Irrgang KD, Govindjee. Concepts in Photobiology: Photosynthesis and Photomorphogenesis. Boston: Kluwer Academic Publishers. p. 13. ISBN 978-0-7923-5519-9.
- Campbell NA, Williamson B, Heyden RJ (2006). Biology Exploring Life. Upper Saddle River, NJ: Prentice Hall. ISBN 0-13-250882-6.
- Raven PH, Evert RF, Eichhorn SE (2005). Biology of Plants, (7th ed.). New York: W. H. Freeman and Company. pp. 124–127. ISBN 0-7167-1007-2.
- Muscatine L, Greene RW (1973). "Chloroplasts and algae as symbionts in molluscs". International Review of Cytology. International Review of Cytology. 36: 137–69. doi:10.1016/S0074-7696(08)60217-X. ISBN 9780123643360.
- Herrero A, Flores E (2008). The Cyanobacteria: Molecular Biology, Genomics and Evolution (1st ed.). Caister Academic Press. ISBN 978-1-904455-15-8.
- Jones HG (2014). Plants and Microclimate: a Quantitative Approach to Environmental Plant Physiology (Third ed.). Cambridge: Cambridge University Press. ISBN 978-0-521-27959-8.
- Harris, Richard W. (1983). Arboriculture: Care of Trees, Shrubs, and Vines in the Landscape. Englewood Cliffs, New Jersey 07632: Prentice-Hall, Inc. pp. 2–3. ISBN 0-13-043935-5.

- Reiley, H. Edward; Shry, Carroll L. (2004). Introductory horticulture. Albany, NY: Delmar/Thomson Learning. p. 54. ISBN 978-0-7668-1567-4.
- Fritz, Robert E.; Simms, Ellen Louise (1992). Plant resistance to herbivores and pathogens: ecology, evolution, and genetics. Chicago: University of Chicago Press. p. 359. ISBN 978-0-226-26554-4.
- Suzie Key; Julian K-C Ma & Pascal MW Drake (1 June 2008). "Genetically modified plants and human health". Journal of the Royal Society of Medecine. pp. 290–298. Retrieved 11 March 2015.
- Oldeman, l (1994). "The global extent of soil degradation" (PDF). Soil resilience and sustainable land use 32 (5967): 818–822. Retrieved 7-11-2013.
- Bänziger (2000). "Breeding for drought and nitrogen stress tolerance in maize: from theory to practice". From Theory to Practice: 7–9. Retrieved 7-11-2013.

5

Plant: An Overview

Plants belong to the kingdom of Plantae. The science that identifies plants and names them is known as plant taxonomy. Plant taxonomy can also be used as a synonym for plant systematics. Bryophyte, embryophyte, spermatophyte, flowering plant and moss are numerous aspects of a plant that have been elaborately explained the chapter. The aspects elucidated in this chapter are of vital importance, and provide a better understanding of botany.

Plant

Plants, also called green plants, are multicellular eukaryotes of the kingdom Plantae. They form an unranked clade Viridiplantae (Latin for green plants) that includes the flowering plants, conifers and other gymnosperms, ferns, clubmosses, hornworts, liverworts, mosses and the green algae. Green plants exclude the red and brown algae, the fungi, archaea, bacteria and animals.

Green plants have cell walls with cellulose and obtain most of their energy from sunlight via photosynthesis by primary chloroplasts, derived from endosymbiosis with cyanobacteria. Their chloroplasts contain chlorophylls a and b, which gives them their green color. Some plants are parasitic and have lost the ability to produce normal amounts of chlorophyll or to photosynthesize. Plants are characterized by sexual reproduction and alternation of generations, although asexual reproduction is also common.

There are about 300–315 thousand species of plants, of which the great majority, some 260–290 thousand, are seed plants (see the table below). Green plants provide most of the world's molecular oxygen and are the basis of most of the earth's ecologies, especially on land. Plants that produce grains, fruits and vegetables form humankind's basic foodstuffs, and have been domesticated for millennia. Plants play many roles in culture. They are used as ornaments and, until recently and in great variety, they have served as the source of most medicines and drugs. The scientific study of plants is known as botany, a branch of biology.

Definition

Plants are one of the two groups into which all living things were traditionally divided; the other is animals. The division goes back at least as far as Aristotle (384 BC – 322 BC), who distinguished between plants, which generally do not move, and animals, which often are mobile to catch their food. Much later, when Linnaeus (1707–1778) created the basis of the modern system of scientific classification, these two groups became the kingdoms Vegetabilia (later Metaphyta or Plantae) and Animalia (also called Metazoa). Since then, it has become clear that the plant kingdom as originally defined included several unrelated groups, and the fungi and several groups of algae were removed to new kingdoms. However, these organisms are still often considered plants, particularly in popular contexts.

Outside of formal scientific contexts, the term "plant" implies an association with certain traits, such as being multicellular, possessing cellulose, and having the ability to carry out photosynthesis.

Current Definitions of Plantae

When the name Plantae or plant is applied to a specific group of organisms or taxon, it usually refers to one of four concepts. From least to most inclusive, these four groupings are:

Name(s)	Scope	Description
Land plants, also known as Embryophyta	Plantae *sensu strictissimo*	This group includes the liverworts, hornworts, mosses, and vascular plants, as well as fossil plants similar to these surviving groups (e.g., Metaphyta Whittaker, 1969, Plantae Margulis, 1971).
Green plants, also known as Viridiplantae, Viridiphyta or Chlorobionta	Plantae *sensu stricto*	This group includes the green algae, and land plants that emerged within them, including stoneworts. The names given to these groups vary considerably as of July 2011. Viridiplantae encompass a group of organisms that have cellulose in their cell walls, possess chlorophylls *a* and *b* and have plastids that are bound by only two membranes that are capable of storing starch. It is this clade that is mainly the subject of this article (e.g., Plantae Copeland, 1956).
Archaeplastida, Plastida or Primoplantae	Plantae *sensu lato*	This group comprises the green plants above plus Rhodophyta (red algae) and Glaucophyta (glaucophyte algae). This clade includes the organisms that eons ago acquired their chloroplasts directly by engulfing cyanobacteria (e.g., Plantae Cavalier-Smith, 1981).
Old definitions of plant	Plantae *sensu amplo*	Old classifications placed diverse algae, fungi or bacteria in Plantae (e.g., Plantae or Vegetabilia Linnaeus, Plantae Haeckel 1866, Metaphyta Haeckel, 1894, Plantae Whittaker, 1969).

Another way of looking at the relationships between the different groups that have been called "plants" is through a cladogram, which shows their evolutionary relationships. The evolutionary history of plants is not yet completely settled, but one accepted relationship between the three groups described above is shown below. Those which have been called "plants" are in bold.

Algae

Green algae from Ernst Haeckel's *Kunstformen der Natur*, 1904.

Algae comprise several different groups of organisms which produce energy through photosynthesis and for that reason have been included in the plant kingdom in the past. Most conspicuous among the algae are the seaweeds, multicellular algae that may roughly resemble land plants, but are classified among the brown, red and green algae. Each of these algal groups also includes various microscopic and single-celled organisms. There is good evidence that some of these algal groups arose independently from separate non-photosynthetic ancestors, with the result that many groups of algae are no longer classified within the plant kingdom as it is defined here.

The Viridiplantae, the green plants – green algae and land plants – form a clade, a group consisting of all the descendants of a common ancestor. With a few exceptions among the green algae, green plants the following features in common; cell walls containing cellulose, chloroplasts containing chlorophylls *a* and *b*, and food stores in the form of starch contained within plastids. They undergo closed mitosis without centrioles, and typically have mitochondria with flat cristae. The chloroplasts of green plants are surrounded by two membranes, suggesting they originated directly from endosymbiotic cyanobacteria.

Two additional groups, the Rhodophyta (red algae) and Glaucophyta (glaucophyte algae), also have chloroplasts which appear to be derived directly from endosymbiotic cyanobacteria, although they differ in the pigments which are used in photosynthesis and so are different in colour. All three groups together are generally believed to have a single common origin, and so are classified together in the taxon Archaeplastida, whose name implies that the chloroplasts or plastids of all the members of the taxon were derived from a single ancient endosymbiotic event. This is the broadest modern definition of the plants.

In contrast, most other algae (e.g. brown algae/diatoms, haptophytes, dinoflagellates, and euglenids) not only have different pigments but also have chloroplasts with three or four surrounding membranes. They are not close relatives of the Archaeplastida, presumably having acquired chloroplasts separately from ingested or symbiotic green and red algae. They are thus not included in even the broadest modern definition of the plant kingdom, although they were in the past.

The green plants or Viridiplantae were traditionally divided into the green algae (including the stoneworts) and the land plants. However, it is now known that the land plants evolved from within a group of green algae, so that the green algae by themselves are a paraphyletic group, i.e. a group that excludes some of the descendants of a common ancestor. Paraphyletic groups are generally avoided in modern classifications, so that in recent treatments the Viridiplantae have been divided into two clades, the Chlorophyta and the Streptophyta (or Charophyta).

The Chlorophyta (a name that has also been used for *all* green algae) are the sister group to the group from which the land plants evolved. There are about 4,300 species of mainly marine organisms, both unicellular and multicellular. The latter include the sea lettuce, *Ulva*.

The other group within the Viridiplantae are the mainly freshwater or terrestrial Streptophyta, which consists of the land plants together with the Charophyta, itself consisting of several groups of green algae such as the desmids and stoneworts. (The names have been used differently, e.g. Streptophyta to mean the group that excludes the land plants and Charophyta for the stoneworts alone or the stoneworts plus the land plants.) Streptophyte algae are either unicellular or form multicellular filaments, branched or unbranched. The genus *Spirogyra* is a filamentous streptophyte alga familiar to many, as it is often used in teaching and is one of the organisms responsible for the algal "scum" that pond-owners so dislike. The freshwater stoneworts strongly resemble land plants and are believed to be their closest relatives. Growing underwater, they consist of a central stalk with whorls of branchlets, giving them a superficial resemblance to horsetails, species of the genus *Equisetum*, which are true land plants.

Fungi

The classification of fungi has been controversial until quite recently in the history of biology. Linnaeus' original classification placed the fungi within the Plantae, since they were unquestionably not animals or minerals and these were the only other alternatives. With later developments in microbiology, in the 19th century Ernst Haeckel felt that another kingdom was required to classify newly discovered micro-organisms. The introduction of the new kingdom Protista in addition to Plantae and Animalia, led to uncertainty as to whether fungi truly were best placed in the Plantae or whether they ought to be reclassified as protists. Haeckel himself found it difficult to decide and it was not until 1969 that a solution was found whereby Robert Whittaker proposed the creation of the kingdom Fungi. Molecular evidence has since shown that the most recent common ancestor (concestor), of the Fungi was probably more similar to that of the Animalia than to that of Plantae or any other kingdom.

Whittaker's original reclassification was based on the fundamental difference in nutrition between the Fungi and the Plantae. Unlike plants, which generally gain carbon through photosynthesis, and so are called autotrophs, fungi generally obtain carbon by breaking down and absorbing surrounding materials, and so are called heterotrophic saprotrophs. In addition, the substructure of multicellular fungi is different from that of plants, taking the form of many chitinous microscopic strands called hyphae, which may be further subdivided into cells or may form a syncytium containing many eukaryotic nuclei. Fruiting bodies, of which mushrooms are the most familiar example, are the reproductive structures of fungi, and are unlike any structures produced by plants.

Diversity

The table below shows some species count estimates of different green plant (Viridiplantae) divisions. It suggests there are about 300,000 species of living Viridiplantae, of which 85–90% are flowering plants. (Note: as these are from different sources and different dates, they are not necessarily comparable, and like all species counts, are subject to a degree of uncertainty in some cases.)

Diversity of living green plant (Viridiplantae) divisions				
Informal group	**Division name**	**Common name**	**No. of living species**	**Approximate No. in informal group**
Green algae	Chlorophyta	green algae (chlorophytes)	3,800–4,300	8,500
	Charophyta	green algae (e.g. desmids & stoneworts)	2,800–6,000	(6,600–10,300)
Bryophytes	Marchantiophyta	liverworts	6,000–8,000	19,000
	Anthocerotophyta	hornworts	100–200	(18,100–20,200)
	Bryophyta	mosses	12,000	
Pteridophytes	Lycopodiophyta	club mosses	1,200	12,000
	Pteridophyta	ferns, whisk ferns & horsetails	11,000	(12,200)

Seed plants	Cycadophyta	cycads	160	260,000
	Ginkgophyta	ginkgo	1	(259,511)
	Pinophyta	conifers	630	
	Gnetophyta	gnetophytes	70	
	Magnoliophyta	flowering plants	258,650	

The naming of plants is governed by the International Code of Nomenclature for algae, fungi, and plants and International Code of Nomenclature for Cultivated Plants (see cultivated plant taxonomy).

Evolution

The evolution of plants has resulted in increasing levels of complexity, from the earliest algal mats, through bryophytes, lycopods, ferns to the complex gymnosperms and angiosperms of today. Plants in all of these groups continue to thrive, especially in the environments in which they evolved.

An algal scum formed on the land 1,200 million years ago, but it was not until the Ordovician Period, around 450 million years ago, that land plants appeared. However, new evidence from the study of carbon isotope ratios in Precambrian rocks has suggested that complex photosynthetic plants developed on the earth over 1000 m.y.a. For more than a century it has been assumed that the ancestors of land plants evolved in aquatic environments and then adapted to a life on land, an idea usually credited to botanist Frederick Orpen Bower in his 1908 book "The Origin of a Land Flora". A more recent alternative view, supported by genetic evidence, is that they evolved from single-celled algae that were already terrestrial. Primitive land plants began to diversify in the late Silurian Period, around 420 million years ago, and the fruits of their diversification are displayed in remarkable detail in an early Devonian fossil assemblage from the Rhynie chert. This chert preserved early plants in cellular detail, petrified in volcanic springs. By the middle of the Devonian Period most of the features recognised in plants today are present, including roots, leaves and secondary wood, and by late Devonian times seeds had evolved. Late Devonian plants had thereby reached a degree of sophistication that allowed them to form forests of tall trees. Evolutionary innovation continued after the Devonian period. Most plant groups were relatively unscathed by the Permo-Triassic extinction event, although the structures of communities changed. This may have set the scene for the evolution of flowering plants in the Triassic (~200 million years ago), which exploded in the Cretaceous and Tertiary. The latest major group of plants to evolve were the grasses, which became important in the mid Tertiary, from around 40 million years ago. The grasses, as well as many other groups, evolved new mechanisms of metabolism to survive the low CO_2 and warm, dry conditions of the tropics over the last 10 million years.

A 1997 proposed phylogenetic tree of Plantae, after Kenrick and Crane, is as follows, with modification to the Pteridophyta from Smith *et al.* The Prasinophyceae are a paraphyletic assemblage of early diverging green algal lineages, but are treated as a group outside the Chlorophyta: later authors have not followed this suggestion

A newer proposed classification follows Leliaert et al. 2011 and modified with Silar 2016 for the green algae clades and Novíkov & Barabaš-Krasni 2015 for the land plants clade. Notice that the Prasinophyceae are here placed inside the Chlorophyta.

Embryophytes

Dicksonia antarctica, a species of tree fern

The plants that are likely most familiar to us are the multicellular land plants, called embryophytes. Embryophytes include the vascular plants, such as ferns, conifers and flowering plants. They also include the *bryophytes*, of which mosses and liverworts are the most common.

All of these plants have eukaryotic cells with cell walls composed of cellulose, and most obtain their energy through photosynthesis, using light, water and carbon dioxide to synthesize food. About three hundred plant species do not photosynthesize but are parasites on other species of photosynthetic plants. Embryophytes are distinguished from green algae, which represent a mode of photosynthetic life similar to the kind modern plants are believed to have evolved from, by having specialized reproductive organs protected by non-reproductive tissues.

Bryophytes first appeared during the early Paleozoic. They can only survive where moisture is available for significant periods, although some species are desiccation-tolerant. Most species of bryophytes remain small throughout their life-cycle. This involves an alternation between two generations: a haploid stage, called the gametophyte, and a diploid stage, called the sporophyte. In bryophytes, the sporophyte is always unbranched and remains nutritionally dependent on its parent gametophyte. The bryophytes have the ability to secrete a cuticle on their outer surface, a waxy layer that confers resistant to desiccation. In the mosses and hornworts a cuticle is usually only produced on the sporophyte. Stomata are absent from liverworts, but occur on the sporangia of mosses and hornworts, allowing gas exchange while controlling water loss.

Vascular plants first appeared during the Silurian period, and by the Devonian had diversified and spread into many different terrestrial environments. They developed a number of adaptations that allowed them to spread into increasingly more arid places, notably the vascular tissues xylem and phloem, that transport water and food throughout the organism. Root systems capable of obtaining soil water and nutrients also evolved during the Devonian. In modern vascular plants, the sporophyte is typically large, branched, nutritionally independent and long-lived, but there is increasing evidence that Paleozoic gametophytes were just as complex as the sporophytes. The gametophytes of all vascular plant groups evolved to become reduced in size and prominence in the life cycle.

The first seed plants, Pteridosperms (seed ferns), now extinct, appeared in the Devonian and diversified through the Carboniferous. In these the microgametophyte is reduced to pollen and the megagametophyte remains inside the megasporangium, attached to the parent plant. A megasporangium invested in protective layer called an integument is known as an ovule. After fertilisation by means of sperm deposited by pollen grains, an embryo develops inside the ovule. The integument becomes a seed coat, and the ovule develops into a seed. Seed plants can survive and reproduce in extremely arid conditions, because they are not dependent on free water for the movement of sperm, or the development of free living gametophytes.

Early seed plants are gymnosperms, as the ovules and subsequent seeds are not enclosed in a protective structure (carpels or fruit), but are found naked, typically on cone scales. Pollen typically lands directly on the ovule. Four surviving groups remain widespread now, particularly the conifers, which are dominant trees in several biomes.

Fossils

A petrified log in Petrified Forest National Park.

Plant fossils include roots, wood, leaves, seeds, fruit, pollen, spores, phytoliths, and amber (the fossilized resin produced by some plants). Fossil land plants are recorded in terrestrial, lacustrine, fluvial and nearshore marine sediments. Pollen, spores and algae (dinoflagellates and acritarchs) are used for dating sedimentary rock sequences. The remains of fossil plants are not as common as fossil animals, although plant fossils are locally abundant in many regions worldwide.

The earliest fossils clearly assignable to Kingdom Plantae are fossil green algae from the Cambrian. These fossils resemble calcified multicellular members of the Dasycladales. Earlier Precambrian fossils are known that resemble single-cell green algae, but definitive identity with that group of algae is uncertain.

The oldest known fossils of embryophytes date from the Ordovician, though such fossils are fragmentary. By the Silurian, fossils of whole plants are preserved, including the lycophyte *Baragwanathia longifolia*. From the Devonian, detailed fossils of rhyniophytes have been found. Early fossils of these ancient plants show the individual cells within the plant tissue. The Devonian period also saw the evolution of what many believe to be the first modern tree, *Archaeopteris*. This fern-like tree combined a woody trunk with the fronds of a fern, but produced no seeds.

The Coal measures are a major source of Paleozoic plant fossils, with many groups of plants in existence at this time. The spoil heaps of coal mines are the best places to collect; coal itself is the remains of fossilised plants, though structural detail of the plant fossils is rarely visible in coal. In the Fossil Grove at Victoria Park in Glasgow, Scotland, the stumps of *Lepidodendron* trees are found in their original growth positions.

The fossilized remains of conifer and angiosperm roots, stems and branches may be locally abundant in lake and inshore sedimentary rocks from the Mesozoic and Cenozoic eras. Sequoia and its allies, magnolia, oak, and palms are often found.

Petrified wood is common in some parts of the world, and is most frequently found in arid or desert areas where it is more readily exposed by erosion. Petrified wood is often heavily silicified (the organic material replaced by silicon dioxide), and the impregnated tissue is often preserved in fine detail. Such specimens may be cut and polished using lapidary equipment. Fossil forests of petrified wood have been found in all continents.

Fossils of seed ferns such as *Glossopteris* are widely distributed throughout several continents of the Southern Hemisphere, a fact that gave support to Alfred Wegener's early ideas regarding Continental drift theory.

The earliest fossils attributed to green algae date from the Precambrian (ca. 1200 mya). The resistant outer walls of prasinophyte cysts (known as phycomata) are well preserved in fossil deposits of the Paleozoic (ca. 250-540 mya). A filamentous fossil (Proterocladus) from middle Neoproterozoic deposits (ca. 750 mya) has been attributed to the Cladophorales, while the oldest reliable records of the Bryopsidales, Dasycladales) and stoneworts are from the Paleozoic.

Structure, Growth and Development

Most of the solid material in a plant is taken from the atmosphere. Through a process known as photosynthesis, most plants use the energy in sunlight to convert carbon dioxide from the atmosphere, plus water, into simple sugars. (Parasitic plants, on the other hand, use the resources of its host to grow.) These sugars are then used as building blocks and form the main structural component of the plant. Chlorophyll, a green-colored, magnesium-containing pigment is essential to this process; it is generally present in plant leaves, and often in other plant parts as well.

Plants usually rely on soil primarily for support and water (in quantitative terms), but also obtain compounds of nitrogen, phosphorus, potassium, magnesium and other elemental nutrients. Epiphytic and lithophytic plants depend on air and nearby debris for nutrients, and carnivorous plants supplement their nutrient requirements with insect prey that they capture. For the majority of plants to grow successfully they also require oxygen in the atmosphere and around their roots (soil gas) for respiration. Plants use oxygen and glucose (which may be produced from stored starch) to provide energy. Some plants grow as submerged aquatics, using oxygen dissolved in the surrounding water, and a few specialized vascular plants, such as mangroves, can grow with their roots in anoxic conditions.

The leaf is usually the primary site of photosynthesis in plants.

There is no photosynthesis in deciduous leaves in autumn.

Factors Affecting Growth

The genotype of a plant affects its growth. For example, selected varieties of wheat grow rapidly, maturing within 110 days, whereas others, in the same environmental conditions, grow more slowly and mature within 155 days.

Growth is also determined by environmental factors, such as temperature, available water, available light, carbon dioxide and available nutrients in the soil. Any change in the availability of these external conditions will be reflected in the plant's growth.

Biotic factors are also capable of affecting plant growth. Plants compete with other plants for space, water, light and nutrients. Plants can be so crowded that no single individual produces normal growth, causing etiolation and chlorosis. Optimal plant growth can be hampered by grazing animals, suboptimal soil composition, lack of mycorrhizal fungi, and attacks by insects or plant diseases, including those caused by bacteria, fungi, viruses, and nematodes.

Simple plants like algae may have short life spans as individuals, but their populations are commonly seasonal. Other plants may be organized according to their seasonal growth pattern: annual plants live and reproduce within one growing season, biennial plants live for two growing seasons and usually reproduce in second year, and perennial plants live for many growing seasons and continue to reproduce once they are mature. These designations often depend on climate and other environmental factors; plants that are annual in alpine or temperate regions can be biennial or perennial in warmer climates. Among the vascular plants, perennials include both evergreens that keep their leaves the entire year, and deciduous plants that lose their leaves for some part of it. In temperate and boreal climates, they generally lose their leaves during the winter; many tropical plants lose their leaves during the dry season.

The growth rate of plants is extremely variable. Some mosses grow less than 0.001 millimeters per hour (mm/h), while most trees grow 0.025-0.250 mm/h. Some climbing species, such as kudzu, which do not need to produce thick supportive tissue, may grow up to 12.5 mm/h.

Dried Dead Plants

Plants protect themselves from frost and dehydration stress with antifreeze proteins, heat-shock proteins and sugars (sucrose is common). LEA (Late Embryogenesis Abundant) protein expression is induced by stresses and protects other proteins from aggregation as a result of desiccation and freezing.

Effects of Cold

When water freezes in plants, the consequences for the plant depend very much on whether the freezing occurs within cells (intracellularly) or outside cells in intercellular spaces (Glerum 1985). Intracellular freezing, which usually kills the cell (Lyons et al. 1979) regardless of the hardiness of the plant and its tissues, seldom occurs in nature because rates of cooling are rarely high enough to support it. Rates of cooling of several degrees Celsius per minute are typically needed to cause intracellular formation of ice (Mazur 1977).

At rates of cooling of a few degrees Celsius per hour, segregation of ice occurs in intercellular spaces, the "extraorgan ice" of Sakai and Larcher (1987) and their coworkers. This may or may not be lethal, depending on the hardiness of the tissue.

The process of intercellular ice formation was described by Glerum (1985). At freezing temperatures, water in the intercellular spaces of plant tissue freezes first, though the water may remain unfrozen until temperatures drop below −7 °C (19 °F). After the initial formation of ice intercellularly, the cells shrink as water is lost to the segregated ice, and the cells undergo freeze-drying. This dehydration is now considered the fundamental cause of freezing injury.

Plant Cell

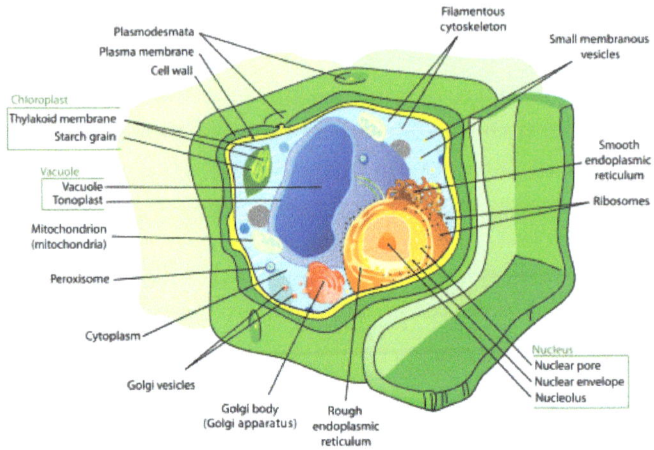

Plant cell structure

Plant cells are typically distinguished by their large water-filled central vacuole, chloroplasts, and rigid cell walls that are made up of cellulose, hemicellulose, and pectin. Cell division is also characterized by the development of a phragmoplast for the construction of a cell plate in the late stages of cytokinesis. Just as in animals, plant cells differentiate and develop into multiple cell types. Totipotent meristematic cells can differentiate into vascular, storage, protective (e.g. epidermal layer), or reproductive tissues, with more primitive plants lacking some tissue types.

Physiology

Photosynthesis

Plants are photosynthetic, which means that they manufacture their own food molecules using energy obtained from light. The primary mechanism plants have for capturing light energy is the pigment chlorophyll. All green plants contain two forms of chlorophyll, chlorophyll *a* and chlorophyll *b*. The latter of these pigments is not found in red or brown algae. The simple equation of photosynthesis is as follows:-

$6CO_2 + 6H_2O \rightarrow$ (in the presence of light and chlorophyll) $C_6H_{12}O_6 + 6O_2$

Immune System

By means of cells that behave like nerves, plants receive and distribute within their systems information about incident light intensity and quality. Incident light that stimulates a chemical reaction in one leaf, will cause a chain reaction of signals to the entire plant via a type of cell termed a *bundle sheath cell*. Researchers, from the Warsaw University of Life Sciences in Poland, found that plants have a specific memory for varying light conditions, which prepares their immune systems against seasonal pathogens. Plants use pattern-recognition receptors to recognize conserved microbial signatures. This recognition triggers an immune response. The first plant receptors of conserved microbial signatures were identified in rice (XA21, 1995) and in *Arabidopsis thaliana* (FLS2, 2000). Plants also carry immune receptors that recognize highly variable pathogen effectors. These include the NBS-LRR class of proteins.

Internal Distribution

Vascular plants differ from other plants in that nutrients are transported between their different parts through specialized structures, called xylem and phloem. They also have roots for taking up water and minerals. The xylem moves water and minerals from the root to the rest of the plant, and the phloem provides the roots with sugars and other nutrient produced by the leaves.

Ecology

The photosynthesis conducted by land plants and algae is the ultimate source of energy and organic material in nearly all ecosystems. Photosynthesis radically changed the composition of the early Earth's atmosphere, which as a result is now 21% oxygen. Animals and most other organisms are aerobic, relying on oxygen; those that do not are confined to relatively rare anaerobic environments. Plants are the primary producers in most terrestrial ecosystems and form the basis of the food web in those ecosystems. Many animals rely on plants for shelter as well as oxygen and food.

Land plants are key components of the water cycle and several other biogeochemical cycles. Some plants have coevolved with nitrogen fixing bacteria, making plants an important part of the nitrogen cycle. Plant roots play an essential role in soil development and prevention of soil erosion.

Distribution

Plants are distributed worldwide in varying numbers. While they inhabit a multitude of biomes and ecoregions, few can be found beyond the tundras at the northernmost regions of continental shelves. At the southern extremes, plants have adapted tenaciously to the prevailing conditions. (See Antarctic flora.)

Plants are often the dominant physical and structural component of habitats where they occur. Many of the Earth's biomes are named for the type of vegetation because plants are the dominant organisms in those biomes, such as grasslands and forests.

Ecological Relationships

Numerous animals have coevolved with plants. Many animals pollinate flowers in exchange for food in the form of pollen or nectar. Many animals disperse seeds, often by eating fruit and passing the seeds in their feces. Myrmecophytes are plants that have coevolved with ants. The plant provides a home, and sometimes food, for the ants. In exchange, the ants defend the plant from herbivores and sometimes competing plants. Ant wastes provide organic fertilizer.

The majority of plant species have various kinds of fungi associated with their root systems in a kind of mutualistic symbiosis known as mycorrhiza. The fungi help the plants gain water and mineral nutrients from the soil, while the plant gives the fungi carbohydrates manufactured in photosynthesis. Some plants serve as homes for endophytic fungi that protect the plant from herbivores by producing toxins. The fungal endophyte, *Neotyphodium coenophialum*, in tall fescue (*Festuca arundinacea*) does tremendous economic damage to the cattle industry in the U.S.

The Venus flytrap, a species of carnivorous plant.

Various forms of parasitism are also fairly common among plants, from the semi-parasitic mistletoe that merely takes some nutrients from its host, but still has photosynthetic leaves, to the fully parasitic broomrape and toothwort that acquire all their nutrients through connections to the roots of other plants, and so have no chlorophyll. Some plants, known as myco-heterotrophs, parasitize mycorrhizal fungi, and hence act as epiparasites on other plants.

Many plants are epiphytes, meaning they grow on other plants, usually trees, without parasitizing them. Epiphytes may indirectly harm their host plant by intercepting mineral nutrients and light that the host would otherwise receive. The weight of large numbers of epiphytes may break tree limbs. Hemiepiphytes like the strangler fig begin as epiphytes but eventually set their own roots and overpower and kill their host. Many orchids, bromeliads, ferns and mosses often grow as epiphytes. Bromeliad epiphytes accumulate water in leaf axils to form phytotelmata that may contain complex aquatic food webs.

Approximately 630 plants are carnivorous, such as the Venus Flytrap (*Dionaea muscipula*) and sundew (*Drosera* species). They trap small animals and digest them to obtain mineral nutrients, especially nitrogen and phosphorus.

Importance

The study of plant uses by people is termed economic botany or ethnobotany; some consider economic botany to focus on modern cultivated plants, while ethnobotany focuses on indigenous plants cultivated and used by native peoples. Human cultivation of plants is part of agriculture, which is the basis of human civilization. Plant agriculture is subdivided into agronomy, horticulture and forestry.

Foods and Beverages

Much of human nutrition depends on plants, either directly through foods and beverages consumed by people, or indirectly as feed for animals or the flavoring of foods. The science of agriculture deals with the planting, raising, nutrition, and harvest of food crops, and has played a key role in the history of world civilizations.

Mechanical harvest of oats.

Human nutrition depends to a large extent on cereals, especially maize (or corn), wheat, rice, oats, and millet. Large areas of many countries are given over to the cultivation of cereals for local consumption or export to other countries. Livestock animals including cows, pigs, sheep, goats and camels are all herbivores; and most feed primarily or entirely on cereal plants. Cereals are staple crops, meaning that they provide calories (in the form of complex carbohydrates such as starch) that are needed to fuel daily activities, and thus form the foundation of a daily diet. Other staple crops include potatoes, cassava, yams, and legumes.

Human food also includes vegetables, which consist principally of leaves and stems eaten as food. Vegetables are important for the vitamins, minerals, and dietary fiber they supply. Fruits provide a higher quantity of sugars and have a sweeter taste than vegetables. However, whether a particular food is considered a "vegetable" or a "fruit" will depend on context, since the word *fruit* has a more precise definition in botany than it does in general use. Nuts and seeds, including foods such as peanuts, walnuts, almonds, and pistachios, contain unsaturated fats that are also necessary for a healthy diet. As with fruits, the terms *nut* and *seed* have stricter definitions in plant science.

Many plants are used to flavor foods. Such plants include herbs (e.g. rosemary and mint), which come from the green leafy parts of plants, and spices (e.g. cumin and cinnamon), which come from other plant parts. Some plants produce edible flowers, which may be added to salads or used to decorate foods. Sweeteners such as sugar and stevia are derived from plants. Sugar is obtained mainly from sugar cane and sugar beet, and honey is created when bees regurgitate the nectar from flowers. Cooking oils and margarine come from maize, soybean, rapeseed, safflower, sunflower, olive and others. Food additives include gum arabic, guar gum, locust bean gum, starch and pectin.

Plants are also the source of beverages produced either by infusion, such as coffee and tea; by fermentation, such as beer and wine; or by distillation, such as whisky, vodka, rum, and other alcoholic spirits.

Nonfood Products

Plants are the source of many natural products such as essential oils, natural dyes, pigments, waxes, resins, tannins, alkaloids, amber and cork. Products derived from plants include soaps, shampoos, perfumes, cosmetics, paint, varnish, turpentine, rubber, latex, lubricants, linoleum, plastics, inks, and gums. Renewable fuels from plants include firewood, peat and many other biofuels. Coal and petroleum are fossil fuels derived from the remains of plants. Olive oil has been used in lamps for centuries to provide illumination.

Timber in storage for later processing at a sawmill.

Structural resources and fibers from plants are used in both the construction of dwellings and the manufacture of clothing. Wood is used not only for buildings, boats, and furniture, but also for smaller items such as musical instruments and sports equipment. Wood also may be pulped for the manufacture of paper and cardboard. Cloth is often made from cotton, flax, ramie or synthetic fibers derived from cellulose, such as rayon and acetate. The thread that is used to sew cloth likewise comes from plant fibers. Hemp and jute are grown for their fibers, which may be woven into rope or rough sacking.

Plants are also a primary source of basic chemicals, both for their medicinal and physiological effects, as well as for the industrial synthesis of a vast array of organic chemicals. Medicines derived from plants include aspirin, taxol, morphine, quinine, reserpine, colchicine, digitalis and vincristine. There are hundreds of herbal supplements such as ginkgo, Echinacea, feverfew, and Saint John's wort. Pesticides derived from plants include nicotine, rotenone, strychnine and pyrethrins. Certain plants contain psychotropic chemicals that are extracted and ingested, including tobacco, cannabis (marijuana), opium, and cocaine. Poisons from plants include ricin, hemlock and curare.

Aesthetic Uses

Thousands of plant species are cultivated for aesthetic purposes as well as to provide shade, modify temperatures, reduce wind, abate noise, provide privacy, and prevent soil erosion. Plants are the basis of a multibillion-dollar per year tourism industry, which includes travel to historic gardens, national parks, rainforests, forests with colorful autumn leaves, and the National Cherry Blossom Festival.

While some gardens are planted with food crops, many are planted for aesthetic, ornamental, or conservation purposes. Arboretums and botanical gardens are public collections of living plants.

In private outdoor gardens, lawn grasses, shade trees, ornamental trees, shrubs, vines, herbaceous perennials and bedding plants are used. Gardens may cultivate the plants in a naturalistic state, or may sculpture their growth, as with topiary or espalier. Gardening is the most popular leisure activity in the U.S., and working with plants or horticulture therapy is beneficial for rehabilitating people with disabilities.

A rose espalier at Niedernhall in Germany.

Plants may also be grown or kept indoors as houseplants, or in specialized buildings such as greenhouses that are designed for the care and cultivation of living plants. Venus Flytrap, sensitive plant and resurrection plant are examples of plants sold as novelties. There are also art forms specializing in the arrangement of cut or living plant, such as bonsai, ikebana, and the arrangement of cut or dried flowers. Ornamental plants have sometimes changed the course of history, as in tulipomania.

Capitals of ancient Egyptian columns decorated to resemble papyrus plants. (at Luxor, Egypt)

Architectural designs resembling plants appear in the capitals of ancient Egyptian columns, which were carved to resemble either the Egyptian white lotus or the papyrus. Images of plants are often used in painting and photography, as well as on textiles, money, stamps, flags and coats of arms.

Scientific and Cultural Uses

Basic biological research has often been done with plants. In genetics, the breeding of pea plants allowed Gregor Mendel to derive the basic laws governing inheritance, and examination of chromosomes in maize allowed Barbara McClintock to demonstrate their connection to inherited traits. The plant *Arabidopsis thaliana* is used in laboratories as a model organism to understand how genes control the growth and development of plant structures. Space stations or space colonies may one day rely on plants for life support.

Barbara McClintock (1902-1992) was a pioneering cytogeneticist who used maize (or corn) to study the mechanism of inheritance of traits.

Ancient trees are revered and many are famous. Tree rings themselves are an important method of dating in archeology, and serve as a record of past climates.

Plants figure prominently in mythology, religion and literature. They are used as national and state emblems, including state trees and state flowers. Plants are often used as memorials, gifts and to mark special occasions such as births, deaths, weddings and holidays. The arrangement of flowers may be used to send hidden messages.

The field of ethnobotany studies plant use by indigenous cultures, which helps to conserve endangered species as well as discover new medicinal plants.

Negative Effects

Weeds are uncultivated and usually unwanted plants growing in managed environments such as farms, urban areas, gardens, lawns, and parks. People have spread plants beyond their native ranges and some of these introduced plants become invasive, damaging existing ecosystems by displacing native species. Invasive plants cause costly damage in crop losses annually by displacing crop plants, they further increase the cost of production and the use of chemicals to control them, which in turn affects the environment.

Plants may cause harm to animals, including people. Plants that produce windblown pollen invoke allergic reactions in people who suffer from hay fever. A wide variety of plants are poisonous. Toxalbumins are plant poisons fatal to most mammals and act as a serious deterrent to consumption. Several plants cause skin irritations when touched, such as poison ivy. Certain plants contain psychotropic chemicals, which are extracted and ingested or smoked, including tobacco, cannabis (marijuana), cocaine and opium. Smoking causes damage to health or even death, while some drugs may also be harmful or fatal to people. Both illegal and legal drugs derived from plants may have negative effects on the economy, affecting worker productivity and law enforcement costs. Some plants cause allergic reactions when ingested, while other plants cause food intolerances that negatively affect health.

Plant Taxonomy

Plant taxonomy is the science that finds, identifies, describes, classifies, and names plants. Thus making it one of the main branches of taxonomy(the science that finds, describes, classifies, and names living things).

Plant taxonomy is closely allied to plant systematics, and there is no sharp boundary between the two. In practice, "Plant systematics" involves relationships between plants and their evolution, especially at the higher levels, whereas "plant taxonomy" deals with the actual handling of plant specimens. The precise relationship between taxonomy and systematics, however, has changed along with the goals and methods employed.

Plant taxonomy is well known for being turbulent, and traditionally not having any close agreement on circumscription and placement of taxa. See the list of systems of plant taxonomy.

Identification, Classification and Description

Three goals of plant taxonomy are the identification, classification and description of plants. The distinction between these three goals is important and often overlooked.

Plant identification is the determination of the identity of an unknown plant by comparison with previously collected specimens or with the aid of books or identification manuals. The process of identification connects the specimen with a published name. Once a plant specimen has been identified, its name and properties are known.

Plant classification is the placing of known plants into groups or categories to show some relationship. Scientific classification follows a system of rules that standardizes the results, and groups successive categories into a hierarchy. For example, the family to which the lilies belong is classified as follows:

- Kingdom: Plantae
- Phylum: Magnoliophyta
- Class: Liliopsida
- Series: Liliales
- Family: Liliaceae
- Genus: ......
- Species: ......

The classification of plants results in an organized system for the naming and cataloging of future specimens, and ideally reflects scientific ideas about inter-relationships between plants. The set of rules and recommendations for formal botanical nomenclature, including plants, is governed by the International Code of Nomenclature for algae, fungi, and plants abbreviated as ICN.

Plant description is a formal description of a newly discovered species, usually in the form of a scientific paper using ICN guidelines. The names of these plants are then registered on the International Plant Names Index along with all other validly published names.

Classification Systems

See List of systems of plant taxonomy

- APG system
- APG II system
- APG III system
- Bessey system
- Cronquist system
- Melchior system

Bryophyte

Marchantia, an example of a liverwort.

Bryophyte is a traditional name used to refer to all embryophytes (land plants) that are non-vascular plants, namely the mosses, hornworts, and liverworts. The defining features of bryophytes are that their life cycles are dominated by the gametophyte stage, that their sporophytes are unbranched and that they do not have a true vascular tissue containing lignin, although some have specialized tissues for the transport of water. Bryophytes are usually considered to be a paraphyletic group and not a natural (monophyletic) group, although some studies have produced contrary results. Regardless of their status, the name is convenient and remains in use as an informal collective term. Bryophytes produce enclosed reproductive structures (gametangia and sporangia), but they produce neither flowers nor seeds, reproducing via spores. The term "bryophyte" comes from Greek βρύον, *bryon*, "tree-moss, oyster-green" + φυτόν – *phyton* "plant".

Life Cycle

Like all land plants (embryophytes), bryophytes have life cycles with alternation of generations. In each cycle, a haploid gametophyte, each of whose cells contains a fixed number of unpaired chromosomes, alternates with a diploid sporophyte, whose cell contain two sets of paired chromosomes. Gametophytes produce haploid sperm and eggs which fuse to form diploid zygotes that grow into sporophytes. Sporophytes produce haploid spores by meiosis, that grow into gametophytes.

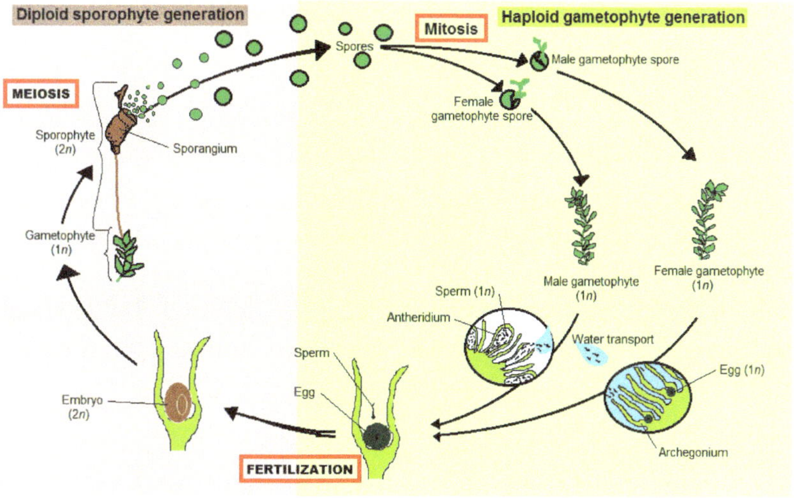

The life cycle of a dioicous bryophyte. The gametophyte (haploid) structures are shown in green, the sporophyte (diploid) in brown.

Bryophytes are gametophyte dominant, meaning that the more prominent, longer-lived plant is the haploid gametophyte. The diploid sporophytes appear only occasionally and remain attached to and nutritionally dependent on the gametophyte. In bryophytes, the sporophytes are always unbranched and produce a single sporangium (spore producing capsule).

Liverworts, mosses and hornworts spend most of their lives as gametophytes. Gametangia (gamete-producing organs), archegonia and antheridia, are produced on the gametophytes, sometimes at the tips of shoots, in the axils of leaves or hidden under thalli. Some bryophytes, such as the liverwort *Marchantia*, create elaborate structures to bear the gametangia that are called gametangiophores. Sperm are flagellated and must swim from the antheridia that produce them to archegonia which may be on a different plant. Arthropods may assist in transfer of sperm.

Fertilized eggs become zygotes, which develop into sporophyte embryos inside the archegonia. Mature sporophytes remain attached to the gametophyte. They consist of a stalk called a seta and a single sporangium or capsule. Inside the sporangium, haploid spores are produced by meiosis. These are dispersed, most commonly by wind, and if they land in a suitable environment can develop into a new gametophyte. Thus bryophytes disperse by a combination of swimming sperm and spores, in a manner similar to lycophytes, ferns and other cryptogams.

Sexuality

The arrangement of antheridia and archegonia on an individual bryophyte plant is usually constant within a species, although in some species it may depend on environmental conditions. The main division is between species in which the antheridia and archegonia occur on the same plant and those in which they occur on different plants. The term monoicous may be used where antheridia and archegonia occur on the same gametophyte and the term dioicous where they occur on different gametophytes.

In seed plants, "monoecious" is used where flowers with anthers (microsporangia) and flowers with ovules (megasporangia) occur on the same sporophyte and "dioecious" where they occur on different sporophytes. These terms occasionally may be used instead of "monoicous" and "dioicous" to describe bryophyte gametophytes. "Monoecious" and "monoicous" are both derived from the Greek for "one house", "dioecious" and "dioicous" from the Greek for two houses. The use of the "oicy" terminology is said to have the advantage of emphasizing the difference between the gametophyte sexuality of bryophytes and the sporophyte sexuality of seed plants.

Monoicous plants are necessarily bisexual (or hermaphroditic), meaning that the same plant has both sexes. The exact arrangement of the antheridia and archegonia in monoicous plants varies. They may be borne on different shoots (autoicous or autoecious), on the same shoot but not together in a common structure (paroicous or paroecious), or together in a common "inflorescence" (synoicous or synoecious). Dioicous plants are unisexual, meaning that the same plant has only one sex. All four patterns (autoicous, paroicous, synoicous and dioicous) occur in species of the moss genus *Bryum*.

Classification and Phylogeny

Hornworts include those bryophytes that are believed to be the closest living relatives of the vascular plants.

Mosses are one group of bryophytes.

Hornworts include those bryophytes that are believed to be the closest living relatives of the vascular plants.

Mosses are one group of bryophytes.

Traditionally, all living land plants without vascular tissues were classified in a single taxonomic group, often a division (or phylum). More recently, phylogenetic research has questioned whether the bryophytes form a monophyletic group and thus whether they should form a single taxon. Although a 2005 study supported the traditional view that the bryophytes form a monophyletic group, by 2010 a broad consensus had emerged among systematists that bryophytes as a whole are not a natural group (i.e., are paraphyletic), although each of the three extant (living) groups is monophyletic.

The three bryophyte clades are the Divisions Marchantiophyta (liverworts), Bryophyta (mosses) and Anthocerotophyta (hornworts). The vascular plants or tracheophytes form a fourth, unranked clade of land plants called the "Polysporangiophyta". In this analysis, hornworts are sister to vascular plants and liverworts are sister to all other land plants, including the hornworts and mosses. Phylogenetic studies continue to produce conflicting results. In particular those based on gene sequences suggest the bryophytes are paraphyletic, whereas those based on the amino acid translations of the same genes suggest they are monophyletic. A 2014 study concluded that composition biases were responsible for these differences and that the bryophytes are monophyletic. The issue remains unresolved.

Paraphyletic View

When extinct plants are taken into account, the picture is slightly altered. Some extinct land plants, such as the horneophytes, are not bryophytes, but also are not vascular plants because, like bryophytes, they do not have true vascular tissue. A different distinction is needed. In bryophytes, the sporophyte is a simple unbranched structure with a single spore-forming organ (sporangium). In all other land plants, the polysporangiophytes, the sporophyte is branched and carries many sporangia. It has been argued that this contrast between bryophytes and other land plants is less misleading than the traditional one of non-vascular versus vascular plant, since many mosses have well-developed water-conducting vessels. The contrast is shown in a slightly different cladogram:

The term "bryophyte" thus refers to a grade of lineages defined primarily by what they lack. Compared to other living land plants, they lack vascular tissue containing lignin and branched sporophytes bearing multiple sporangia. The prominence of the gametophyte in the life cycle is also a shared feature of the three bryophyte lineages (extant vascular plants are all sporophyte dominant).

Other Views

An alternative phylogeny, based on amino acids rather than genes, shows bryophytes as a monophyletic group:

If this phylogeny proves correct, then the complex sporophyte of living vascular plants might have evolved independently of the simpler unbranched sporophyte present in bryophytes. Other studies suggest a monophyletic group comprising liverworts and mosses, with hornworts being sister to vascular plants.

Evolution

Bryophytes are the oldest of all lineages of land plants and are believed to be a vital link in the migration of plants from aquatic environments onto land. A number of physical features link bryophytes to both land plants and aquatic plants. Two distinct adaptations have helped to make the move from water to land possible and forged the way for plants to colonize the Earth's terrestrial environments. A waxy cuticle covering the soft tissue of the plant provides protection and prevents desiccation of the plant's tissues. The development of gametangia provided further protection specifically for gametes.

They also have embryonic development which is a significant adaptation seen in land plants and not green algae. Connections to their aquatic ancestry are also evident through their dependence on water for reproduction and survival. A thin layer of water is required on the surface of the plant to enable the movement of sperm between gametophytes and the fertilization of an egg.

Comparative Morphology

Summary of the morphological characteristics of the gametophytes of the three groups of bryophytes:

	Liverworts	Mosses	Hornworts
Structure	Thalloid or Foliose	Foliose	Thalloid
Symmetry	Dorsiventral or radial	Radial	Dorsiventral
Rhizoids	Unicellular	Pluricellular	Unicellular
Chloroplasts/cell	Many	Many	One
Protonemata	Reduced	Present	Absent
Gametangia (antheridia and archegonia)	Superficial	Superficial	Immersed

Summary of the morphological characteristics of the sporophytes of the three groups of bryophytes:

	Liverworts	Mosses	Hornworts
Stomata	Absent	Present	Present
Structure	Small, without chlorophyll	Large, with chlorophyll	Large, with chlorophyll
Persistence	Ephemeral	Persistent	Persistent
Growth	Defined	Defined	Continuous
Seta	Present	Present	Absent
Capsule form	Simple	Differentiated (operculum, peristome)	Elongated
Maturation of spores	Simultaneous	Simultaneous	Graduate
Dispersion of spores	Elaters	Peristome teeth	Pseudo-elaters
Columella	Absent	Present	Present
Dehiscence	Longitudinal or irregular	Transverse	Longitudinal

Embryophyte

The Embryophyta are the most familiar group of green plants that form vegetation on earth. Living embryophytes include hornworts, liverworts, mosses, ferns, lycophytes, gymnosperms and flowering plants, and emerged from Charophyte green algae. The Embryophyta are informally called land plants because they live primarily in terrestrial habitats, while the related green algae are primarily aquatic. All are complex multicellular eukaryotes with specialized reproductive organs. The name derives from their innovative characteristic of nurturing the young embryo sporophyte during the early stages of its multicellular development within the tissues of the parent gametophyte. With very few exceptions, embryophytes obtain their energy by photosynthesis, that is by using the energy of sunlight to synthesize their food from carbon dioxide and water.

Description

The evolutionary origins of the embryophytes are discussed further below, but they are believed to have evolved from within a group of complex green algae during the Paleozoic era (which started around 540 million years ago). Charales or the stoneworts may be the best living illustration of that developmental step. Embryophytes are primarily adapted for life on land, although some are secondarily aquatic. Accordingly, they are often called *land plants* or *terrestrial plants.*

On a microscopic level, the cells of embryophytes are broadly similar to those of green algae, but differ in that in cell division the daughter nuclei are separated by a phragmoplast. They are eukaryotic, with a cell wall composed of cellulose and plastids surrounded by two membranes. The latter include chloroplasts, which conduct photosynthesis and store food in the form of starch, and are characteristically pigmented with chlorophylls *a* and *b*, generally giving them a bright green color. Embryophyte cells also generally have an enlarged central vacuole enclosed by a vacuolar membrane or tonoplast, which maintains cell turgor and keeps the plant rigid.

In common with all groups of multicellular algae they have a life cycle which involves 'alternation of generations'. A multicellular generation with a single set of chromosomes – the haploid gametophyte – produces sperm and eggs which fuse and grow into a multicellular generation with twice the number of chromosomes – the diploid sporophyte. The mature sporophyte produces haploid spores which grow into a gametophyte, thus completing the cycle. Embryophytes have two features related to their reproductive cycles which distinguish them from all other plant lineages. Firstly, their gametophytes produce sperm and eggs in multicellular structures (called 'antheridia' and 'archegonia'), and fertilization of the ovum takes place within the archegonium rather than in the external environment. Secondly, and most importantly, the initial stage of development of the fertilized egg (the zygote) into a diploid multicellular sporophyte, take place within the archegonium where it is both protected and provided with nutrition. This second feature is the origin of the term 'embryophyte' – the fertilized egg develops into a protected embryo, rather than dispersing as a single cell. In the bryophytes the sporophyte remains dependent on the gametophyte, while in all other embryophytes the sporophyte generation is dominant and capable of independent existence.

Embryophytes also differ from algae by having metamers. Metamers are repeated units of development, in which each unit derives from a single cell, but the resulting product tissue or part is largely the same for each cell. The whole organism is thus constructed from similar, repeating parts or *metamers*. Accordingly, these plants are sometimes termed 'metaphytes' and classified as the group Metaphyta (but Haeckel's definition of Metaphyta places some algae in this group). In all land plants a disc-like structure called a phragmoplast forms where the cell will divide, a trait only found in the land plants in the streptophyte lineage, some species within their relatives Coleochaetales, Charales and Zygnematales, as well as within subaerial species of the algae order Trentepohliales, and appears to be essential in the adaptation towards a terrestrial life style.

Phylogeny and Classification

All green algae and land plants are now known to form a single evolutionary lineage or clade, one name for which is Viridiplantae (i.e. 'green plants'). According to several molecular clock estimates the Viridiplantae split 1,200 million years ago to 725 million years ago into two clades: chlorophytes and streptophytes. The chlorophytes are considerably more diverse (with around 700 genera) and were originally marine, although some groups have since spread into fresh water. The streptophyte algae (i.e. the streptophyte clade minus the land plants) are less diverse (with around 122 genera) and adapted to fresh water very early in their evolutionary history. They have not spread into marine environments (only a few stoneworts, which belong to this group, tolerate brackish water). Some time during the Ordovician period (which started around 490 million years ago) one or more streptophytes invaded the land and began the evolution of the embryophyte land plants.

Becker and Marin speculate that land plants evolved from streptophytes rather than any other group of algae because streptophytes were adapted to living in fresh water. This prepared them to tolerate a range of environmental conditions found on land. Fresh water living made them tolerant of exposure to rain; living in shallow pools required tolerance to temperature variation, high levels of ultra-violet light and seasonal dehydration.

Relationships between the groups making up Viridiplantae are still being elucidated. Views have changed considerably since 2000 and classifications have not yet caught up. However, the division between chlorophytes and streptophytes and the evolution of embryophytes from within the latter group, as shown in the cladogram below, are well established. Three approaches to classification are shown. Older classifications, as on the left, treated all green algae as a single division of the plant kingdom under the name Chlorophyta. Land plants were then placed in separate divisions. All the streptophyte algae can be grouped into one paraphyletic taxon, as in the middle, allowing the embryophytes to form a taxon at the same level. Alternatively, the embryophytes can be sunk into a monophyletic taxon comprising all the streptophytes, as shown below. A variety of names have been used for the different groups which result from these approaches; those used below are only one of a number of possibilities. The higher-level classification of the Viridiplantae varies considerably, resulting in widely different ranks being assigned to the embryophytes, from kingdom to class.

The precise relationships within the streptophytes are less clear as of March 2012. The stoneworts (Charales) have traditionally been identified as closest to the embryophytes, but recent work suggests that either the Zygnematales or a clade consisting of the Zygnematales and the Coleochaetales may be the sister group to the land plants. That the Zygnematales (or Zygnematophyceae) are the closest algal relatives to land plants was underpinned by an exhaustive phylogenetic analysis (phylogenomics) performed in 2014, which is supported by both plastid genome phylogenies as well as plastid gene content and properties.

The preponderance of currently available molecular evidence suggests that the groups making up the embryophytes are related as shown in the cladogram below (based on Qiu *et al.* 2006 with additional names from Crane *et al.* 2004).

Studies based on morphology rather than on genes and proteins have regularly reached different conclusions; for example that neither the monilophytes (ferns and horsetails) nor the gymnosperms are a natural or monophyletic group.

There is considerable variation in how these relationships are converted into a formal classification. Consider the angiosperms or flowering plants. Many botanists, following Lindley in 1830, have treated the angiosperms as a division. Researchers concerned with fossil plants have usually followed Banks in treating the tracheophytes or vascular plants as a division, so that the angiosperms become a class or even a subclass. Two very different systems are shown below. The classification on the left is a traditional one, in which ten living groups are treated as separate divisions; the classification on the right (based on Kenrick and Crane's 1997 treatment) sharply reduces the rank of groups such as the flowering plants. (More complex classifications are needed if extinct plants are included.)

Two contrasting classifications of living land plants		
Liverworts	Marchiantiophyta	Marchiantiophyta
Mosses	Bryophyta	Bryophyta
Hornworts	Anthocerotophyta	Anthocerotophyta
		Tracheophyta
Lycophytes	Lycopodiophyta	Lycophytina

		Euphyllophytina
Ferns and horsetails	Pteridophyta	Moniliformopses
		Radiatopses
Cycads	Cycadophyta	Cycadatae
Conifers	Pinophyta	Coniferophytatae
Ginkgo	Ginkgophyta	Ginkgoatae
Gnetophytes	Gnetophyta	Anthophytatae
Flowering plants	Magnoliophyta	

An updated phylogeny of Embryophyta based on the work by Novíkov & Barabaš-Krasni 2015 with plant taxon authors from Anderson, Anderson & Cleal 2007 and some clade names from Pelletier 2012.

Diversity

Bryophytes

Most bryophytes, such as these mosses, produce stalked sporophytes from which their spores are released.

Bryophytes consist of all nonvascular land plants (embryophytes without vascular tissue). All are relatively small and are usually confined to environments that are humid or at least seasonally moist. They are limited by their reliance on water needed to disperse their gametes, although only a few bryophytes are truly aquatic. Most species are tropical, but there are many arctic species as well. They may locally dominate the ground cover in tundra and Arctic–alpine habitats or the epiphyte flora in rain forest habitats.

The three living divisions are the mosses (Bryophyta), hornworts (Anthocerotophyta), and liverworts (Marchantiophyta). Originally, these three groups were included together as classes within the single division Bryophyta. However, they now are placed separately into three divisions since the bryophytes as a whole are known to be a paraphyletic (artificial) group instead of a single lineage. Instead, the three bryophyte groups form an evolutionary grade of those land plants that are not vascular. Some closely related green algae are also non-vascular, but are not considered "land plants."

- Marchantiophyta (liverworts)
- Bryophyta (mosses)
- Anthocerotophyta (hornworts)

Despite the fact that they are no longer classified as a single group, the bryophytes are still studied together because of their many biological similarities as non-vascular land plants. All three bryophyte groups share a haploid-dominant life cycle and unbranched sporophytes. These are traits that appear to be plesiotypic within the land plants, and thus were common to all early diverging lineages of plants on the land. The fact that the bryophytes have a life cycle in common is thus an artefact of being the oldest extant lineages of land plant, and not the result of close shared ancestry. (See the phylogeny above.)

The bryophyte life-cycle is strongly dominated by the haploid gametophyte generation. The sporophyte remains small and dependent on the parent gametophyte for its entire brief life. All other living groups of land plants have a life cycle dominated by the diploid sporophyte generation. It is in the diploid sporophyte that vascular tissue develops. Although some mosses have quite complex water-conducting vessels, bryophytes lack true vascular tissue.

Like the vascular plants, bryophytes do have differentiated stems, and although these are most often no more than a few centimeters tall, they do provide mechanical support. Most bryophytes also have leaves, although these typically are one cell thick and lack veins. Unlike the vascular plants, bryophytes lack true roots or any deep anchoring structures. Some species do grow a filamentous network of horizontal stems, but these have a primary function of mechanical attachment rather than extraction of soil nutrients (Palaeos 2008).

Rise of Vascular Plants

Reconstruction of a plant of *Rhynia*

During the Silurian and Devonian periods (around 440 to 360 million years ago), plants evolved which possessed true vascular tissue, including cells with walls strengthened by lignin (tracheids). Some extinct early plants appear to be between the grade of organization bryophytes and that of true vascular plants (eutracheophytes). Genera such as *Horneophyton* have water-conducting tissue more like that of mosses, but a different life-cycle in which the sporophyte is more developed than the gametophyte. Genera such as *Rhynia* have a similar life-cycle but have simple tracheids and so are a kind of vascular plant.

During the Devonian period, vascular plants diversified and spread to many different land environments. In addition to vascular tissues which transport water throughout the body, tracheophytes have an outer layer or cuticle that resists drying out. The sporophyte is the dominant generation, and in modern species develops leaves, stems and roots, while the gametophyte remains very small.

Lycophytes and Euphyllophytes

Lycopodiella inundata, a lycophyte

All the vascular plants which disperse through spores were once thought to be related (and were often grouped as 'ferns and allies'). However, recent research suggests that leaves evolved quite separately in two different lineages. The lycophytes or lycopodiophytes – modern clubmosses, spikemosses and quillworts – make up less than 1% of living vascular plants. They have small leaves, often called 'microphylls' or 'lycophylls', which are borne all along the stems in the clubmosses and spikemosses, and which effectively grow from the base, via an intercalary meristem. It is believed that microphylls evolved from outgrowths on stems, such as spines, which later acquired veins (vascular traces).

Although the living lycophytes are all relatively small and inconspicuous plants, more common in the moist tropics than in temperate regions, during the Carboniferous period tree-like lycophytes (such as *Lepidodendron*) formed huge forests that dominated the landscape.

The euphyllophytes, making up more than 99% of living vascular plant species, have large 'true' leaves (megaphylls), which effectively grow from the sides or the apex, via marginal or apical meristems. One theory is that megaphylls developed from three-dimensional branching systems by first 'planation' – flattening to produce a two dimensional branched structure – and then 'webbing' – tissue growing out between the flattened branches. Others have questioned whether megaphylls developed in the same way in different groups.

Ferns and Horsetails

Euphyllophytes are divided into two lineages: the ferns and horsetails (monilophytes) and the seed plants (spermatophytes). Like all the preceding groups, the monilophytes continue to use spores as their main method of dispersal. Traditionally, whisk ferns and horsetails were treated as distinct from 'true' ferns. Recent research suggests that they all belong together, although there are differences of opinion on the exact classification to be used. Living whisk ferns and horsetails do not have the large leaves (megaphylls) which would be expected of euphyllophytes. However, this has probably resulted from reduction, as evidenced by early fossil horsetails, in which the leaves are broad with branching veins.

Athyrium filix-femina, unrolling young frond

Ferns are a large and diverse group, with some 12,000 species. A stereotypical fern has broad, much divided leaves, which grow by unrolling.

Seed Plants

Conifer forest in Northern California

Seed plants, which first appeared in the fossil record towards the end of the Paleozoic era, reproduce using desiccation-resistant capsules called seeds. Starting from a plant which disperses by spores, highly complex changes are needed to produce seeds. The sporophyte has two kinds of spore-forming organs (sporangia). One kind, the megasporangium, produces only a single large spore (a megaspore). This sporangium is surrounded by one or more sheathing layers (integuments) which form the seed coat. Within the seed coat, the megaspore develops into a tiny gametophyte, which in turn produces one or more egg cells. Before fertilization, the sporangium and its contents plus its coat is called an 'ovule'; after fertilization a 'seed'. In parallel to these developments, the other kind of sporangium, the microsporangium, produces microspores. A tiny gametophyte develops inside the wall of a microspore, producing a pollen grain. Pollen grains are physically transferred between plants by the wind or animals, most commonly insects. When a pollen grain reaches an ovule, it enters via a microscopic gap in the coat (the micropyle). The tiny gametophyte inside the pollen grain then produces sperm cells which move to the egg cell and fertilize it. Seed plants include two groups with living members, the gymnosperms and the angiosperms or flowering plants. In gymnosperms, the ovules or seeds are not further enclosed. In angiosperms, they are enclosed in ovaries. A split ovary with a visible seed can be seen in the image to the right. Angiosperms typically also have other, secondary structures, such as petals, which together form a flower.

Large seed of a horse chestnut, *Aesculus hippocastanum*

Extant seed plants are divided into five groups:

Gymnosperms

- Pinophyta - conifers
- Cycadophyta - cycads
- Ginkgophyta - ginkgo
- Gnetophyta - gnetophytes

Angiosperms

- Magnoliophyta – flowering plants

Spermatophyte

The spermatophytes (from the Greek word Σπερματόφυτα), also known as phanerogams or phenogamae, comprise those plants that produce seeds, hence the alternative name seed plants. They are a subset of the embryophytes or land plants. The term *phanerogams* or *phanerogamae* is derived from the Greek φανερός, *phanerós* meaning "visible", in contrast to the cryptogamae from Greek κρυπτός *kryptós* = "hidden" together with the suffix γαμέω, *gameein*, "to marry". These terms distinguished those plants with hidden sexual organs (cryptogamae) from those with visible sexual organs (phanerogamae).

Description

The living spermatophytes form five groups:

- cycads, a subtropical and tropical group of plants with a large crown of compound leaves and a stout trunk,
- *ginkgo*, a single living species of tree,
- conifers, cone-bearing trees and shrubs,
- gnetophytes, woody plants in the genera *Ephedra*, *Gnetum*, and *Welwitschia*
- angiosperms, (or magnoliophyta) the flowering plants, a large group including many familiar plants in a wide variety of habitats.

In addition to the taxa listed above, the fossil (old creature) record contains evidence of many extinct taxa of seed plants. The so-called "seed ferns" (Pteridospermae) were one of the earliest successful groups of land plants, and forests dominated by seed ferns were prevalent in the late Paleozoic. *Glossopteris* was the most prominent tree genus in the ancient southern supercontinent of Gondwana during the Permian period. By the Triassic period, seed ferns had declined in ecological importance, and representatives of modern gymnosperm groups were abundant and dominant through the end of the Cretaceous, when angiosperms radiated.

Evolution

A whole genome duplication event in the ancestor of seed plants occurred about 319 million years ago. This gave rise to a series of evolutionary changes that resulted in the origin of seed plants.

A middle Devonian precursor to seed plants from Belgium has been identified predating the earliest seed plants by about 20 million years. *Runcaria*, small and radially symmetrical, is an integumented megasporangium surrounded by a cupule. The megasporangium bears an unopened distal extension protruding above the mutlilobed integument. It is suspected that the extension was involved in anemophilous pollination. *Runcaria* sheds new light on the sequence of character acquisition leading to the seed. *Runcaria* has all of the qualities of seed plants except for a solid seed coat and a system to guide the pollen to the seed.

Relationships and Nomenclature

Seed-bearing plants were traditionally divided into angiosperms, or flowering plants, and gymnosperms, which includes the gnetophytes, cycads, ginkgo, and conifers. Older morphological studies believed in a close relationship between the gnetophytes and the angiosperms, in particular based on vessel elements. However, molecular studies (and some more recent morphological and fossil papers) have generally shown a clade of gymnosperms, with the gnetophytes in or near the conifers. For example, one common proposed set of relationships is known as the *gne-pine hypothesis* and looks like:

However, the relationships between these groups should not be considered settled.

Other classifications group all the seed plants in a single division, with classes for our five groups:

- Division Spermatophyta
    - Cycadopsida, the cycads
    - Ginkgoopsida, the ginkgo
    - Pinopsida, the conifers, ("Coniferopsida")
    - Gnetopsida, the gnetophytes
    - Magnoliopsida, the flowering plants, or Angiospermopsida

A more modern classification ranks these groups as separate divisions (sometimes under the Superdivision Spermatophyta):

- Cycadophyta, the cycads
- Ginkgophyta, the ginkgo
- Pinophyta, the conifers
- Gnetophyta, the gnetophytes
- Magnoliophyta, the flowering plants

An alternative phylogeny of spermatophytes based on the work by Novíkov & Barabaš-Krasni 2015 with plant taxon authors from Anderson, Anderson & Cleal 2007 showing the relationship of extinct clades.

Flowering Plant

The flowering plants (angiosperms), also known as Angiospermae or Magnoliophyta, are the most diverse group of land plants, with 416 families, approx. 13,164 known genera and a total of c. 295,383 known species. Like gymnosperms, angiosperms are seed-producing plants; they are distinguished from gymnosperms by characteristics including flowers, endosperm within the seeds, and the production of fruits that contain the seeds. Etymologically, angiosperm means a plant that produces seeds within an enclosure, in other words, a fruiting plant. The term "angiosperm" comes from the Greek composite word (*angeion*, "case" or "casing", and *sperma*, "seed") meaning "enclosed seeds", after the enclosed condition of the seeds.

The ancestors of flowering plants diverged from gymnosperms in the Triassic Period, during the range 245 to 202 million years ago (mya), and the first flowering plants are known from 160 mya. They diversified extensively during the Lower Cretaceous, became widespread by 120 mya, and replaced conifers as the dominant trees during 100 to 60 mya.

Angiosperm Derived Characteristics

Angiosperms differ from other seed plants in several ways, described in the table. These distinguishing characteristics taken together have made the angiosperms the most diverse and numerous land plants and the most commercially important group to humans.

Bud of a pink rose

Distinctive features of Angiosperms	
Feature	**Description**
Flowering organs	Flowers, the reproductive organs of flowering plants, are the most remarkable feature distinguishing them from the other seed plants. Flowers provided angiosperms with the means to have a more species-specific breeding system, and hence a way to evolve more readily into different species without the risk of crossing back with related species. Faster speciation enabled the Angiosperms to adapt to a wider range of ecological niches. This has allowed flowering plants to largely dominate terrestrial ecosystems.
Stamens with two pairs of pollen sacs	Stamens are much lighter than the corresponding organs of gymnosperms and have contributed to the diversification of angiosperms through time with adaptations to specialized pollination syndromes, such as particular pollinators. Stamens have also become modified through time to prevent self-fertilization, which has permitted further diversification, allowing angiosperms eventually to fill more niches.
Reduced male parts, three cells	The male gametophyte in angiosperms is significantly reduced in size compared to those of gymnosperm seed plants. The smaller size of the pollen reduces the amount of time between pollination — the pollen grain reaching the female plant — and fertilization. In gymnosperms, fertilization can occur up to a year after pollination, whereas in angiosperms, fertilization begins very soon after pollination. The shorter amount of time between pollination and fertilization allows angiosperms to produce seeds earlier after pollination than gymnosperms, providing angiosperms a distinct evolutionary advantage.
Closed carpel enclosing the ovules (carpel or carpels and accessory parts may become the fruit)	The closed carpel of angiosperms also allows adaptations to specialized pollination syndromes and controls. This helps to prevent self-fertilization, thereby maintaining increased diversity. Once the ovary is fertilized, the carpel and some surrounding tissues develop into a fruit. This fruit often serves as an attractant to seed-dispersing animals. The resulting cooperative relationship presents another advantage to angiosperms in the process of dispersal.
Reduced female gametophyte, seven cells with eight nuclei	The reduced female gametophyte, like the reduced male gametophyte, may be an adaptation allowing for more rapid seed set, eventually leading to such flowering plant adaptations as annual herbaceous life-cycles, allowing the flowering plants to fill even more niches.
Endosperm	In general, endosperm formation begins after fertilization and before the first division of the zygote. Endosperm is a highly nutritive tissue that can provide food for the developing embryo, the cotyledons, and sometimes the seedling when it first appears.

Evolution

Fossilized spores suggest that higher plants (embryophytes) have lived on land for at least 475 million years. Early land plants reproduced sexually with flagellated, swimming sperm, like the green algae from which they evolved. An adaptation to terrestrialization was the development of upright meiosporangia for dispersal by spores to new habitats. This feature is lacking in the descendants of their nearest algal relatives, the Charophycean green algae. A later terrestrial adaptation took place with retention of the delicate, avascular sexual stage, the gametophyte, within the tissues of the vascular sporophyte. This occurred by spore germination within sporangia rather than spore release, as in non-seed plants. A current example of how this might have happened can be seen in the precocious spore germination in *Selaginella*, the spike-moss. The result for the ancestors of angiosperms was enclosing them in a case, the seed. The first seed bearing plants, like the ginkgo, and conifers (such as pines and firs), did not produce flowers. The pollen grains (males) of *Ginkgo* and cycads produce a pair of flagellated, mobile sperm cells that "swim" down the developing pollen tube to the female and her eggs.

Flowers of *Malus sylvestris* (crab apple)

The apparently sudden appearance of nearly modern flowers in the fossil record initially posed such a problem for the theory of evolution that Charles Darwin called it an "*abominable mystery*". However, the fossil record has considerably grown since the time of Darwin, and recently discovered angiosperm fossils such as *Archaefructus*, along with further discoveries of fossil gymnosperms, suggest how angiosperm characteristics may have been acquired in a series of steps. Several groups of extinct gymnosperms, in particular seed ferns, have been proposed as the ancestors of flowering plants, but there is no continuous fossil evidence showing exactly how flowers evolved. Some older fossils, such as the upper Triassic *Sanmiguelia*, have been suggested. Based on current evidence, some propose that the ancestors of the angiosperms diverged from an unknown group of gymnosperms in the Triassic period (245–202 million years ago). Fossil angiosperm-like pollen from the Middle Triassic (247.2–242.0 Ma) suggests an older date for their origin. A close relationship between angiosperms and gnetophytes, proposed on the basis of morphological evidence, has more recently been disputed on the basis of molecular evidence that suggest gnetophytes are instead more closely related to other gymnosperms.

The evolution of seed plants and later angiosperms appears to be the result of two distinct rounds of whole genome duplication events. These occurred at 319 million years ago and 192 million years ago. Another possible whole genome duplication event at 160 million years ago perhaps created the ancestral line that led to all modern flowering plants. That event was studied by sequencing the genome of an ancient flowering plant, *Amborella trichopoda*, and directly addresses Darwin's "*abominable mystery*."

Flowers and leaves of *Oxalis pes-caprae* (Bermuda buttercup)

The earliest known macrofossil confidently identified as an angiosperm, *Archaefructus liaoningensis*, is dated to about 125 million years BP (the Cretaceous period), whereas pollen considered to be of angiosperm origin takes the fossil record back to about 130 million years BP. However, one study has suggested that the early-middle Jurassic plant *Schmeissneria*, traditionally considered a type of ginkgo, may be the earliest known angiosperm, or at least a close relative. In addition, circumstantial chemical evidence has been found for the existence of angiosperms as early as 250 million years ago. Oleanane, a secondary metabolite produced by many flowering plants, has been found in Permian deposits of that age together with fossils of gigantopterids. Gigantopterids are a group of extinct seed plants that share many morphological traits with flowering plants, although they are not known to have been flowering plants themselves.

In 2013 flowers encased in amber were found and dated 100 million years before present. The amber had frozen the act of sexual reproduction in the process of taking place. Microscopic images showed tubes growing out of pollen and penetrating the flower's stigma. The pollen was sticky, suggesting it was carried by insects.

Recent DNA analysis based on molecular systematics showed that *Amborella trichopoda*, found on the Pacific island of New Caledonia, belongs to a sister group of the other flowering plants, and morphological studies suggest that it has features that may have been characteristic of the earliest flowering plants.

The orders Amborellales, Nymphaeales, and Austrobaileyales diverged as separate lineages from the remaining angiosperm clade at a very early stage in flowering plant evolution.

The great angiosperm radiation, when a great diversity of angiosperms appears in the fossil record, occurred in the mid-Cretaceous (approximately 100 million years ago). However, a study in 2007 estimated that the division of the five most recent (the genus *Ceratophyllum*, the family Chloranthaceae, the eudicots, the magnoliids, and the monocots) of the eight main groups occurred around 140 million years ago. By the late Cretaceous, angiosperms appear to have dominated environments formerly occupied by ferns and cycadophytes, but large canopy-forming trees replaced conifers as the dominant trees only close to the end of the Cretaceous 66 million years ago or even later, at the beginning of the Tertiary. The radiation of herbaceous angiosperms occurred much later. Yet, many fossil plants recognizable as belonging to modern families (including beech, oak, maple, and magnolia) had already appeared by the late Cretaceous.

Two bees on a flower head of Creeping Thistle, *Cirsium arvense*

It is generally assumed that the function of flowers, from the start, was to involve mobile animals in their reproduction processes. That is, pollen can be scattered even if the flower is not brightly colored or oddly shaped in a way that attracts animals; however, by expending the energy required to create such traits, angiosperms can enlist the aid of animals and, thus, reproduce more efficiently.

Island genetics provides one proposed explanation for the sudden, fully developed appearance of flowering plants. Island genetics is believed to be a common source of speciation in general, especially when it comes to radical adaptations that seem to have required inferior transitional forms. Flowering plants may have evolved in an isolated setting like an island or island chain, where the plants bearing them were able to develop a highly specialized relationship with some specific animal (a wasp, for example). Such a relationship, with a hypothetical wasp carrying pollen from one plant to another much the way fig wasps do today, could result in the development of a high degree of specialization in both the plant(s) and their partners. Note that the wasp example is not incidental; bees, which, it is postulated, evolved specifically due to mutualistic plant relationships, are descended from wasps.

Animals are also involved in the distribution of seeds. Fruit, which is formed by the enlargement of flower parts, is frequently a seed-dispersal tool that attracts animals to eat or otherwise disturb it, incidentally scattering the seeds it contains (see frugivory). Although many such mutualistic relationships remain too fragile to survive competition and to spread widely, flowering proved to be an unusually effective means of reproduction, spreading (whatever its origin) to become the dominant form of land plant life.

Flower ontogeny uses a combination of genes normally responsible for forming new shoots. The most primitive flowers probably had a variable number of flower parts, often separate from (but in contact with) each other. The flowers tended to grow in a spiral pattern, to be bisexual (in plants, this means both male and female parts on the same flower), and to be dominated by the ovary (female part). As flowers evolved, some variations developed parts fused together, with a much more specific number and design, and with either specific sexes per flower or plant or at least "ovary-inferior".

Flower evolution continues to the present day; modern flowers have been so profoundly influenced by humans that some of them cannot be pollinated in nature. Many modern domesticated flower species were formerly simple weeds, which sprouted only when the ground was disturbed. Some of them tended to grow with human crops, perhaps already having symbiotic companion plant relationships with them, and the prettiest did not get plucked because of their beauty, developing a dependence upon and special adaptation to human affection.

A few paleontologists have also proposed that flowering plants, or angiosperms, might have evolved due to interactions with dinosaurs. One of the idea's strongest proponents is Robert T. Bakker. He proposes that herbivorous dinosaurs, with their eating habits, provided a selective pressure on plants, for which adaptations either succeeded in deterring or coping with predation by herbivores.

Classification

There are eight groups of living angiosperms:

- *Amborella*, a single species of shrub from New Caledonia;
- Nymphaeales, about 80 species, water lilies and Hydatellaceae;
- Austrobaileyales, about 100 species of woody plants from various parts of the world;
- Chloranthales, several dozen species of aromatic plants with toothed leaves;
- Magnoliids, about 9,000 species, characterized by trimerous flowers, pollen with one pore, and usually branching-veined leaves—for example magnolias, bay laurel, and black pepper;
- Monocots, about 70,000 species, characterized by trimerous flowers, a single cotyledon, pollen with one pore, and usually parallel-veined leaves—for example grasses, orchids, and palms;
- *Ceratophyllum*, about 6 species of aquatic plants, perhaps most familiar as aquarium plants;
- Eudicots, about 175,000 species, characterized by 4- or 5-merous flowers, pollen with three pores, and usually branching-veined leaves—for example sunflowers, petunia, buttercup, apples, and oaks.

The exact relationship between these eight groups is not yet clear, although there is agreement that the first three groups to diverge from the ancestral angiosperm were Amborellales, Nymphaeales, and Austrobaileyales. The term basal angiosperms refers to these three groups. Among the rest, the relationship between the three broadest of these groups (magnoliids, monocots, and eudicots) remains unclear. Some analyses make the magnoliids the first to diverge, others the monocots. *Ceratophyllum* seems to group with the eudicots rather than with the monocots.

APG IV Classification

Based on the 4th version of the Angiosperm Phylogeny Group classification.

History of Classification

From 1736, an illustration of Linnaean classification

The botanical term "Angiosperm", from the Ancient Greek αγγεἰον, *angeíon* (bottle, vessel) and σπέρμα, (seed), was coined in the form Angiospermae by Paul Hermann in 1690, as the name of one of his primary divisions of the plant kingdom. This included flowering plants possessing seeds enclosed in capsules, distinguished from his Gymnospermae, or flowering plants with achenial or schizo-carpic fruits, the whole fruit or each of its pieces being here regarded as a seed and naked. The term and its antonym were maintained by Carl Linnaeus with the same sense, but with restricted application, in the names of the orders of his class Didynamia. Its use with any approach to its modern scope became possible only after 1827, when Robert Brown established the existence of truly naked ovules in the Cycadeae and Coniferae, and applied to them the name Gymnosperms. From that time onward, as long as these Gymnosperms were, as was usual, reckoned as dicotyledonous flowering plants, the term Angiosperm was used antithetically by botanical writers, with varying scope, as a group-name for other dicotyledonous plants.

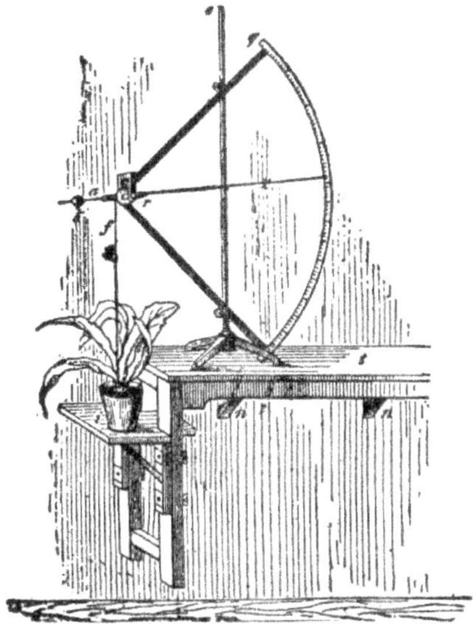

Auxanometer: Device for measuring increase or rate of growth in plants

In 1851, Hofmeister discovered the changes occurring in the embryo-sac of flowering plants, and determined the correct relationships of these to the Cryptogamia. This fixed the position of Gymnosperms as a class distinct from Dicotyledons, and the term Angiosperm then gradually came to be accepted as the suitable designation for the whole of the flowering plants other than Gymnosperms, including the classes of Dicotyledons and Monocotyledons. This is the sense in which the term is used today.

In most taxonomies, the flowering plants are treated as a coherent group. The most popular descriptive name has been Angiospermae (Angiosperms), with Anthophyta ("flowering plants") a second choice. These names are not linked to any rank. The Wettstein system and the Engler system use the name Angiospermae, at the assigned rank of subdivision. The Reveal system treated flowering plants as subdivision Magnoliophytina (Frohne & U. Jensen ex Reveal, Phytologia 79: 70 1996), but later split it to Magnoliopsida, Liliopsida, and Rosopsida. The Takhtajan system and Cronquist system treat this group at the rank of division, leading to the name Magnoliophyta (from the family name Magnoliaceae). The Dahlgren system and Thorne system (1992) treat this group at the rank of class, leading to the name Magnoliopsida. The APG system of 1998, and the later 2003 and 2009 revisions, treat the flowering plants as a clade called angiosperms without a formal botanical name. However, a formal classification was published alongside the 2009 revision in which the flowering plants form the Subclass Magnoliidae.

The internal classification of this group has undergone considerable revision. The Cronquist system, proposed by Arthur Cronquist in 1968 and published in its full form in 1981, is still widely used but is no longer believed to accurately reflect phylogeny. A consensus about how the flowering plants should be arranged has recently begun to emerge through the work of the Angiosperm Phylogeny Group (APG), which published an influential reclassification of the angiosperms in 1998. Updates incorporating more recent research were published as APG II in 2003 and as APG III in 2009.

Traditionally, the flowering plants are divided into two groups, which in the Cronquist system are called Magnoliopsida (at the rank of class, formed from the family name Magnoliaceae) and Liliopsida (at the rank of class, formed from the family name Liliaceae). Other descriptive names allowed by Article 16 of the ICBN include Dicotyledones or Dicotyledoneae, and Monocotyledones or Monocotyledoneae, which have a long history of use. In English a member of either group may be called a dicotyledon (plural dicotyledons) and monocotyledon (plural monocotyledons), or abbreviated, as dicot (plural dicots) and monocot (plural monocots). These names derive from the observation that the dicots most often have two cotyledons, or embryonic leaves, within each seed. The monocots usually have only one, but the rule is not absolute either way. From a diagnostic point of view, the number of cotyledons is neither a particularly handy nor a reliable character.

Monocot (left) and dicot seedlings

Recent studies, as by the APG, show that the monocots form a monophyletic group (clade) but that the dicots do not (they are paraphyletic). Nevertheless, the majority of dicot species do form a monophyletic group, called the eudicots or tricolpates. Of the remaining dicot species, most belong to a third major clade known as the magnoliids, containing about 9,000 species. The rest include a paraphyletic grouping of primitive species known collectively as the basal angiosperms, plus the families Ceratophyllaceae and Chloranthaceae.

Flowering Plant Diversity

A poster of twelve different species of flowers of the *Asteraceae* family

Lupinus pilosus

The number of species of flowering plants is estimated to be in the range of 250,000 to 400,000. This compares to around 12,000 species of moss or 11,000 species of pteridophytes, showing that the flowering plants are much more diverse. The number of families in APG (1998) was 462. In APG II (2003) it is not settled; at maximum it is 457, but within this number there are 55 optional segregates, so that the minimum number of families in this system is 402. In APG III (2009) there are 415 families.

The diversity of flowering plants is not evenly distributed. Nearly all species belong to the eudicot (75%), monocot (23%), and magnoliid (2%) clades. The remaining 5 clades contain a little over 250 species in total; i.e. less than 0.1% of flowering plant diversity, divided among 9 families. The 42 most-diverse of 443 families of flowering plants by species, in their APG circumscriptions, are

1. Asteraceae or Compositae (daisy family): 22,750 species;
2. Orchidaceae (orchid family): 21,950;
3. Fabaceae or Leguminosae (bean family): 19,400;
4. Rubiaceae (madder family): 13,150;
5. Poaceae or Gramineae (grass family): 10,035;
6. Lamiaceae or Labiatae (mint family): 7,175;
7. Euphorbiaceae (spurge family): 5,735;
8. Melastomataceae or Melastomaceae (melastome family): 5,005;
9. Myrtaceae (myrtle family): 4,625;
10. Apocynaceae (dogbane family): 4,555;
11. Cyperaceae (sedge family): 4,350;

12. Malvaceae (mallow family): 4,225;
13. Araceae (arum family): 4,025;
14. Ericaceae (heath family): 3,995;
15. Gesneriaceae (gesneriad family): 3,870;
16. Apiaceae or Umbelliferae (parsley family): 3,780;
17. Brassicaceae or Cruciferae (cabbage family): 3,710:
18. Piperaceae (pepper family): 3,600;
19. Acanthaceae (acanthus family): 3,500;
20. Rosaceae (rose family): 2,830;
21. Boraginaceae (borage family): 2,740;
22. Urticaceae (nettle family): 2,625;
23. Ranunculaceae (buttercup family): 2,525;
24. Lauraceae (laurel family): 2,500;
25. Solanaceae (nightshade family): 2,460;
26. Campanulaceae (bellflower family): 2,380;
27. Arecaceae (palm family): 2,361;
28. Annonaceae (custard apple family): 2,220;
29. Caryophyllaceae (pink family): 2,200;
30. Orobanchaceae (broomrape family): 2,060;
31. Amaranthaceae (amaranth family): 2,050;
32. Iridaceae (iris family): 2,025;
33. Aizoaceae or Ficoidaceae (ice plant family): 2,020;
34. Rutaceae (rue family): 1,815;
35. Phyllanthaceae (phyllanthus family): 1,745;
36. Scrophulariaceae (figwort family): 1,700;
37. Gentianaceae (gentian family): 1,650;
38. Convolvulaceae (bindweed family): 1,600;
39. Proteaceae (protea family): 1,600;

40. Sapindaceae (soapberry family): 1,580;

41. Cactaceae (cactus family): 1,500;

42. Araliaceae (*Aralia* or ivy family): 1,450.

Of these, the Orchidaceae, Poaceae, Cyperaceae, Arecaceae, and Iridaceae are monocot families; Piperaceae, Lauraceae, and Annonaceae are magnoliid (acot); the others are eudicot.

Vascular Anatomy

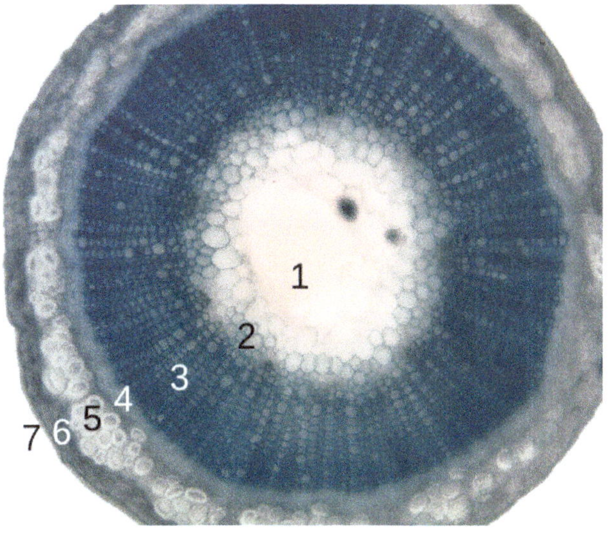

Cross-section of a stem of the angiosperm flax:
1. Pith,
2. Protoxylem,
3. Xylem I,
4. Phloem I,
5. Sclerenchyma (bast fibre),
6. Cortex,
7. Epidermis

The amount and complexity of tissue-formation in flowering plants exceeds that of gymnosperms. The vascular bundles of the stem are arranged such that the xylem and phloem form concentric rings.

In the dicotyledons, the bundles in the very young stem are arranged in an open ring, separating a central pith from an outer cortex. In each bundle, separating the xylem and phloem, is a layer of meristem or active formative tissue known as cambium. By the formation of a layer of cambium between the bundles (interfascicular cambium), a complete ring is formed, and a regular periodical increase in thickness results from the development of xylem on the inside and phloem on the outside. The soft phloem becomes crushed, but the hard wood persists and forms the bulk of the stem and branches of the woody perennial. Owing to differences in the character of the elements produced at the beginning and end of the season, the wood is marked out in transverse section into concentric rings, one for each season of growth, called annual rings.

Among the monocotyledons, the bundles are more numerous in the young stem and are scattered through the ground tissue. They contain no cambium and once formed the stem increases in diameter only in exceptional cases.

The Flower, Fruit, and Seed

Flowers

The characteristic feature of angiosperms is the flower. Flowers show remarkable variation in form and elaboration, and provide the most trustworthy external characteristics for establishing relationships among angiosperm species. The function of the flower is to ensure fertilization of the ovule and development of fruit containing seeds. The floral apparatus may arise terminally on a shoot or from the axil of a leaf (where the petiole attaches to the stem). Occasionally, as in violets, a flower arises singly in the axil of an ordinary foliage-leaf. More typically, the flower-bearing portion of the plant is sharply distinguished from the foliage-bearing or vegetative portion, and forms a more or less elaborate branch-system called an inflorescence.

A collection of flowers forming an inflorescence

There are two kinds of reproductive cells produced by flowers. Microspores, which will divide to become pollen grains, are the "male" cells and are borne in the stamens (or microsporophylls). The "female" cells called megaspores, which will divide to become the egg cell (megagametogenesis), are contained in the ovule and enclosed in the carpel (or megasporophyll).

The flower may consist only of these parts, as in willow, where each flower comprises only a few stamens or two carpels. Usually, other structures are present and serve to protect the sporophylls and to form an envelope attractive to pollinators. The individual members of these surrounding structures are known as sepals and petals (or tepals in flowers such as *Magnolia* where sepals and petals are not distinguishable from each other). The outer series (calyx of sepals) is usually green and leaf-like, and functions to protect the rest of the flower, especially the bud. The inner series (corolla of petals) is, in general, white or brightly colored, and is more delicate in structure. It functions to attract insect or bird pollinators. Attraction is effected by color, scent, and nectar, which may be secreted in some part of the flower. The characteristics that attract pollinators account for the popularity of flowers and flowering plants among humans.

While the majority of flowers are perfect or hermaphrodite (having both pollen and ovule producing parts in the same flower structure), flowering plants have developed numerous morphological and physiological mechanisms to reduce or prevent self-fertilization. Heteromorphic flowers have short carpels and long stamens, or vice versa, so animal pollinators cannot easily transfer pollen to the pistil (receptive part of the carpel). Homomorphic flowers may employ a biochemical (physiological) mechanism called self-incompatibility to discriminate between self and non-self pollen grains. In other species, the male and female parts are morphologically separated, developing on different flowers.

Fertilization and Embryogenesis

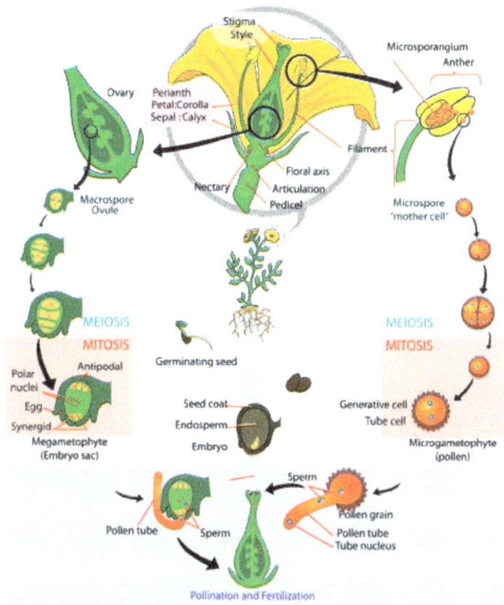

Angiosperm life cycle

Double fertilization refers to a process in which two sperm cells fertilize cells in the ovary. This process begins when a pollen grain adheres to the stigma of the pistil (female reproductive structure), germinates, and grows a long pollen tube. While this pollen tube is growing, a haploid generative cell travels down the tube behind the tube nucleus. The generative cell divides by mitosis to produce two haploid (n) sperm cells. As the pollen tube grows, it makes its way from the stigma, down the style and into the ovary. Here the pollen tube reaches the micropyle of the ovule and digests its way into one of the synergids, releasing its contents (which include the sperm cells). The synergid that the cells were released into degenerates and one sperm makes its way to fertilize the egg cell, producing a diploid ($2n$) zygote. The second sperm cell fuses with both central cell nuclei, producing a triploid ($3n$) cell. As the zygote develops into an embryo, the triploid cell develops into the endosperm, which serves as the embryo's food supply. The ovary now will develop into fruit and the ovule will develop into seed.

Fruit and Seed

As the development of embryo and endosperm proceeds within the embryo sac, the sac wall enlarges and combines with the nucellus (which is likewise enlarging) and the integument to form the *seed coat*. The ovary wall develops to form the fruit or pericarp, whose form is closely associated with the manner of distribution of the seed.

The fruit of the *Aesculus* or Horse Chestnut tree

Frequently, the influence of fertilization is felt beyond the ovary, and other parts of the flower take part in the formation of the fruit, e.g., the floral receptacle in the apple, strawberry, and others.

The character of the seed coat bears a definite relation to that of the fruit. They protect the embryo and aid in dissemination; they may also directly promote germination. Among plants with indehiscent fruits, in general, the fruit provides protection for the embryo and secures dissemination. In this case, the seed coat is only slightly developed. If the fruit is dehiscent and the seed is exposed, in general, the seed-coat is well developed, and must discharge the functions otherwise executed by the fruit.

Economic Importance

Agriculture is almost entirely dependent on angiosperms, which provide virtually all plant-based food, and also provide a significant amount of livestock feed. Of all the families of plants, the Poaceae, or grass family (grains), is by far the most important, providing the bulk of all feedstocks (rice, corn — maize, wheat, barley, rye, oats, pearl millet, sugar cane, sorghum). The Fabaceae, or legume family, comes in second place. Also of high importance are the Solanaceae, or nightshade family (potatoes, tomatoes, and peppers, among others), the Cucurbitaceae, or gourd family (also including pumpkins and melons), the Brassicaceae, or mustard plant family (including rapeseed and the innumerable varieties of the cabbage species *Brassica oleracea*), and the Apiaceae, or parsley family. Many of our fruits come from the Rutaceae, or rue family (including oranges, lemons, grapefruits, etc.), and the Rosaceae, or rose family (including apples, pears, cherries, apricots, plums, etc.).

In some parts of the world, certain single species assume paramount importance because of their variety of uses, for example the coconut (*Cocos nucifera*) on Pacific atolls, and the olive (*Olea europaea*) in the Mediterranean region.

Flowering plants also provide economic resources in the form of wood, paper, fiber (cotton, flax, and hemp, among others), medicines (digitalis, camphor), decorative and landscaping plants, and many other uses. The main area in which they are surpassed by other plants — namely, coniferous trees (Pinales), which are non-flowering (gymnosperms) — is timber and paper production.

Perennial Plant

A perennial plant or simply perennial is a plant that lives for more than two years. The term is often used to differentiate a plant from shorter-lived annuals and biennials. The term is also widely used to distinguish plants with little or no woody growth from trees and shrubs, which are also technically perennials.

Common chicory, *Cichorium intybus*, is a herbaceous perennial plant

Perennials, especially small flowering plants, that grow and bloom over the spring and summer, die back every autumn and winter, and then return in the spring from their rootstock, are known as *herbaceous perennials*. However, depending on the rigors of local climate, a plant that is a perennial in its native habitat, or in a milder garden, may be treated by a gardener as an annual and planted out every year, from seed, from cuttings or from divisions. Tomato vines, for example, live several years in their natural tropical/subtropical habitat but are grown as annuals in temperate regions because they don't survive the winter.

There is also a class of evergreen, or non-herbaceous, perennials, including plants like *Bergenia* which retain a mantle of leaves throughout the year. An intermediate class of plants is known as subshrubs, which retain a vestigial woody structure in winter, e.g. *Penstemon*. The local climate may dictate whether plants are treated as shrubs or perennials. For instance, many varieties of *Fuchsia* are shrubs in warm regions, but in colder temperate climates may be cut to the ground every year as a result of winter frosts.

The symbol for a perennial plant, based on *Species Plantarum* by Linnaeus, which is also the astronomical symbol for the planet Jupiter.

Life Cycle and Structure

Perennial plants can be short-lived (only a few years) or they can be long-lived, as are some woody plants like trees. They include a wide assortment of plant groups from ferns and liverworts to the highly diverse flowering plants like orchids and grasses.

Plants that flower and fruit only once and then die are termed monocarpic or semelparous. However, most perennials are polycarpic (or iteroparous), flowering over many seasons in their lifetime.

Perennials typically grow structures that allow them to adapt to living from one year to the next through a form of vegetative reproduction rather than seeding. These structures include bulbs, tubers, woody crowns, rhizomes plus others. They might have specialized stems or crowns that allow them to survive periods of dormancy over cold or dry seasons during the year. Annuals produce seeds to continue the species as a new generation while the growing season is suitable, and the seeds survive over the cold or dry period to begin growth when the conditions are again suitable.

Many perennials have developed specialized features that allow them to survive extreme climatic and environmental conditions. Some have adapted to survive hot and dry conditions or cold temperatures. Those plants tend to invest a lot of resource into their adaptations and often do not flower and set seed until after a few years of growth. Many perennials produce relatively large seeds, which can have an advantage, with larger seedlings produced after germination that can better compete with other plants. Some annuals produce many more seeds per plant in one season, while some (polycarpic) perennials are not under the same pressure to produce large numbers of seeds but can produce seeds over many years.

Dahlia plants are perennial.

Growth

In warmer and more favorable climates, perennials grow continuously. In seasonal climates, their growth is limited to the growing season.

In some species, perennials retain their foliage all year round; these are evergreen perennials. Other plants are deciduous perennials, for example, in temperate regions a perennial plant may grow and bloom during the warm part of the year, with the foliage dying back in the winter. In many parts of the world, seasonality is expressed as wet and dry periods rather than warm and cold periods, and deciduous perennials lose their leaves in the dry season.

With their roots protected below ground in the soil layer, perennial plants are notably tolerant of wildfire. Herbaceous perennials are also able to tolerate the extremes of cold in temperate and Arctic winters, with less sensitivity than trees or shrubs.

Benefits in Agriculture

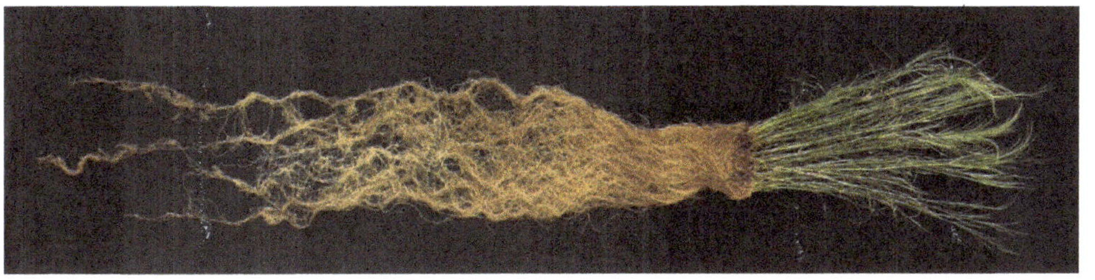

Switchgrass is a deep-rooted perennial. These roots are more than 3 meters long.

Although most of humanity is fed by the re-sowing of the seeds of annual grain crops, (either naturally or by the manual efforts of man), perennial crops provide numerous benefits. Perennial plants often have deep, extensive root systems which can hold soil to prevent erosion, capture dissolved nitrogen before it can contaminate ground and surface water, and out-compete weeds (reducing the need for herbicides). These potential benefits of perennials have resulted in new attempts to increase the seed yield of perennial species, which could result in the creation of new perennial grain crops. Some examples of new perennial crops being developed are perennial rice and intermediate wheatgrass. The Land Institute estimates that profitable, productive perennial grain crops will take at least 25 years to achieve.

Location

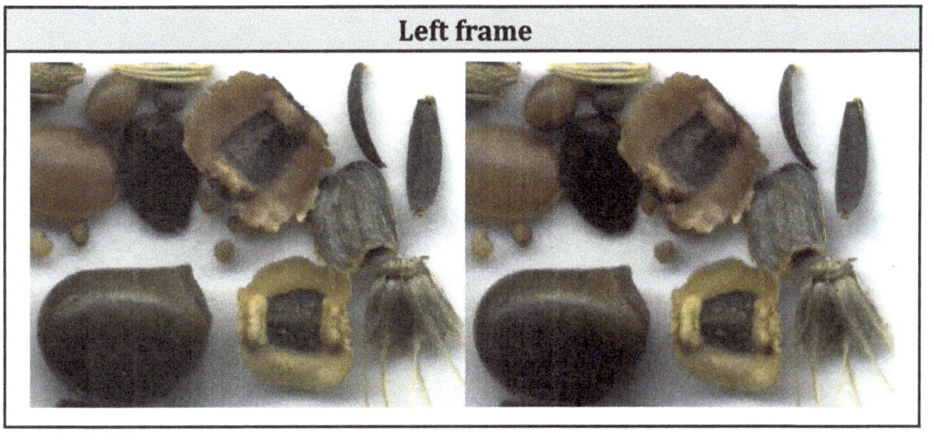

Seeds from various perennial flowers

Perennial plants dominate many natural ecosystems on land and in fresh water, with only a very few (e.g. *Zostera*) occurring in shallow sea water. Herbaceous perennial plants are particularly dominant in conditions too fire-prone for trees and shrubs, e.g., most plants on prairies and steppes are perennials; they are also dominant on tundra too cold for tree growth. Nearly all forest plants are perennials, including the trees and shrubs.

Perennial plants are usually better competitors than annual plants, especially under stable, resource-poor conditions. This is due to the development of larger root systems which can access water and soil nutrients deeper in the soil and to earlier emergence in the spring.

Types

- Examples of evergreen perennials include *Begonia* and banana.
- Examples of deciduous perennials include goldenrod and mint.
- Examples of monocarpic perennials include *Agave* and some species of *Streptocarpus*.
- Examples of woody perennials include maple, pine, and apple trees.
- Examples of herbaceous perennials used in agriculture include alfalfa, *Thinopyrum intermedium*, and Red clover.

List of Perennials

Perennial Flowers

- Gulmohur
- Hibiscus

Perennial Fruits

- Apple
- Apricot
- Avocado
- Banana
- Blackcurrant
- Blueberry
- Currant
- Feijoa
- Grape
- Kiwi Fruit
- Japanese Wineberry
- Pear
- Persimmon
- Pineapple
- Plum
- Raspberries
- Strawberry

- Strawberry Tree
- Tomato

Perennial Herbs

The following perennial plants are used as herbs:

- *Agastache*
- Alfalfa
- *Althaea officinalis* (marshmallow)
- Basil, many varieties: African Blue, East Indian
- Chives
- Fennel
- *Ferula*
- Garlic
- Ginger
- Hops - *Humulus*
- Hyssop
- Horseradish
- Lavender
- Lemon Balm
- Mint
- Onions, many varieties: Potato onions, Shallots, Egyptian onions, Japanese bunching onions, Welsh onions, Chinese leeks
- Oregano
- *Piper nigrum* (black pepper)
- Rosemary
- Sage
- Thyme
- Valerian
- White Horehound - *Marrubium vulgare*
- Yarrow - *Achillea millefolium*

Perennial Vegetables

Many vegetable plants can grow as perennials in tropical climates, but die in cold weather. Some of the more completely perennial vegetables are:

- Allium tricoccum
- Asparagus
- Broccoli: Nine Star
- Chives
- *Colocasia esculenta*
- Globe artichoke
- *Apios americana* ground Nut
- Jerusalem Artichoke
- Konjac
- Leek
- Milkweed (*Asclepias*)
- New Zealand Spinach
- Potato
- Radicchio or a.k.a. Italian Chicory
- Rhubarb
- Siberian Pea Tree (*Caragana arborescens*)
- Sorrel
- Rakkyo
- Sea Kale
- Collard greens
- Mustard greens
- Turnip greens
- Kale
- Sweet potato
- Taro
- Watercress

Moss

Mosses are small flowerless plants that typically grow in dense green clumps or mats, often in damp or shady locations. The individual plants are usually composed of simple, one-cell thick leaves, attached to a stem that may be branched or unbranched and has only a limited role in conducting water and nutrients. Although some species have vascular tissue this is generally poorly developed and structurally different from similar tissue found in other plants. They do not have seeds and after fertilisation develop sporophytes (unbranched stalks topped with single capsules containing spores). They are typically 0.2–10 cm (0.1–3.9 in) tall, though some species are much larger, like *Dawsonia*, the tallest moss in the world, which can grow to 50 cm (20 in) in height.

Mosses are commonly confused with lichens, hornworts, and liverworts. Lichens may superficially look like mosses, and have common names that include the word "moss" (e.g., "reindeer moss" or "iceland moss"), but are not related to mosses.:3 Mosses used to be grouped together with the hornworts and liverworts as "non-vascular" plants in the former division "bryophytes", all of them having the haploid gametophyte generation as the dominant phase of the life cycle. This contrasts with the pattern in all "vascular" plants (seed plants and pteridophytes), where the diploid sporophyte generation is dominant.

Mosses are now classified on their own as the division Bryophyta. There are approximately 12,000 species.

The main commercial significance of mosses is as the main constituent of peat (mostly the genus *Sphagnum*), although they are also used for decorative purposes, such as in gardens and in the florist trade. Traditional uses of mosses included as insulation and for the ability to absorb liquids up to 20 times their weight.

Physical Characteristics

Description

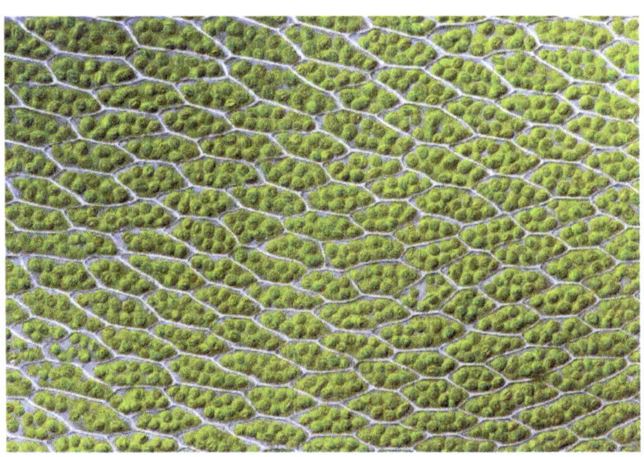

Chloroplasts and accumulated starch granules in *Bryum capillare*

Botanically, mosses are non-vascular plants in the land plant division Bryophyta. They are small (a few centimeters tall) herbaceous (non-woody) plants that absorb water and nutrients mainly through their leaves and harvest carbon dioxide and sunlight to create food by photosynthesis.

They differ from vascular plants in lacking water-bearing xylem tracheids or vessels. As in liverworts and hornworts, the haploid gametophyte generation is the dominant phase of the life cycle. This contrasts with the pattern in all vascular plants (seed plants and pteridophytes), where the diploid sporophyte generation is dominant. Mosses reproduce using spores, not seeds, and have no flowers.

Moss gametophytes have stems which may be simple or branched and upright or prostrate. Their leaves are simple, usually only a single layer of cells with no internal air spaces, often with thicker midribs. They do not have proper roots, but have threadlike rhizoids that anchor them to their substrate. Mosses do not absorb water or nutrients from their substrate through their rhizoids. They can be distinguished from liverworts (Marchantiophyta or Hepaticae) by their multi-cellular rhizoids. Spore-bearing capsules or sporangia of mosses are borne singly on long, unbranched stems, thereby distinguishing them from the polysporangiophytes, which include all vascular plants. The spore-bearing sporophytes (i.e. the diploid multicellular generation) are short-lived and dependent on the gametophyte for water supply and nutrition. Also, in most mosses, the spore-bearing capsule enlarges and matures after its stalk elongates, while in liverworts the capsule enlarges and matures before its stalk elongates. Other differences are not universal for all mosses and all liverworts, but the presence of clearly differentiated stem with simple-shaped, ribbed leaves, without deeply lobed or segmented leaves and not arranged in three ranks, all point to the plant being a moss.

Life Cycle

Vascular plants have two sets of chromosomes in their vegetative cells and are said to be diploid, i.e. each chromosome has a partner that contains the same, or similar, genetic information. By contrast, mosses and other bryophytes have only a single set of chromosomes and so are haploid (i.e. each chromosome exists in a unique copy within the cell). There is a period in the moss life cycle when they do have a double set of paired chromosomes, but this happens only during the sporophyte stage.

The moss life-cycle starts with a haploid spore that germinates to produce a protonema (*pl.* protonemata), which is either a mass of thread-like filaments or thalloid (flat and thallus-like). Massed moss protonemata typically look like a thin green felt, and may grow on damp soil, tree bark, rocks, concrete, or almost any other reasonably stable surface. This is a transitory stage in the life of a moss, but from the protonema grows the gametophore ("gamete-bearer") that is structurally differentiated into stems and leaves. A single mat of protonemata may develop several gametophore shoots, resulting in a clump of moss.

From the tips of the gametophore stems or branches develop the sex organs of the mosses. The female organs are known as archegonia (*sing.* archegonium) and are protected by a group of modified leaves known as the perichaetum (plural, perichaeta). The archegonia are small flask-shaped clumps of cells with an open neck (venter) down which the male sperm swim. The male organs are known as antheridia (*sing.* antheridium) and are enclosed by modified leaves called the perigonium (*pl.* perigonia). The surrounding leaves in some mosses form a splash cup, allowing the sperm contained in the cup to be splashed to neighboring stalks by falling water droplets.

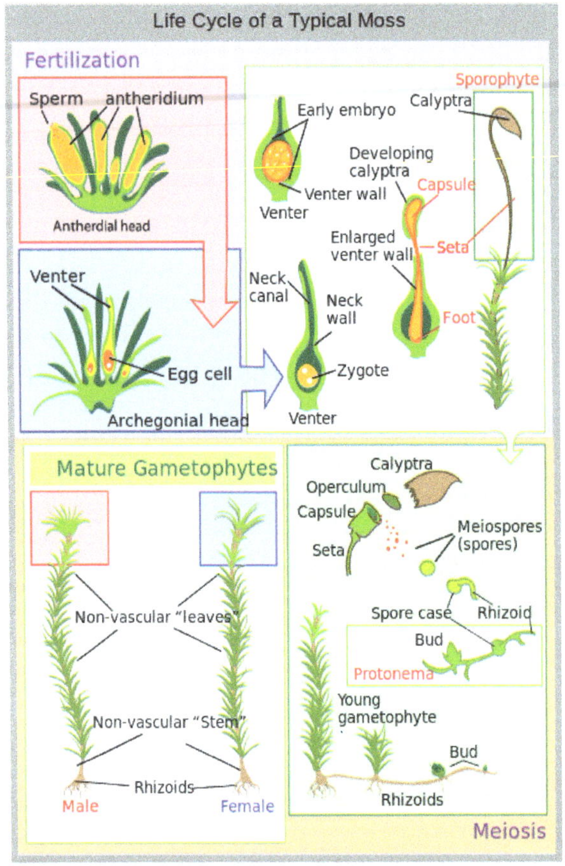

Life cycle of a typical moss (*Polytrichum commune*)

Mosses can be either dioicous (compare dioecious in seed plants) or monoicous (compare monoecious). In dioicous mosses, male and female sex organs are borne on different gametophyte plants. In monoicous (also called autoicous) mosses, both are borne on the same plant. In the presence of water, sperm from the antheridia swim to the archegonia and fertilisation occurs, leading to the production of a diploid sporophyte. The sperm of mosses is biflagellate, i.e. they have two flagellae that aid in propulsion. Since the sperm must swim to the archegonium, fertilisation cannot occur without water. Some species (for example *Mnium hornum* or several species of *Polytrichum*) keep their antheridia in so called 'splash cups', bowl-like structures on the shoot tips that propel the sperm several decimeters when water droplets hit it, increasing the fertilization distance.

After fertilisation, the immature sporophyte pushes its way out of the archegonial venter. It takes about a quarter to half a year for the sporophyte to mature. The sporophyte body comprises a long stalk, called a seta, and a capsule capped by a cap called the operculum. The capsule and operculum are in turn sheathed by a haploid calyptra which is the remains of the archegonial venter. The calyptra usually falls off when the capsule is mature. Within the capsule, spore-producing cells undergo meiosis to form haploid spores, upon which the cycle can start again. The mouth of the capsule is usually ringed by a set of teeth called peristome. This may be absent in some mosses.

Most mosses rely on the wind to disperse the spores. In the genus *Sphagnum* the spores are projected about 10–20 cm (4–8 in) off the ground by compressed air contained in the capsules; the spores are accelerated to about 36,000 times the earth's gravitational acceleration g.

A patch of moss showing both gametophytes (the low, leaf-like forms) and sporophytes (the tall, stalk-like forms)

It has recently been found that microarthropods, such as springtails and mites, can effect moss fertilization and that this process is mediated by moss-emitted scents. Male and female fire moss, for example emit different and complex volatile organic scents. Female plants emit more compounds than male plants. Springtails were found to choose female plants preferentially, and one study found that springtails enhance moss fertilization, suggesting a scent-mediated relationship analogous to the plant-pollinator relationship found in many seed plants. The stinkmoss species *Splachnum sphaericum* develops insect pollination further by attracting flies to its sporangia with a strong smell of carrion, and providing a strong visual cue in the form of red-coloured swollen collars beneath each spore capsule. Flies attracted to the moss carry its spores to fresh herbivore dung, which is the favoured habitat of the species of this genus.

In many mosses, e.g. *Ulota phyllantha*, green vegetative structures called gemmae are produced on leaves or branches, which can break off and form new plants without the need to go through the cycle of fertilization. This is a means of asexual reproduction, and the genetically identical units can lead to the formation of clonal populations.

Dwarf Males

Moss dwarf males (also known as nannandry or phyllodioicy) originate from wind-dispersed male spores that settle and germinate on the female shoot where their growth is restricted to a few millimeters. In some species, dwarfness is genetically determined in that all male spores become dwarf. More often however, it is environmentally determined in that male spores that land on a female become dwarf, while those that land elsewhere develop into large, female-sized males. In the latter case, dwarf males that are transplanted from females to another substrate develop into large shoots, suggesting that the females emit a substance which inhibits the growth of germinating males and possibly also quickens their onset of sexual maturation. The nature of such a substance is unknown, but the phytohormone auxin may be involved

Having the males growing as dwarfs on the female is expected to increase the fertilization efficiency by minimizing the distance between male and female reproductive organs. Accordingly, it has been observed that fertilization frequency is positively associated with the presence of dwarf males in several phyllodioicous species.

Dwarf males occur in several unrelated lineages and is showing to be more common than previously thought. For example, it is estimated that between one quarter and half of all dioicous pleurocarps have dwarf males.

DNA Repair

The moss *Phycomitrella patens* has proven to be an excellent model organism for analyzing repair of DNA damages in plants, especially repair by the homologous recombination pathway. Loss of the ability to repair double-strand breaks and other DNA damages by homologous recombination in somatic cells leads to cell dysfunction or death, and loss of this ability during meiosis leads to infertility. The determination of the genome sequence of *P. patens* has allowed identification of numerous genes that encode proteins necessary for removal of DNA damages by homologous recombination and by other DNA repair pathways. Studies of *P. patens* mutants defective in key steps of homologous recombination have provided insight into the functions of this process in plants. For example, a study of *P. patens* mutants defective in *Rp*RAD51, a gene that encodes a protein at the core of the recombinational repair reaction, indicated that homologous recombination is essential for repairing DNA double-strand breaks in this plant. Similarly, studies of mutants defective in *Ppmre11* or *Pprad50* (that encode key proteins of the MRN complex, the principal sensor of DNA double-strand breaks) showed that these genes are necessary for repair of DNA damages as well as for normal growth and development.

Classification

Traditionally, mosses were grouped with the liverworts and hornworts in the division Bryophyta (bryophytes, or Bryophyta sensu lato), within which the mosses made up the class Musci. This definition of Bryophyta, however, is paraphyletic and is now split into three divisions, Bryophyta, Marchantiophyta and Anthocerotophyta. Currently, the division Bryophyta contains only the mosses.

"Muscinae" from Ernst Haeckel's *Kunstformen der Natur*, 1904

Six of the eight classes contain only one or two genera each. Polytrichopsida includes 23 genera, and Bryopsida includes the majority of moss diversity with over 95% of moss species belonging to this class.

The Sphagnopsida, the peat-mosses, comprise the two living genera *Ambuchanania* and *Sphagnum*, as well as fossil taxa. However, the genus *Sphagnum* is a diverse, widespread, and economically important one. These large mosses form extensive acidic bogs in peat swamps. The leaves of *Sphagnum* have large dead cells alternating with living photosynthetic cells. The dead cells help to store water. Aside from this character, the unique branching, thallose (flat and expanded) protonema, and explosively rupturing sporangium place it apart from other mosses.

Andreaeopsida and Andreaeobryopsida are distinguished by the biseriate (two rows of cells) rhizoids, multiseriate (many rows of cells) protonema, and sporangium that splits along longitudinal lines. Most mosses have capsules that open at the top.

Polytrichopsida have leaves with sets of parallel lamellae, flaps of chloroplast-containing cells that look like the fins on a heat sink. These carry out photosynthesis and may help to conserve moisture by partially enclosing the gas exchange surfaces. The Polytrichopsida differ from other mosses in other details of their development and anatomy too, and can also become larger than most other mosses, with e.g. *Polytrichum commune* forming cushions up to 40 cm (16 in) high. The tallest land moss, a member of the Polytrichidae is probably *Dawsonia superba*, a native to New Zealand and other parts of Australasia.

They appear to be the closest living relatives of the vascular plants.

Geological History

Red moss capsules, a winter native of the Yorkshire Dales moorland

The fossil record of moss is sparse, due to their soft-walled and fragile nature. Unambiguous moss fossils have been recovered from as early as the Permian of Antarctica and Russia, and a case is put forwards for Carboniferous mosses. It has further been claimed that tube-like fossils from the Silurian are the macerated remains of moss calyptræ. Mosses also appear to evolve 2–3 times slower than ferns, gymnosperms and angiosperms.

Recent research show that ancient moss could explain why the Ordovician ice ages occurred. When the ancestors of today's moss started to spread on land 470 million years ago, they absorbed CO_2

from the atmosphere and extracted minerals by secreting organic acids that dissolved the rocks they were growing on. These chemically altered rocks in turn reacted with the atmospheric CO_2 and formed new carbonate rocks in the ocean through the weathering of calcium and magnesium ions from silicate rocks. The weathered rocks also released a lot of phosphorus and iron which ended up in the oceans, where it caused massive algal blooms, resulting in organic carbon burial, extracting more carbon dioxide from the atmosphere. Small organisms feeding on the nutrients created large areas without oxygen, which caused a mass extinction of marine species, while the levels of CO_2 dropped all over the world, allowing the formation of ice caps on the poles.

Habitat

Dense moss colonies in a cool coastal forest

Moss with sporophytes on brick

A closeup of moss on a rock

Young sporophytes of the common moss *Tortula muralis* (wall screw-moss)

Retaining wall covered in moss

A small clump of moss.

The moss garden at the Bloedel Reserve, Bainbridge Island, Washington State.

Since moss gametophytes have no vascular system to transport water through the plant or water-proofing systems to prevent tissue water from evaporating, they must have a damp environment in which to grow, and a surrounding of liquid water to reproduce. Since mosses are autotrophic they require enough sunlight to conduct photosynthesis. Shade tolerance varies by species, just as it does with higher plants. In most areas, mosses grow chiefly in areas of dampness and shade, such as wooded areas and at the edges of streams; but they can grow anywhere in cool damp cloudy climates, and some species are adapted to sunny, seasonally dry areas like alpine rocks or stabilized sand dunes.

Choice of substrate varies by species as well. Moss species can be classed as growing on: rocks, exposed mineral soil, disturbed soils, acid soil, calcareous soil, cliff seeps and waterfall spray areas, streamsides, shaded humusy soil, downed logs, burnt stumps, tree trunk bases, upper tree trunks, and tree branches or in bogs. Moss species growing on or under trees are often specific about the species of trees they grow on, such as preferring conifers to broadleaf trees, oaks to alders, or vice versa. While mosses often grow on trees as epiphytes, they are never parasitic on the tree.

Mosses are also found in cracks between paving stones in damp city streets, and on roofs. Some species adapted to disturbed, sunny areas are well adapted to urban conditions and are commonly found in cities. Examples would be *Rhytidiadelphus squarrosus*, a garden weed in Vancouver and Seattle areas; *Bryum argenteum*, the cosmopolitan sidewalk moss, and *Ceratodon purpureus*, red roof moss, another cosmopolitan species. A few species are wholly aquatic, such as *Fontinalis antipyretica*, common water moss; and others such as *Sphagnum* inhabit bogs, marshes and very slow-moving waterways. Such aquatic or semi-aquatic mosses can greatly exceed the normal range of lengths seen in terrestrial mosses. Individual plants 20–30 cm (8–12 in) or more long are common in *Sphagnum* species for example.

Wherever they occur, mosses require high levels of moisture to survive because of the lack of a vascular system, and the need for liquid water to complete fertilisation. Many mosses can survive desiccation, sometimes for months, returning to life within a few hours of rehydration.

It is generally believed that in northern latitudes, the north side of trees and rocks will generally have more luxuriant moss growth on average than other sides. This is assumed to be because the sun on the south side creates a dry environment. South of the equator the reverse would be true. However, naturalists feel that mosses grow on the damper side of trees and rocks. In some cases, such as sunny climates in temperate northern latitudes, this will be the shaded north side of the tree or rock. On steep slopes it may be the uphill side. For mosses that grow on tree branches, this is generally the upper side of the branch on horizontally growing sections or near the crotch. In cool damp cloudy climates, all sides of tree trunks and rocks may be equally damp enough for mosses. And different species of mosses have different moisture and sun requirements so will grow on different sections of the same tree or rock.

Cultivation

Moss is often considered a weed in grass lawns, but is deliberately encouraged to grow under aesthetic principles exemplified by Japanese gardening. In old temple gardens, moss can carpet a forest scene. Moss is thought to add a sense of calm, age, and stillness to a garden scene. Moss is also used in bonsai to cover the soil and enhance the impression of age. Rules of cultivation are not widely established. Moss collections are quite often begun using samples transplanted from the wild in a water-retaining bag. However, specific species of moss can be extremely difficult to maintain away from their natural sites with their unique requirements of combinations of light, humidity, substrate chemistry, shelter from wind, etc.

Growing moss from spores is even less controlled. Moss spores fall in a constant rain on exposed surfaces; those surfaces which are hospitable to a certain species of moss will typically be colonised by that moss within a few years of exposure to wind and rain. Materials which are porous and moisture retentive, such as brick, wood, and certain coarse concrete mixtures are hospitable to moss. Surfaces can also be prepared with acidic substances, including buttermilk, yogurt, urine, and gently puréed mixtures of moss samples, water and ericaceous compost.

In the cool cloudy damp Pacific Northwest, moss is sometimes allowed to grow naturally as a lawn substitute, one that needs little or no mowing, fertilizing or watering. In this case, grass is considered to be the weed. Landscapers in the Seattle area sometimes collect boulders and downed logs growing mosses for installation in gardens and landscapes. Woodland gardens in many parts of the world can include a carpet of natural mosses. The Bloedel Reserve on Bainbridge Island, Washington State, is famous for its moss garden. The moss garden was created by removing shrubby underbrush and herbaceous groundcovers, thinning trees, and allowing mosses to fill in naturally.

Mosses are sometimes used in green roofs. Advantages of mosses over higher plants in green roofs include reduced weight loads, increased water absorption, no fertilizer requirements, and high drought tolerance. Since mosses do not have true roots, they require less planting medium than higher plants with extensive root systems. With proper species selection for the local climate, mosses in green roofs require no irrigation once established and are low maintenance.

Inhibiting Moss Growth

Moss can be a troublesome weed in containerized nursery operations and greenhouses. Vigorous moss growth can inhibit seedling emergence and penetration of water and fertilizer to the plant roots.

Moss growth can be inhibited by a number of methods:

- Decreasing availability of water through drainage.
- Increasing direct sunlight.
- Increasing number and resources available for competitive plants like grasses.
- Increasing the soil pH with the application of lime.
- Heavy traffic or manually disturbing the moss bed with a rake
- Application of chemicals such as ferrous sulfate (e.g. in lawns) or bleach (e.g. on solid surfaces).
- In containerized nursery operations, coarse mineral materials such as sand, gravel, and rock chips are used as a fast-draining top dressing in plant containers to discourage moss growth.

The application of products containing ferrous sulfate or ferrous ammonium sulfate will kill moss; these ingredients are typically in commercial moss control products and fertilizers. Sulfur and Iron are essential nutrients for some competing plants like grasses. Killing moss will not prevent regrowth unless conditions favorable to their growth are changed.

Mossery

A passing fad for moss-collecting in the late 19th century led to the establishment of mosseries in many British and American gardens. The mossery is typically constructed out of slatted wood, with a flat roof, open to the north side (maintaining shade). Samples of moss were installed in the cracks between wood slats. The whole mossery would then be regularly moistened to maintain growth.

Traditional Uses

Preindustrial societies made use of the mosses growing in their areas.

- Laplanders and other circumpolar people used mosses for bedding.
- North American tribal people used mosses for basketry, bedding, wound dressing, diapers, and menstrual fluid absorption.
- Circumpolar and alpine people used mosses as insulation in boots and mittens. Ötzi the Iceman had moss-packed boots.
- Tribes of the northeastern United States and southeastern Canada used moss to fill chinks in wooden longhouses.
- Tribes of the Pacific Northwest in the United States and Canada used mosses to clean salmon prior to drying, and packed wet moss into pit ovens for steaming camas bulbs. Food storage baskets and boiling baskets were also packed with mosses.

Commercial Use

There is a substantial market in mosses gathered from the wild. The uses for intact moss are principally in the florist trade and for home decoration. Decaying moss in the genus *Sphagnum* is also the major component of peat, which is "mined" for use as a fuel, as a horticultural soil additive, and in smoking malt in the production of Scotch whisky.

Sphagnum moss, generally the species *cristatum* and *subnitens*, is harvested while still growing and is dried out to be used in nurseries and horticulture as a plant growing medium. *Peat moss* can be harvested on a sustainable basis and managed so that regrowth is allowed, whereas the harvest-ing of *moss peat* is generally considered to cause significant environmental damage as the peat is stripped with little or no chance of recovery.

Some sphagnum mosses can absorb up to 20 times their own weight in water. In World War I, *Sphagnum* mosses were used as first-aid dressings on soldiers' wounds, as these mosses said to absorb liquids three times faster than cotton, retain liquids better, better distribute liquids uniformly throughout themselves, and are cooler, softer, and be less irritating. It was also claimed that they have mild antibacterial properties. Native Americans were one of the peoples to use *Sphagnum* for diapers and napkins, which is still done in Canada.

In rural UK, *Fontinalis antipyretica* was traditionally used to extinguish fires as it could be found in substantial quantities in slow-moving rivers and the moss retained large volumes of water which helped extinguish the flames. This historical use is reflected in its specific Latin/Greek name, the approximate meaning of which is "against fire".

In Finland, peat mosses have been used to make bread during famines. In Mexico, Moss is used as a Christmas decoration.

Moss bioreactor cultivating the moss *Physcomitrella patens*

Physcomitrella patens is increasingly used in biotechnology. Prominent examples are the identification of moss genes with implications for crop improvement or human health and the safe production of complex biopharmaceuticals in the moss bioreactor, developed by Ralf Reski and his co-workers.

References

- Kenrick, Paul; Crane, Peter R. (1997). The origin and early diversification of land plants: A cladistic study. Washington, D. C.: Smithsonian Institution Press. ISBN 1-56098-730-8.
- Raven, Peter H.; Evert, Ray F.; Eichhorn, Susan E. (2005). Biology of Plants (7th ed.). New York: W. H. Freeman and Company. ISBN 0-7167-1007-2.
- Van den Hoek, C., D. G. Mann, & H. M. Jahns, 1995. Algae: An Introduction to Phycology. pages 343, 350, 392, 413, 425, 439, & 448 (Cambridge: Cambridge University Press). ISBN 0-521-30419-9
- Van den Hoek, C., D. G. Mann, & H. M. Jahns, 1995. Algae: An Introduction to Phycology. pages 457, 463, & 476. (Cambridge: Cambridge University Press). ISBN 0-521-30419-9
- Crandall-Stotler, Barbara. & Stotler, Raymond E., 2000. "Morphology and classification of the Marchantiophyta". page 21 in A. Jonathan Shaw & Bernard Goffinet (Eds.), Bryophyte Biology. (Cambridge: Cambridge University Press). ISBN 0-521-66097-1
- Gifford, Ernest M.; Foster, Adriance S. (1988). Morphology and Evolution of Vascular Plants (3rd ed.). New York: W. H. Freeman and Company. p. 358. ISBN 0-7167-1946-0.
- Taylor, Thomas N.; Taylor, Edith L. (1993). The Biology and Evolution of Fossil Plants. New Jersey: Prentice-Hall. p. 636. ISBN 0-13-651589-4.
- Kenrick, Paul & Peter R. Crane. 1997. The Origin and Early Diversification of Land Plants: A Cladistic Study. (Washington, D.C., Smithsonian Institution Press.) ISBN 1-56098-730-8.
- Novíkov & Barabaš-Krasni (2015). "Modern plant systematics". Liga-Pres: 685. doi:10.13140/RG.2.1.4745.6164. ISBN 978-966-397-276-3.

- Kenrick, Paul & Crane, Peter R. (1997a). "The Origin and Early Diversification of Land Plants: A Cladistic Study". Washington, D.C.: Smithsonian Institution Press. ISBN 978-1-56098-730-7.
- Taylor, T.N.; Taylor, E.L. & Krings, M. (2009), Paleobotany, The Biology and Evolution of Fossil Plants (2nd ed.), Amsterdam; Boston: Academic Press, ISBN 978-0-12-373972-8.
- Kenrick, P. & Crane, P.R. (1997), The Origin and Early Diversification of Land Plants: A Cladistic Study, Washington, D.C.: Smithsonian Institution Press, ISBN 978-1-56098-730-7.
- Anderson, Anderson & Cleal (2007). "Brief history of the gymnosperms: classification, biodiversity, phytogeography and ecology". Strelitzia. SANBI. 20: 280. ISBN 978-1-919976-39-6.
- Boyce, C.K. (2005), "The evolutionary history of roots and leaves", in Holbrook, N.M. & Zwieniecki, M.A., Vascular Transport in Plants, Burlington: Academic Press, doi:10.1016/B978-012088457-5/50025-3, ISBN 978-0-12-088457-5, retrieved 2011-02-06.
- Taylor, T.N.; Taylor, E.L. & Krings, M. (2009), Paleobotany, The Biology and Evolution of Fossil Plants (2nd ed.), Amsterdam; Boston: Academic Press, ISBN 978-0-12-373972-8.
- Anderson, Anderson & Cleal (2007). "Brief history of the gymnosperms: classification, biodiversity, phytogeography and ecology". Strelitzia. SANBI. 20: 280. ISBN 978-1-919976-39-6.
- Stewart, Wilson Nichols; Rothwell, Gar W. (1993). Paleobotany and the evolution of plants (2nd ed.). Cambridge Univ. Press. p. 498. ISBN 0-521-23315-1.
- Raven, Peter H., Ray F. Evert, & Susan E. Eichhorn, 2005. Biology of Plants, 7th edition. (New York: W. H. Freeman and Company). ISBN 0-7167-1007-2.
- Mathews, Daniel (1994). Cascade-Olympic Natural History. Portland, Oregon: Audubon Society of Portland/Raven Editions. ISBN 0-9620782-0-4.
- Pojar and MacKinnon (1994). Plants of the Pacific Northwest Coast. Vancouver, British Columbia: Lone Pine Publishing. ISBN 1-55105-040-4.
- Buck, William R. & Bernard Goffinet. (2000). "Morphology and classification of mosses", pages 71-123 in A. Jonathan Shaw & Bernard Goffinet (Eds.), Bryophyte Biology. (Cambridge: Cambridge University Press). ISBN 0-521-66097-1.

6

Plant Anatomy: A Comprehensive Study

Plant anatomy is the study of the structure of a plant. Plant anatomy investigates plant cell, pollen and meristem. Plant anatomy is an emerging field of study. Advances in technology have allowed great progress to be made in this field. The text helps the reader in developing a profound understanding on the topic.

Plant Anatomy

Plant anatomy or phytotomy is the general term for the study of the internal structure of plants. While originally it included plant morphology, which is the description of the physical form and external structure of plants, since the mid-20th century the investigations of plant anatomy are considered a separate, distinct field, and plant anatomy refers to *just* the internal plant structures. Plant anatomy is now frequently investigated at the cellular level, and often involves the sectioning of tissues and microscopy.

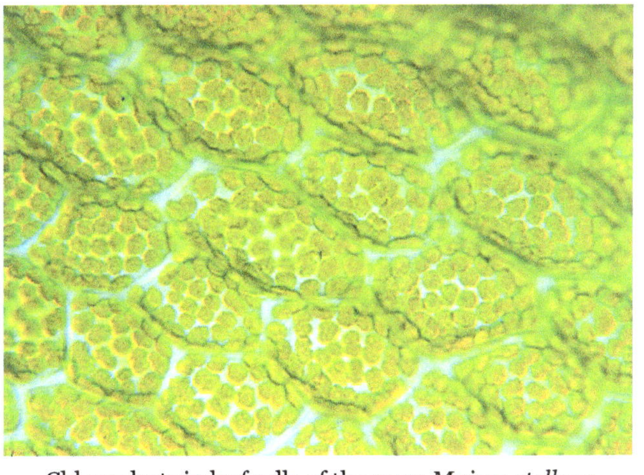

Chloroplasts in leaf cells of the moss *Mnium stellare*

Structural Divisions

Plant anatomy is often divided into the following categories:

- Flower anatomy
- Calyx
- Corolla
- Androecium

Gynoecium

Leaf anatomy

Stem anatomy

Stem structure

Fruit/Seed anatomy

Ovule

Seed structure

Pericarp

Accessory fruit

Wood anatomy

Bark

Cork

Phloem

Vascular cambium

Heartwood and sapwood

branch collar

Root anatomy

Root structure

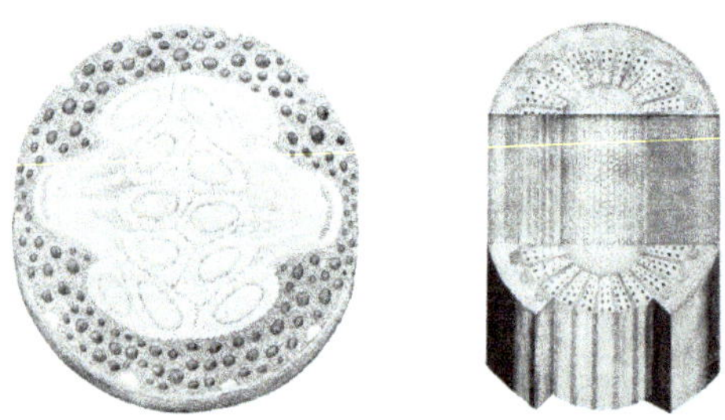

Vascular tissue of a gooseberry vine from Grew's *Anatomy of Plants*

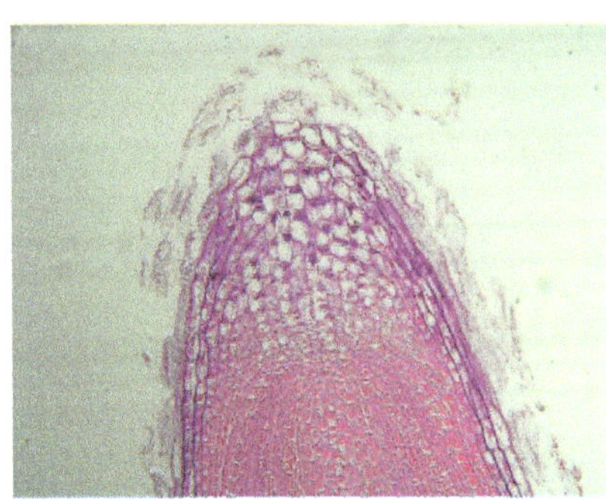

Root tip

History

About 300 BC Theophrastus wrote a number of plant treatises, only two of which survive, *Enquiry into Plants* (Περὶ φυτῶν ἱστορία), and *On the Causes of Plants* (Περὶ φυτῶν αἰτιῶν). He developed concepts of plant morphology and classification, which did not withstand the scientific scrutiny of the Renaissance.

A Swiss physician and botanist, Gaspard Bauhin, introduced binomial nomenclature into plant taxonomy. He published *Pinax theatri botanici* in 1596, which was the first to use this convention for naming of species. His criteria for classification included natural relationships, or 'affinities', which in many cases were structural.

It was in the late 1600s that plant anatomy became refined into a modern science. Italian doctor and microscopist, Marcello Malpighi, was one of the two founders of plant anatomy. In 1671 he published his *Anatomia Plantarum*, the first major advance in plant physiogamy since Aristotle. The other founder was the British doctor Nehemiah Grew. He published *An Idea of a Philosophical History of Plants* in 1672 and *The Anatomy of Plants* in 1682. Grew is credited with the recognition of plant cells, although he called them 'vesicles' and 'bladders'. He correctly identified and described the sexual organs of plants (flowers) and their parts.

In the Eighteenth Century, Carolus Linnaeus established taxonomy based on structure, and his early work was with plant anatomy. While the exact structural level which is to be considered to be scientifically valid for comparison and differentiation has changed with the growth of knowledge, the basic principles were established by Linnaeus. He published his master work, *Species Plantarum* in 1753.

In 1802, French botanist Charles-François Brisseau de Mirbel, published *Traité d'anatomie et de physiologie végétale* (*Treatise on Plant Anatomy and Physiology*) establishing the beginnings of the science of plant cytology.

In 1812, Johann Jacob Paul Moldenhawer published *Beyträge zur Anatomie der Pflanzen*, describing microscopic studies of plant tissues.

In 1813 a Swiss botanist, Augustin Pyrame de Candolle, published *Théorie élémentaire de la botanique*, in which he argued that plant anatomy, not physiology, ought to be the sole basis for plant classification. Using a scientific basis, he established structural criteria for defining and separating plant genera.

In 1830, Franz Meyen published *Phytotomie*, the first comprehensive review of plant anatomy.

In 1838 German botanist Matthias Jakob Schleiden, published *Contributions to Phytogenesis*, stating, "the lower plants all consist of one cell, while the higher plants are composed of (many) individual cells" thus confirming and continuing Mirbel's work.

A German-Polish botanist, Eduard Strasburger, described the mitotic process in plant cells and further demonstrated that new cell nuclei can only arise from the division of other pre-existing nuclei. His *Studien über Protoplasma* was published in 1876.

Gottlieb Haberlandt, a German botanist, studied plant physiology and classified plant tissue based upon function. On this basis, in 1884 he published *Physiologische Pflanzenanatomie* (*Physiological Plant Anatomy*) in which he described twelve types of tissue systems (absorptive, mechanical, photosynthetic, etc.).

British paleobotanists Dunkinfield Henry Scott and William Crawford Williamson described the structures of fossilized plants at the end of the Nineteenth Century. Scott's *Studies in Fossil Botany* was published in 1900.

Following Charles Darwin's *Origin of Species* a Canadian botanist, Edward Charles Jeffrey, who was studying the comparative anatomy and phylogeny of different vascular plant groups, applied the theory to plants using the form and structure of plants to establish a number of evolutionary lines. He published his *The Anatomy of Woody Plants* in 1917.

The growth of comparative plant anatomy was spearheaded by British botanist Agnes Arber. She published *Water Plants: A Study of Aquatic Angiosperms* in 1920, *Monocotyledons: A Morphological Study* in 1925, and *The Gramineae: A Study of Cereal, Bamboo and Grass* in 1934.

Following World War II, Katherine Esau published, *Plant Anatomy* (1953), which became the definitive textbook on plant structure in North American universities and elsewhere, it was still in print as of 2006. She followed up with her *Anatomy of seed plants* in 1960.

Plant Cell

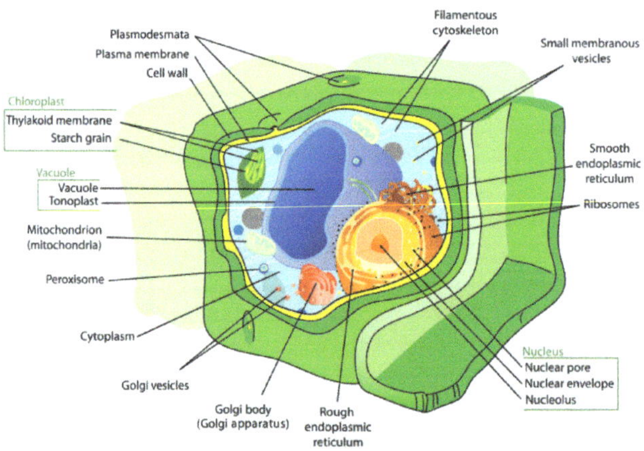

Plant cell structure

Plant cells are eukaryotic cells that differ in several key aspects from the cells of other eukaryotic organisms. Their distinctive features include:

- A large central vacuole, a water-filled volume enclosed by a membrane known as the *tonoplast* that maintains the cell's turgor, controls movement of molecules between the cytosol and sap, stores useful material and digests waste proteins and organelles.
- A cell wall composed of cellulose and hemicellulose, pectin and in many cases lignin, is

secreted by the protoplast on the outside of the cell membrane. This contrasts with the cell walls of fungi (which are made of chitin), and of bacteria, which are made of peptidoglycan. Cell walls perform many essential functions: they provide shape to form the tissue and organs of the plant, and play an important role in intercellular communication and plant-microbe interactions

- Specialized cell-to-cell communication pathways known as plasmodesmata, pores in the primary cell wall through which the plasmalemma and endoplasmic reticulum of adjacent cells are continuous.
- Plastids, the most notable being the chloroplast, which contains chlorophyll, a green-colored pigment that absorbs sunlight, and allows the plant to make its own food in the process known as photosynthesis. Other types of plastids are the amyloplasts, specialized for starch storage, elaioplasts specialized for fat storage, and chromoplasts specialized for synthesis and storage of pigments. As in mitochondria, which have a genome encoding 37 genes, plastids have their own genomes of about 100–120 unique genes and, it is presumed, arose as prokaryotic endosymbionts living in the cells of an early eukaryotic ancestor of the land plants and algae.
- Cell division by construction of a phragmoplast as a template for building a cell plate late in cytokinesis is characteristic of land plants and a few groups of algae, notably the Charophytes and the Order Trentepohliales
- The sperm of bryophytes and pteridophytes, cycads and *Ginkgo* have flagella similar to those in animal cells, but conifers and flowering plants lack flagella and centrioles.

Cell Types

- Parenchyma cells are living cells that have functions ranging from storage and support to photosynthesis and phloem loading (transfer cells). Apart from the xylem and phloem in their vascular bundles, leaves are composed mainly of parenchyma cells. Some parenchyma cells, as in the epidermis, are specialized for light penetration and focusing or regulation of gas exchange, but others are among the least specialized cells in plant tissue, and may remain totipotent, capable of dividing to produce new populations of undifferentiated cells, throughout their lives. Parenchyma cells have thin, permeable primary walls enabling the transport of small molecules between them, and their cytoplasm is responsible for a wide range of biochemical functions such as nectar secretion, or the manufacture of secondary products that discourage herbivory. Parenchyma cells that contain many chloroplasts and are concerned primarily with photosynthesis are called chlorenchyma cells. Others, such as the majority of the parenchyma cells in potato tubers and the seed cotyledons of legumes, have a storage function.
- Collenchyma cells – collenchyma cells are alive at maturity and have only a primary wall. These cells mature from meristem derivatives that initially resemble parenchyma, but differences quickly become apparent. Plastids do not develop, and the secretory apparatus (ER and Golgi) proliferates to secrete additional primary wall. The wall is most commonly thickest at the corners, where three or more cells come in contact, and thinnest where only two cells come in contact, though other arrangements of the wall thickening are possible.

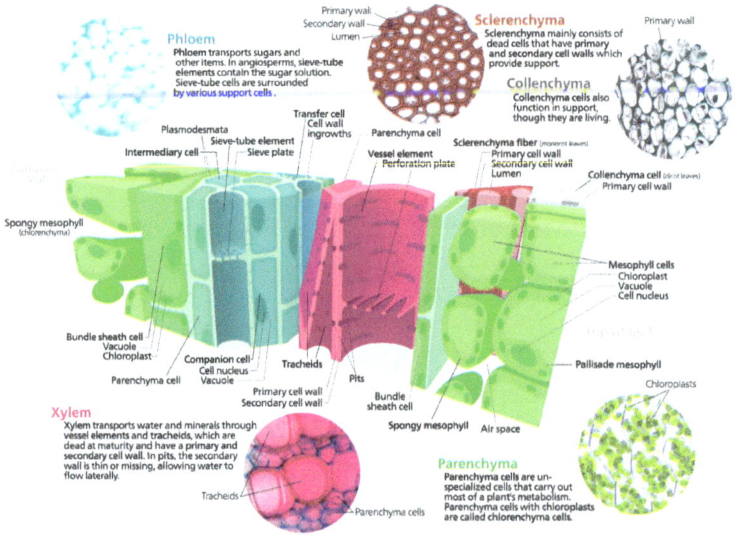

Cross section of a leaf showing various plant cell types

Pectin and hemicellulose are the dominant constituents of collenchyma cell walls of dicotyledon angiosperms, which may contain as little as 20% of cellulose in *Petasites*. Collenchyma cells are typically quite elongated, and may divide transversely to give a septate appearance. The role of this cell type is to support the plant in axes still growing in length, and to confer flexibility and tensile strength on tissues. The primary wall lacks lignin that would make it tough and rigid, so this cell type provides what could be called plastic support – support that can hold a young stem or petiole into the air, but in cells that can be stretched as the cells around them elongate. Stretchable support (without elastic snap-back) is a good way to describe what collenchyma does. Parts of the strings in celery are collenchyma.

- Sclerenchyma cells – Sclerenchyma cells (from the Greek skleros, *hard*) are hard and tough cells with a function in mechanical support. They are of two broad types – sclereids or stone cells and fibres. The cells develop an extensive secondary cell wall that is laid down on the inside of the primary cell wall. The secondary wall is impregnated with lignin, making it hard and impermeable to water. Thus, these cells cannot survive for long' as they cannot exchange sufficient material to maintain active metabolism. Sclerenchyma cells are typically dead at functional maturity, and the cytoplasm is missing, leaving an empty central cavity.

Functions for sclereid cells (hard cells that give leaves or fruits a gritty texture) include discouraging herbivory, by damaging digestive passages in small insect larval stages, and physical protection (a solid tissue of hard sclereid cells form the pit wall in a peach and many other fruits). Functions of fibres include provision of load-bearing support and tensile strength to the leaves and stems of herbaceous plants. Sclerenchyma fibres are not involved in conduction, either of water and nutrients (as in the xylem) or of carbon compounds (as in the phloem), but it is likely that they may have evolved as modifications of xylem and phloem initials in early land plants.

Tissue Types

The major classes of cells differentiate from undifferentiated meristematic cells (analogous to the stem cells of animals) to form the tissue structures of roots, stems, leaves, flowers, and reproductive structures.

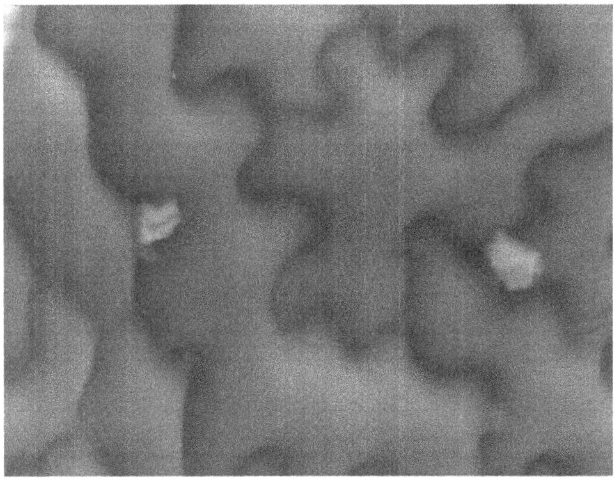

cells of *Arabidopsis thaliana* epidermis

Xylem cells are elongated cells with lignified secondary thickening of the cell walls. Xylem cells are specialised for conduction of water, and first appeared in plants during their transition to land in the Silurian period more than 425 million years ago (see *Cooksonia*). The possession of xylem defines the vascular plants or Tracheophytes. Xylem tracheids are pointed, elongated xylem cells, the simplest of which have continuous primary cell walls and lignified secondary wall thickenings in the form of rings, hoops, or reticulate networks. More complex tracheids with valve-like perforations called bordered pits characterise the gymnosperms. The ferns and other pteridophytes and the gymnosperms have only xylem tracheids, while the angiosperms also have xylem vessels. Vessel members are hollow xylem cells without end walls that are aligned end-to-end so as to form long continuous tubes. The bryophytes lack true xylem cells, but their sporophytes have a water-conducting tissue known as the hydrome that is composed of elongated cells of simpler construction.

Phloem is a specialised tissue for food transport in higher plants. Phloem cells mainly transport sucrose along pressure gradients generated by osmosis. This phenomenon is called translocation. Phloem consists of two cell types, the sieve tubes and the intimately associated companion cells. The sieve tube elements lack nuclei and ribosomes, and their metabolism and functions are regulated by the adjacent nucleate companion cells. Sieve tubes are joined end-to-end with perforate end-plates between known as *sieve plates*, which allow transport of photosynthate between the sieve elements. The companion cells, connected to the sieve tubes via plasmodesmata, are responsible for loading the phloem with sugars. The bryophytes lack phloem, but moss sporophytes have a simpler tissue with analogous function known as the leptome.

Plant epidermal cells are specialised parenchyma cells covering the external surfaces of leaves, stems and roots. The epidermal cells of aerial organs arise from the superficial layer of cells known as the *tunica* (L1 and L2 layers) that covers the plant shoot apex, whereas the cortex and vascular tissues arise from innermost layer of the shoot apex known as the *corpus* (L3 layer). The epidermis of roots originates from the layer of cells immediately beneath the root cap.

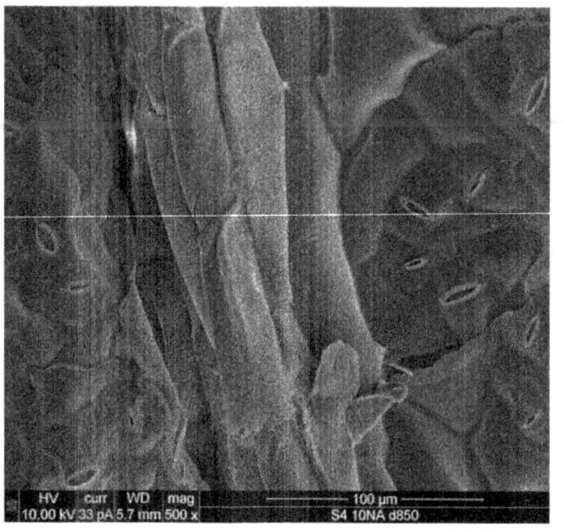

This is an electron micrograph of the epidermal cells of a Brassica chinensis leaf.
The stomates are also visible.

The epidermis of all aerial organs, but not roots, is covered with a cuticle made of the polyester cutin and/or the hydrocarbon polymer cutan with a superficial layer of epicuticular waxes. The epidermal cells of the primary shoot are thought to be the only plant cells with the biochemical capacity to synthesize cutin. Several cell types may be present in the epidermis. Notable among these are the stomatal guard cells, glandular and clothing hairs or trichomes, and the root hairs of primary roots. In the shoot epidermis of most plants, only the guard cells have chloroplasts. Chloroplasts contain the green pigment chlorophyll which is needed for photosynthesis.

Organelles

- Cell membrane
- Cell wall
- Nuclear membrane
- Plasmodesma
- Vacuole
- Plastids
- Chloroplast
- Leucoplast
- Chromoplast
- Golgi Bodies
- Cytoplasm
- Nucleus
- Chromatin
- Cytoskeleton

- Nucleolus
- Mitochondrion

Leaf

A leaf is an organ of a vascular plant and is the principal lateral appendage of the stem. The leaves and stem together form the shoot. Leaves are collectively referred to as foliage, as in "autumn foliage."

Leaf of *Tilia tomentosa* (Silver lime tree)

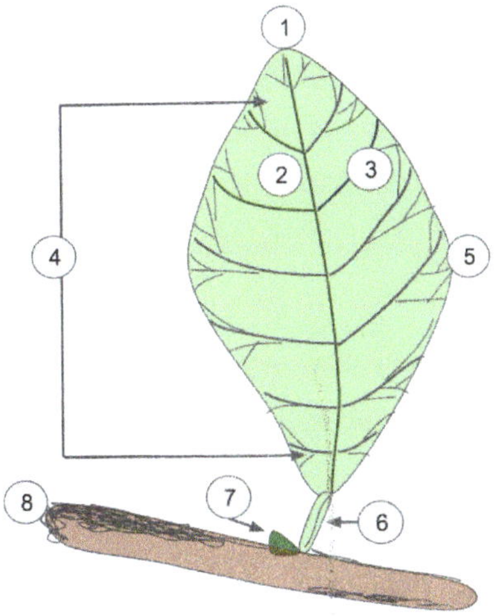

Diagram of a simple leaf.

1. Apex
2. Midvein (Primary vein)
3. Secondary vein.

4. Lamina.
5. Leaf margin
6. Petiole
7. Bud
8. Stem

Although leaves can be seen in many different textures and sizes, typically a leaf is a thin, dorsiventrally flattened organ, borne above ground and specialized for photosynthesis. In most leaves, the primary photosynthetic tissue, the (palisade mesophyll), is located on the upper side of the blade or lamina of the leaf but in some species, including the mature foliage of *Eucalyptus*, palisade mesophyll is present on both sides and the leaves are said to be isobilateral. Most leaves have distinctive upper (adaxial) and lower (abaxial) surfaces that differ in colour, hairiness, the number of stomata (pores that intake and output gases), epicuticular wax amount and structure and other features.

Broad, flat leaves with complex venation are known as megaphylls and the species that bear them, the majority, as broad-leaved or megaphyllous plants. In others, such as the clubmosses, with different evolutionary origins, the leaves are simple, with only a single vein and are known as microphylls.

Some leaves, such as bulb scales are not above ground, and in many aquatic species the leaves are submerged in water. Succulent plants often have thick juicy leaves, but some leaves are without major photosynthetic function and may be dead at maturity, as in some cataphylls, and spines). Furthermore, several kinds of leaf-like structures found in vascular plants are not totally homologous with them. Examples include flattened plant stems called phylloclades and cladodes, and flattened leaf stems called phyllodes which differ from leaves both in their structure and origin. Many structures of non-vascular plants, such as the phyllids of mosses and liverworts and even of some foliose lichens, which are not plants at all (in the sense of being members of the kingdom Plantae), look and function much like leaves.

General Characteristics

3D rendering of a computed tomography scan of a leaf

Typically leaves are broad, flat and thin (dorsiventrally flattened), thereby maximising the surface area directly exposed to light and enabling the light to penetrate the tissues and reach the chloroplasts, thus promoting photosynthesis. They are arranged on the plant so as to expose their surfaces to light as efficiently as possible without shading each other, but there are many exceptions and complications. For instance plants adapted to windy conditions may have pendent leaves, such as in many willows and eucalypts. Also conifers, whose leaves are needle shaped. The flat, or laminar, shape also maximises thermal contact with the surrounding air. The leaf shape also minimises damage from wind, by creating turbulence rather than resistance. Functionally, in addition to photosynthesis the leaf is the principal site of transpiration and guttation. Leaves also function to store chemical energy and water (especially succulents) and may become specialised organs serving other functions.

Vein skeleton of a leaf. Veins contain lignin that make them harder to degrade for microorganisms.

The internal organisation of most kinds of leaves has evolved to maximise exposure of the photosynthetic organelles, the chloroplasts, to light and to increase the absorption of carbon dioxide. Their surfaces are waterproofed by the plant cuticle and gas exchange between the mesophyll cells and the atmosphere is controlled by minute openings called stomata, about 10 μm which open or close to regulate the rate exchange of carbon dioxide, oxygen, and water vapour into and out of the internal intercellular space system. Stomatal opening is controlled by the turgor pressure in a pair of guard cells that surround the stomatal aperture. In any square centimeter of a plant leaf there may be from 1,000 to 100,000 stomata.

Many gymnosperms have thin needle-like or scale-like leaves that can be advantageous in cold climates with frequent snow and frost. These are interpreted as reduced from megaphyllous leaves of their Devonian ancestors. Some leaf forms are adapted to modulate the amount of light they absorb to avoid or mitigate excessive heat, ultraviolet damage, or desiccation, or to sacrifice light-absorption efficiency in favour of protection from herbivory. For xerophytes the major constraint is not light flux or intensity, but drought. Some window plants such as *Fenestraria* species and some *Haworthia* species such as *Haworthia tesselata* and *Haworthia truncata* are examples of xerophytes. and *Bulbine mesembryanthemoides*.

The shape and structure of leaves vary considerably from species to species of plant, depending largely on their adaptation to climate and available light, but also to other factors such as grazing animals (such as deer), available nutrients, and ecological competition from other plants. Considerable changes in leaf type occur within species too, for example as a plant matures; as a case in point *Eucalyptus* species commonly have isobilateral, pendent leaves when mature and dominating their neighbours; however, such trees tend to have erect or horizontal dorsiventral leaves as seedlings, when their growth is limited by the available light. Other factors include the need to balance water loss at high temperature and low humidity against the need to absorb atmospheric carbon dioxide. In most plants leaves also are the primary organs responsible for transpiration and guttation (beads of fluid forming at leaf margins).

Near the ground these *Eucalyptus* saplings have juvenile dorsiventral foliage from the previous year, but this season their newly sprouting foliage is isobilateral, like the mature foliage on the adult trees above

Leaves can also store food and water, and are modified accordingly to meet these functions, for example in the leaves of succulent plants and in bulb scales. The concentration of photosynthetic structures in leaves requires that they be richer in protein, minerals, and sugars than, say, woody stem tissues. Accordingly, leaves are prominent in the diet of many animals.

A leaf shed in autumn.

Correspondingly, leaves represent heavy investment on the part of the plants bearing them, and their retention or disposition are the subject of elaborate strategies for dealing with pest pressures, seasonal conditions, and protective measures such as the growth of thorns and the production of phytoliths, lignins, tannins and poisons.

Deciduous plants in frigid or cold temperate regions typically shed their leaves in autumn, whereas in areas with a severe dry season, some plants may shed their leaves until the dry season ends. In either case the shed leaves may be expected to contribute their retained nutrients to the soil where they fall.

In contrast, many other non-seasonal plants, such as palms and conifers, retain their leaves for long periods; *Welwitschia* retains its two main leaves throughout a lifetime that may exceed a thousand years.

The leaf-like organs of Bryophytes (e.g., mosses and liverworts), known as phyllids, differ morphologically from the leaves of vascular plants in that they lack vascular tissue, are usually only a single cell thick and have no cuticle stomata or internal system of intercellular spaces.

Simple, vascularised leaves (microphylls) first evolved as enations, extensions of the stem, in clubmosses such as *Baragwanathia* during the Silurian period. True leaves or euphylls of larger size and with more complex venation did not become widespread in other groups until the Devonian period, by which time the carbon dioxide concentration in the atmosphere had dropped significantly. This occurred independently in several separate lineages of vascular plants, in progymnosperms like *Archaeopteris*, in Sphenopsida, ferns and later in the gymnosperms and angiosperms. Euphylls are also referred to as macrophylls or megaphylls (large leaves).

Morphology (Large-scale Features)

A structurally complete leaf of an angiosperm consists of a petiole (leaf stalk), a lamina (leaf blade), and stipules (small structures located to either side of the base of the petiole). Not every species produces leaves with all of these structural components. In certain species, paired stipules are not obvious or are absent altogether. A petiole may be absent, or the blade may not be laminar (flattened). The tremendous variety shown in leaf structure (anatomy) from species to species is presented in detail below under morphology. The petiole mechanically links the leaf to the plant and provides the route for transfer of water and sugars to and from the leaf. The lamina is typically the location of the majority of photosynthesis. The upper (adaxial) angle between a leaf and a stem is known as the axil of the leaf. It is often the location of a bud. Structures located there are called "axillary".

Translucent glands in *Citrus* leaves

External leaf characteristics, such as shape, margin, hairs, the petiole, and the presence of stipules, are important for identifying plant species, and botanists have developed a rich terminology for describing leaf characteristics. Leaves have determinate growth. They grow to a specific pattern and shape and then stop. Other plant parts like stems or roots have non-determinate growth, and will usually continue to grow as long as they have the resources to do so.

The type of leaf is usually characteristic of a species (monomorphic), although some species produce more than one type of leaf (dimorphic or polymorphic). The longest leaves are those of the Raffia palm, *R. regalis* which may be up to 25 m (82 ft) long and 3 m (9.8 ft) wide.

Prostrate leaves in *Crossyne guttata*

Where leaves are basal, and lie on the ground, they are referred to as prostrate.

Basic Leaf Types

Leaves of the White Spruce (*Picea glauca*) are needle-shaped and their arrangement is spiral

- Ferns have fronds
- Conifer leaves are typically needle- or awl-shaped or scale-like
- Angiosperm (flowering plant) leaves: the standard form includes stipules, a petiole, and a lamina
- Lycophytes have microphyll leaves.
- Sheath leaves (type found in most grasses and many other monocots)
- Other specialized leaves (such as those of *Nepenthes*, a pitcher plant)

Arrangement on the Stem

Different terms are usually used to describe the arrangement of leaves on the stem (phyllotaxis):

The leaves on this plant are arranged in pairs opposite one another, with successive pairs at right angles to each other (*decussate*) along the red stem. Note the developing buds in the axils of these leaves.

Alternate

One leaf, branch, or flower part attaches at each point or node on the stem, and leaves alternate direction, to a greater or lesser degree, along the stem.

Basal

Arising from the base of the stem.

Cauline

Arising from the aerial stem.

Opposite

Two leaves, branches, or flower parts attach at each point or node on the stem. Leaf attachments are paired at each node and decussate if, as typical, each successive pair is rotated 90° progressing along the stem.

Whorled, or Verticillate

Three or more leaves, branches, or flower parts attach at each point or node on the stem. As with opposite leaves, successive whorls may or may not be decussate, rotated by half the angle between the leaves in the whorl (i.e., successive whorls of three rotated 60°, whorls of four rotated 45°, etc.). Opposite leaves may appear whorled near the tip of the stem. Pseudoverticillate describes an arrangement only appearing whorled, but not actually so.

Rosulate

Leaves form a rosette.

Rows

The term, *distichous*, literally means *two rows*. Leaves in this arrangement may be alternate or opposite in their attachment. The term, *2-ranked*, is equivalent. The terms, *tristichous* and *tetrastichous*, are sometimes encountered. For example, the "leaves" (actually microphylls) of most species of *Selaginella* are tetrastichous, but not decussate.

As a *stem* grows, leaves tend to appear arranged around the stem in a way that optimizes yield of light. In essence, leaves form a helix pattern centered around the stem, either clockwise or counterclockwise, with (depending upon the species) the same angle of divergence. There is a regularity in these angles and they follow the numbers in a Fibonacci sequence: 1/2, 2/3, 3/5, 5/8, 8/13, 13/21, 21/34, 34/55, 55/89. This series tends to the golden angle, which is approximately $360° \times 34/89 \approx 137.52° \approx 137° 30'$. In the series, the numerator indicates the number of complete turns or "gyres" until a leaf arrives at the initial position and the denominator indicates the number of leaves in the arrangement. This can be demonstrated by the following:

- Alternate leaves have an angle of 180° (or 1/2)
- 120° (or 1/3): 3 leaves in 1 circle
- 144° (or 2/5): 5 leaves in 2 gyres
- 135° (or 3/8): 8 leaves in 3 gyres.

Divisions of the Blade

A leaf with laminar structure and pinnate venation

Two basic forms of leaves can be described considering the way the blade (lamina) is divided. A simple leaf has an undivided blade. However, the leaf shape may be formed of lobes, but the gaps between lobes do not reach to the main vein. A compound leaf has a fully subdivided blade, each leaflet of the blade being separated along a main or secondary vein. Because each leaflet can appear to be a simple leaf, it is important to recognize where the petiole occurs to identify a compound leaf. Compound leaves are a characteristic of some families of higher plants, such as the Fabaceae. The middle vein of a compound leaf or a frond, when it is present, is called a rachis.

Palmately Compound

Leaves have the leaflets radiating from the end of the petiole, like fingers of the palm of a hand; e.g., *Cannabis* (hemp) and *Aesculus* (buckeyes).

Pinnately Compound

Leaves have the leaflets arranged along the main or mid-vein.

Odd pinnate

With a terminal leaflet; e.g., *Fraxinus* (ash).

Even pinnate

lacking a terminal leaflet; e.g., *Swietenia* (mahogany).

Bipinnately Compound

Leaves are twice divided: the leaflets are arranged along a secondary vein that is one of several branching off the rachis. Each leaflet is called a "pinnule". The group of pinnules on each secondary vein forms a "pinna"; e.g., *Albizia* (silk tree).

Trifoliate (or Trifoliolate)

A pinnate leaf with just three leaflets; e.g., *Trifolium* (clover), *Laburnum* (laburnum).

Pinnatifid

Pinnately dissected to the central vein, but with the leaflets not entirely separate; e.g., *Polypodium*, some *Sorbus* (whitebeams). In pinnately veined leaves the central vein in known as the *midrib*.

Characteristics of the Petiole

The overgrown petioles of rhubarb (*Rheum rhabarbarum*) are edible.

Petiolated leaves have a petiole (leaf stalk), and are said to be petiolate.

Sessile (epetiolate) leaves have no petiole and the blade attaches directly to the stem. Subpetiolate leaves are nearly petiolate or have an extremely short petiole and may appear to be sessile.

In clasping or decurrent leaves, the blade partially surrounds the stem.

When the leaf base completely surrounds the stem, the leaves are said to be perfoliate, such as in *Eupatorium perfoliatum*.

In peltate leaves, the petiole attaches to the blade inside the blade margin.

In some *Acacia* species, such as the koa tree (*Acacia koa*), the petioles are expanded or broadened and function like leaf blades; these are called phyllodes. There may or may not be normal pinnate leaves at the tip of the phyllode.

A stipule, present on the leaves of many dicotyledons, is an appendage on each side at the base of the petiole, resembling a small leaf. Stipules may be lasting and not be shed (a stipulate leaf, such as in roses and beans), or be shed as the leaf expands, leaving a stipule scar on the twig (an exstipulate leaf). The situation, arrangement, and structure of the stipules is called the "stipulation".

Free, Lateral

As in *Hibiscus*.

Adnate

Fused to the petiole base, as in *Rosa*.

Ochreate

Provided with ochrea, or sheath-formed stipules, as in Polygonaceae; e.g., rhubarb.

Encircling the Petiole Base

Interpetiolar

Between the petioles of two opposite leaves, as in Rubiaceae.

Intrapetiolar

Between the petiole and the subtending stem, as in Malpighiaceae.

Venation

Branching veins on underside of taro leaf

The venation within the bract of a lime tree.

There are two subtypes of venation, namely, *craspedodromous*, where the major veins stretch up to the margin of the leaf, and *camptodromous*, when major veins extend close to the margin, but bend before they intersect with the margin.

Feather-veined, reticulate (also called pinnate-netted, penniribbed, penninerved, or penniveined)

The veins arise pinnately from a single mid-vein and subdivide into veinlets. These, in turn, form a complicated network. This type of venation is typical for (but by no means limited to) dicotyledons.

Plinervy

Three main veins branch at the base of the lamina and run essentially parallel subsequently, as in *Ceanothus* and in *Celtis*. A similar pattern (with 3-7 veins) is especially conspicuous in Melastomataceae.

Palmate-netted, palmate-veined, fan-veined

Several main veins diverge from near the leaf base where the petiole attaches, and radiate toward the edge of the leaf; e.g., most *Acer* (maples).

Parallel-veined, parallel-ribbed, parallel-nerved, penniparallel

Veins run parallel for the length of the leaf, from the base to the apex. Commissural veins (small veins) connect the major parallel veins. Typical for most monocotyledons, such as grasses.

Dichotomous

There are no dominant bundles, with the veins forking regularly by pairs; found in *Ginkgo* and some pteridophytes.

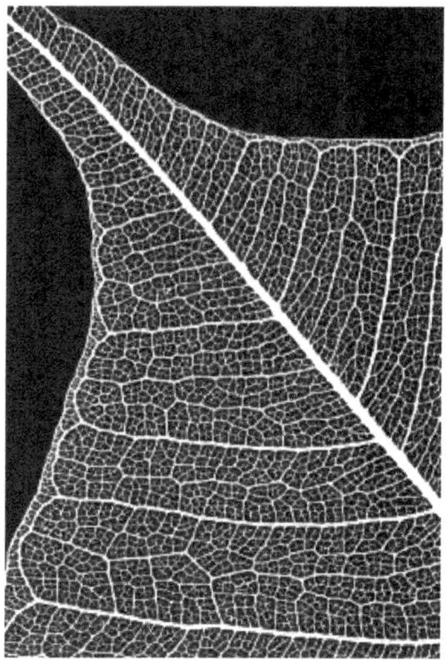

Micrograph of a leaf skeleton

Although it is the more complex pattern, branching veins appear to be plesiomorphic and in some form were present in ancient seed plants as long as 250 million years ago. A pseudo-reticulate venation that is actually a highly modified penniparallel one is an autapomorphy of some Melanthiaceae, which are monocots; e.g., *Paris quadrifolia* (True-lover's Knot).

Morphology Changes within a Single Plant

Homoblasty

Characteristic in which a plant has small changes in leaf size, shape, and growth habit between juvenile and adult stages.

Heteroblasty

Characteristic in which a plant has marked changes in leaf size, shape, and growth habit between juvenile and adult stages.

Terminology

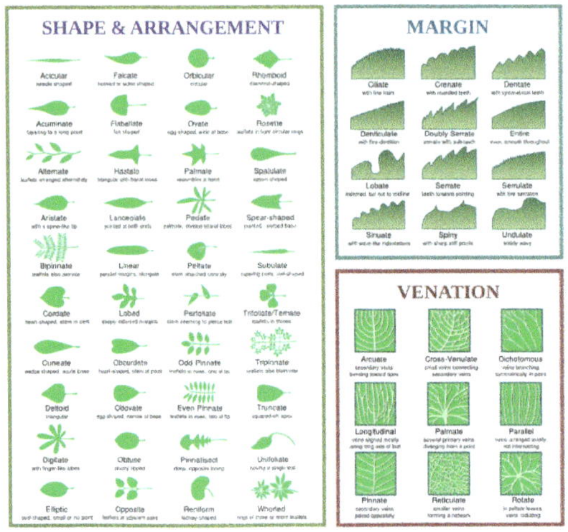

Leaf morphology terms

Shape

Leaves showing various morphologies. Clockwise from upper left: tripartite lobation, elliptic with serrulate margin, palmate venation, acuminate odd-pinnate (center), pinnatisect, lobed, elliptic with entire margin

Edge (Margin)

Image	Term	Latin	Description
	Entire	*Forma integra*	Even; with a smooth margin; without toothing
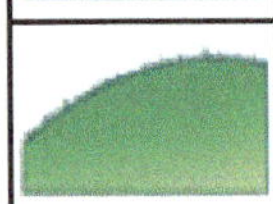	Ciliate	*Ciliata*	Fringed with hairs

	Crenate	*Crenata*	Wavy-toothed; dentate with rounded teeth
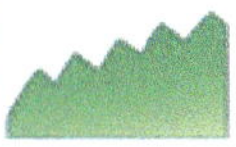	Dentate	*Dentata*	Toothed May be coarsely dentate, having large teeth or glandular dentate, having teeth which bear glands
	Denticulate	*Denticulata*	Finely toothed
	Doubly serrate	*Duplicato-dentata*	Each tooth bearing smaller teeth
	Serrate	*Serrata*	Saw-toothed; with asymmetrical teeth pointing forward
	Serrulate	*Serrulata*	Finely serrate
	Sinuate	*Sinuosa*	With deep, wave-like indentations; coarsely crenate
	Lobate	*Lobata*	Indented, with the indentations not reaching the center
	Undulate	*Undulata*	With a wavy edge, shallower than sinuate
	Spiny or pungent	*Spiculata*	With stiff, sharp points such as thistles

Tip

Acuminate

Long-pointed, prolonged into a narrow, tapering point in a concave manner.

Acute

Ending in a sharp, but not prolonged point

Cuspidate

With a sharp, elongated, rigid tip; tipped with a cusp.

Emarginate

Indented, with a shallow notch at the tip.

Mucronate

Abruptly tipped with a small short point, as a continuation of the midrib; tipped with a mucro.

Mucronulate

Mucronate, but with a noticeably diminutive spine, a mucronule.

Obcordate

Inversely heart-shaped, deeply notched at the top.

Obtuse

Rounded or blunt

Truncate

Ending abruptly with a flat end, that looks cut off.

Haworthia truncata, a classic example of truncate leaves

Base

Acuminate

Coming to a sharp, narrow, prolonged point.

Acute

Coming to a sharp, but not prolonged point.

Auriculate

Ear-shaped.

Cordate

Heart-shaped with the notch towards the stalk.

Cuneate

Wedge-shaped.

Hastate

Shaped like an halberd and with the basal lobes pointing outward.

Oblique

Slanting.

Reniform

Kidney-shaped but rounder and broader than long.

Rounded

Curving shape.

Sagittate

Shaped like an arrowhead and with the acute basal lobes pointing downward.

Truncate

Ending abruptly with a flat end, that looks cut off.

Surface

Scale-shaped leaves of a Norfolk Island Pine, *Araucaria heterophylla*.

Coriaceous

Leathery; stiff and tough, but somewhat flexible.

Farinose

Bearing farina; mealy, covered with a waxy, whitish powder.

Glabrous

Smooth, not hairy.

Glaucous

With a whitish bloom; covered with a very fine, bluish-white powder.

Glutinous

Sticky, viscid.

Lepidote

Coated with small scales (thus elepidote, without such scales).

Maculate

Stained, spotted, compare immaculate.

Papillate, or Papillose

Bearing papillae (minute, nipple-shaped protuberances).

Pubescent

Covered with erect hairs (especially soft and short ones).

Punctate

Marked with dots; dotted with depressions or with translucent glands or colored dots.

Rugose

Deeply wrinkled; with veins clearly visible.

Scurfy

Covered with tiny, broad scalelike particles.

Tuberculate

Covered with tubercles; covered with warty prominences.

Verrucose

Warted, with warty outgrowths.

Viscid, or Viscous

Covered with thick, sticky secretions.

The leaf surface is also host to a large variety of microorganisms; in this context it is referred to as the phyllosphere.

The parallel veins within an iris leaf

Hairiness

Common mullein (*Verbascum thapsus*) leaves are covered in dense, stellate trichomes.

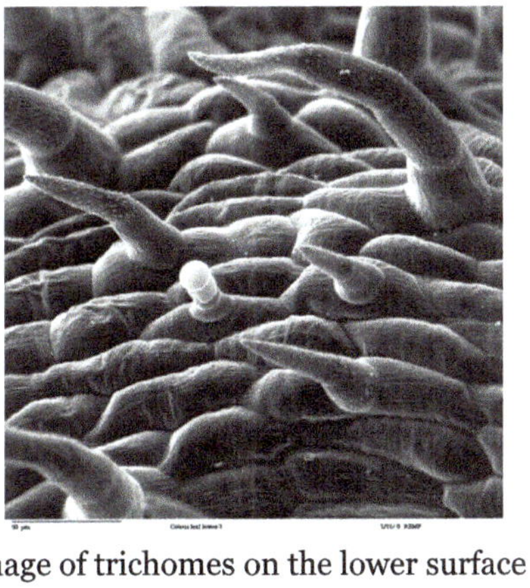

Scanning electron microscope image of trichomes on the lower surface of a *Coleus blumei* (coleus) leaf

"Hairs" on plants are properly called trichomes. Leaves can show several degrees of hairiness. The meaning of several of the following terms can overlap.

Arachnoid, or Arachnose

With many fine, entangled hairs giving a cobwebby appearance.

Barbellate

With finely barbed hairs (barbellae).

Bearded

With long, stiff hairs.

Bristly

With stiff hair-like prickles.

Canescent

Hoary with dense grayish-white pubescence.

Ciliate

Marginally fringed with short hairs (cilia).

Ciliolate

Minutely ciliate.

Floccose

With flocks of soft, woolly hairs, which tend to rub off.

Glabrescent

Losing hairs with age.

Glabrous

No hairs of any kind present.

Glandular

With a gland at the tip of the hair.

Hirsute

With rather rough or stiff hairs.

Hispid

With rigid, bristly hairs.

Hispidulous

Minutely hispid.

Hoary

With a fine, close grayish-white pubescence.

Lanate, or Lanose

With woolly hairs.

Pilose

With soft, clearly separated hairs.

Puberulent, or Puberulous

With fine, minute hairs.

Pubescent

With soft, short and erect hairs.

Scabrous, or Scabrid

Rough to the touch.

Sericeous

Silky appearance through fine, straight and appressed (lying close and flat) hairs.

Silky

With adpressed, soft and straight pubescence.

Stellate, or Stelliform

With star-shaped hairs.

Strigose

With appressed, sharp, straight and stiff hairs.

Tomentose

Densely pubescent with matted, soft white woolly hairs.

Cano-tomentose

Between canescent and tomentose.

Felted-tomentose

Woolly and matted with curly hairs.

Tomentulose

Minutely or only slightly tomentose.

Villous

With long and soft hairs, usually curved.

Woolly

With long, soft and tortuous or matted hairs.

Timing

Hysteranthous (Hysteranthy)

Developing after the flowers

Synanthous (Synanthy)

Developing at the same time as the flowers

Vein Patterning

Channelled

Sunken below the surface, resulting in a rounded channel

Image	Term	Description
	Arcuate	Secondary arching toward the apex
	Dichotomous	Veins splitting in two
	Longitudinal	All veins aligned mostly with the midvein
	Palmate	Veins coming from one point like fingers come from the palm of the hand
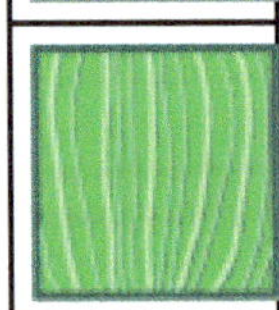	Parallel	All veins parallel and not intersecting
	Pinnate	Secondary veins borne from midrib

	Reticulate	All veins branching repeatedly, net veined
	Rotate	Veins coming from the center of the leaf and radiating toward the edges
	Transverse	Nervure tertiaire reliant les nervures secondaires

Size

The terms megaphyll, macrophyll, mesophyll, notophyll, microphyll, nanophyll and leptophyll are used to describe leaf sizes (in descending order), in a classification devised in 1934 by Christen C. Raunkiær and since modified by others.

Anatomy (Medium and Small Scale)

Medium-scale Features

Leaves are normally extensively vascularised and typically have networks of vascular bundles containing xylem, which supplies water for photosynthesis, and phloem, which transports the sugars produced by photosynthesis. Many leaves are covered in trichomes (small hairs) which have diverse structures and functions.

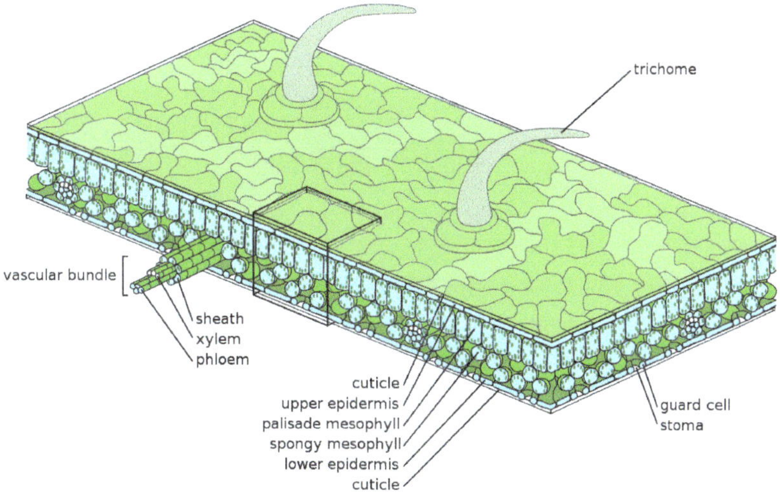

Small-scale features

The major tissue systems present are

1. The epidermis, which covers the upper and lower surfaces

2. The mesophyll tissue inside the leaf, which is rich in chloroplasts (also called chlorenchyma)

3. The arrangement of veins (the vascular tissue)

These three tissue systems typically form a regular organisation at the cellular scale. Specialised cells that differ markedly from surrounding cells, and which often synthesise specialised products such as crystals, are termed idioblasts.

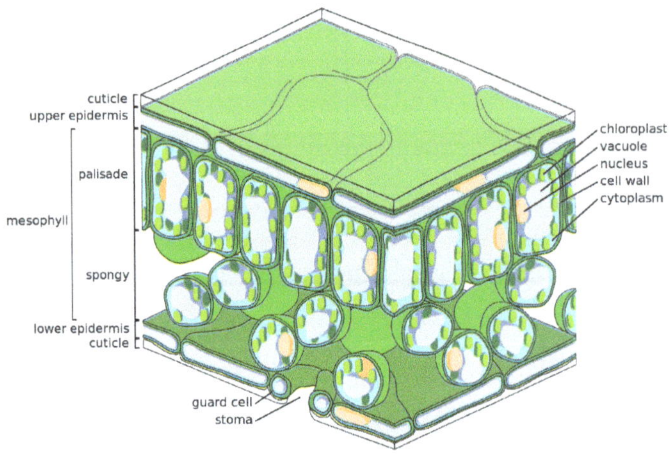

Major leaf tissues

Cross-section of a leaf

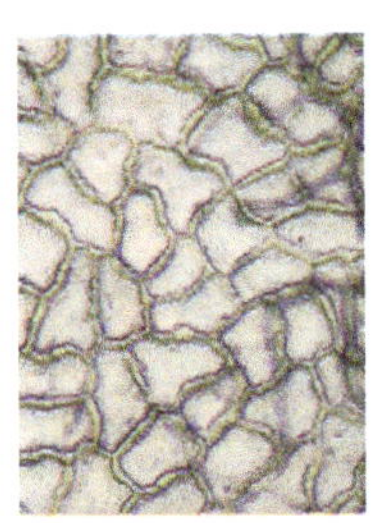

Epidermal cells

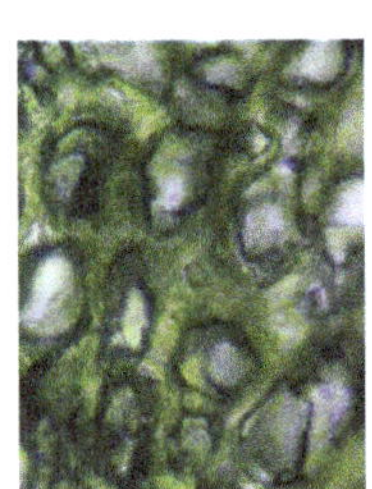

Spongy mesophyll cells

Epidermis

The epidermis is the outer layer of cells covering the leaf. It is covered with a waxy cuticle which is impermeable to liquid water and water vapor and forms the boundary separating the plant's inner cells from the external world. The cuticle is in some cases thinner on the lower epidermis than on the upper epidermis, and is generally thicker on leaves from dry climates as compared with those from wet climates. The epidermis serves several functions: protection against water loss by way of transpiration, regulation of gas exchange, secretion of metabolic compounds, and (in some species) absorption of water. Most leaves show dorsoventral anatomy: The upper (adaxial) and lower (abaxial) surfaces have somewhat different construction and may serve different functions.

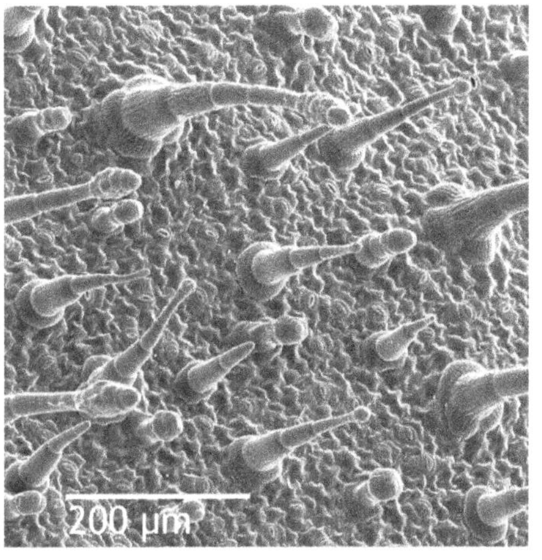

SEM image of the leaf epidermis of *Nicotiana alata*, showing trichomes (hair-like appendages) and stomata (eye-shaped slits, visible at full resolution).

The epidermis tissue includes several differentiated cell types; epidermal cells, epidermal hair cells (trichomes), cells in the stomatal complex; guard cells and subsidiary cells. The epidermal cells are the most numerous, largest, and least specialized and form the majority of the epidermis. These are typically more elongated in the leaves of monocots than in those of dicots.

Chloroplasts are generally absent in epidermal cells, the exception being the guard cells of the stomata. The stomatal pores perforate the epidermis and are surrounded on each side by chloroplast-containing guard cells, and two to four subsidiary cells that lack chloroplasts, forming a specialized cell group known as the stomatal complex. The opening and closing of the stomatal aperture is controlled by the stomatal complex and regulates the exchange of gases and water vapor between the outside air and the interior of the leaf. Stomata therefore play the important role in allowing photosynthesis without letting the leaf dry out. In a typical leaf, the stomata are more numerous over the abaxial (lower) epidermis than the adaxial (upper) epidermis and are more numerous in plants from cooler climates.

Mesophyll

Most of the interior of the leaf between the upper and lower layers of epidermis is a *parenchyma* (ground tissue) or *chlorenchyma* tissue called the mesophyll (Greek for "middle leaf"). This assimilation tissue is the primary location of photosynthesis in the plant. The products of photosynthesis are called "assimilates".

In ferns and most flowering plants, the mesophyll is divided into two layers:

- An upper palisade layer of vertically elongated cells, one to two cells thick, directly beneath the adaxial epidermis, with intercellular air spaces between them. Its cells contain many more chloroplasts than the spongy layer. These long cylindrical cells are regularly arranged in one to five rows. Cylindrical cells, with the *chloroplasts* close to the walls of the cell, can take optimal advantage of light. The slight separation of the cells provides maximum absorption of carbon dioxide. This separation must be minimal to afford capillary action for water distribution. In order to adapt to their different environment (such as sun or shade), plants had to adapt this structure to obtain optimal result. Sun leaves have a multi-layered palisade layer, while shade leaves or older leaves closer to the soil are single-layered.
- Beneath the palisade layer is the spongy layer. The cells of the spongy layer are more branched and not so tightly packed, so that there are large intercellular air spaces between them for oxygen and carbon dioxide to diffuse in and out of during respiration and photosynthesis. These cells contain fewer chloroplasts than those of the palisade layer. The pores or *stomata* of the epidermis open into substomatal chambers, which are connected to the air spaces between the spongy layer cells.

Leaves are normally green, due to chlorophyll in chloroplasts in the chlorenchyma cells. Plants that lack chlorophyll cannot photosynthesize.

Veins

The veins of a bramble leaf

The veins are the vascular tissue of the leaf and are located in the spongy layer of the mesophyll. The pattern of the veins is called venation. In angiosperms the venation is typically parallel in monocotyledons and forms an interconnecting network in broad-leaved plants. They were once thought to be typical examples of pattern formation through ramification, but they may instead exemplify a pattern formed in a stress tensor field.

A vein is made up of a vascular bundle. At the core of each bundle are clusters of two distinct types of conducting cells:

- Xylem: cells that bring water and minerals from the roots into the leaf.
- Phloem: cells that usually move sap, with dissolved sucrose, produced by photosynthesis in the leaf, out of the leaf.
- A sheath of ground tissue made of lignin surrounding the vascular tissue. This sheath has a mechanical role in strengthening the rigidity of the leaf.

The xylem typically lies on the adaxial side of the vascular bundle and the phloem typically lies on the abaxial side. Both are embedded in a dense parenchyma tissue, called the sheath, which usually includes some structural collenchyma tissue.

Leaf Development

According to Agnes Arber's partial-shoot theory of the leaf, leaves are partial shoots, being derived from leaf primordia of the shoot apex. Compound leaves are closer to shoots than simple leaves. Developmental studies have shown that compound leaves, like shoots, may branch in three dimensions. On the basis of molecular genetics, Eckardt and Baum (2010) concluded that "it is now generally accepted that compound leaves express both leaf and shoot properties."

Adaptations

Poinsettia bracts are leaves which have evolved red pigmentation in order to attract insects and birds to the central flowers, an adaptive function normally served by petals (which are themselves leaves highly modified by evolution).

In the course of evolution, leaves have adapted to different environments in the following ways:

- A certain surface structure avoids moistening by rain and contamination (*See Lotus effect*).
- Sliced leaves reduce wind resistance.
- Hairs on the leaf surface trap humidity in dry climates and create a boundary layer reducing water loss.
- Waxy leaf surfaces reduce water loss.
- Large surface area provides large area for sunlight and shade for plant to minimize heating and reduce water loss.

- In harmful levels of sunlight, specialised leaves, opaque or partly buried, admit light through a translucent leaf window for photosynthesis at inner leaf surfaces (e.g. *Fenestraria*).
- Succulent leaves store water and organic acids for use in CAM photosynthesis.
- Aromatic oils, poisons or pheromones produced by leaf borne glands deter herbivores (e.g. eucalypts).
- Inclusions of crystalline minerals deter herbivores (e.g. silica phytoliths in grasses, raphides in Araceae).
- Petals attract pollinators.
- Spines protect the plants (*e.g.* cacti).
- Special leaves on carnivorous plants are adapted to trapping food, mainly invertebrate prey, though some species trap small vertebrates as well (see carnivorous plants).
- Bulbs store food and water (*e.g.* onions).
- Tendrils allow the plant to climb (*e.g.* peas).
- Bracts and pseudanthia (false flowers) replace normal flower structures when the true flowers are greatly reduced (*e.g.* Spurges).
- Spathe.

Seasonal Leaf Loss

Leaves shifting color in autumn/fall

A girl playing with leaves

Leaves in temperate, boreal, and seasonally dry zones may be seasonally deciduous (falling off or dying for the inclement season). This mechanism to shed leaves is called abscission. When the leaf is shed, it leaves a leaf scar on the twig. In cold autumns, they sometimes change color, and turn yellow, bright-orange, or red, as various accessory pigments (carotenoids and xanthophylls) are revealed when the tree responds to cold and reduced sunlight by curtailing chlorophyll production. Red anthocyanin pigments are now thought to be produced in the leaf as it dies, possibly to mask the yellow hue left when the chlorophyll is lost—yellow leaves appear to attract herbivores such as aphids. Optical masking of chlorophyll by anthocyanins reduces risk of photo-oxidative damage to leaf cells as they senesce, which otherwise may lower the efficiency of nutrient retrieval from senescing autumn leaves.

Interactions with other Organisms

Some insects, like *Kallima inachus*, mimic leaves

Leaf damaged by insects with chewing mouthparts, probably weevils or katydids

Although not as nutritious as other organs such as fruit, leaves provide a food source for many organisms. The leaf is a vital source of energy production for the plant, and plants have evolved protection against animals that consume leaves, such as tannins, chemicals which hinder the digestion of proteins and have an unpleasant taste. Animals that are specialized to eat leaves are known as folivores.

Some species have cryptic adaptations by which they use leaves in avoiding predators. For example, the caterpillars of some leaf-roller moths will create a small home in the leaf by folding it over themselves. Some sawflies similarly roll the leaves of their food plants into tubes. Females of the Attelabidae, so-called leaf-rolling weevils, lay their eggs into leaves that they then roll up as means of protection. Other herbivores and their predators mimic the appearance of the leaf. Reptiles such as some chameleons, and insects such as some katydids, also mimic the oscillating movements of leaves in the wind, moving from side to side or back and forth while evading a possible threat.

Meristem

A meristem is the tissue in most plants containing undifferentiated cells (meristematic cells), found in zones of the plant where growth can take place.

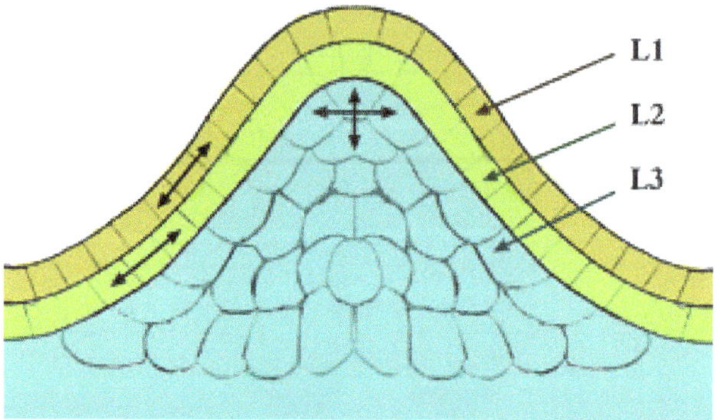

Tunica-Corpus model of the apical meristem (growing tip). The epidermal (L1) and subepidermal (L2) layers form the outer layers called the tunica. The inner L3 layer is called the corpus. Cells in the L1 and L2 layers divide in a sideways fashion, which keeps these layers distinct, whereas the L3 layer divides in a more random fashion.

Meristematic cells give rise to various organs of the plant and keep the plant growing. The *shoot apical meristem* (SAM) gives rise to organs like the leaves and flowers, while the *root apical meristem* (RAM) provides the meristematic cells for the future root growth. SAM and RAM cells divide rapidly and are considered indeterminate, in that they do not possess any defined end status. In that sense, the meristematic cells are frequently compared to the stem cells in animals, which have an analogous behavior and function.

The term *meristem* was first used in 1858 by Karl Wilhelm von Nägeli (1817–1891) in his book *Beiträge zur Wissenschaftlichen Botanik* ("Contributions to Scientific Botany"). It is derived from the Greek word *merizein* (μερίζειν), meaning to divide, in recognition of its inherent function.

In general, differentiated plant cells cannot divide or produce cells of a different type. Therefore, cell division in the meristem is required to provide new cells for expansion and differentiation of tissues and initiation of new organs, providing the basic structure of the plant body.

Meristematic cells are incompletely or not at all differentiated, and are capable of continued cellular division (youthful). Furthermore, the cells are small and protoplasm fills the cell completely. The vacuoles are extremely small. The cytoplasm does not contain differentiated plastids (chloroplasts or chromoplasts), although they are present in rudimentary form (proplastids). Meristematic cells are packed closely together without intercellular cavities. The cell wall is a very thin *primary cell wall.*

Maintenance of the cells requires a balance between two antagonistic processes: organ initiation and stem cell population renewal.

Apical meristems are the completely undifferentiated (indeterminate) meristems in a plant. These differentiate into three kinds of primary meristems. The primary meristems in turn produce the two secondary meristem types. These secondary meristems are also known as lateral meristems because they are involved in lateral growth.

At the meristem summit, there is a small group of slowly dividing cells, which is commonly called the central zone. Cells of this zone have a stem cell function and are essential for meristem maintenance. The proliferation and growth rates at the meristem summit usually differ considerably from those at the periphery.

Meristems also are induced in the roots of legumes such as soybean, *Lotus japonicus*, pea, and *Medicago truncatula* after infection with soil bacteria commonly called Rhizobium. Cells of the inner or outer cortex in the so-called "window of nodulation" just behind the developing root tip are induced to divide. The critical signal substance is the lipo-oligosaccharide Nod-factor, decorated with side groups to allow specificity of interaction. The Nod factor receptor proteins NFR1 and NFR5 were cloned from several legumes including *Lotus japonicus*, *Medicago truncatula* and soybean (*Glycine max*). Regulation of nodule meristems utilizes long distance regulation commonly called "Autoregulation of Nodulation" (AON). This process involves a leaf-vascular tissue located LRR receptor kinases (LjHAR1, GmNARK and MtSUNN), CLE peptide signalling, and KAPP interaction, similar to that seen in the CLV1,2,3 system. LjKLAVIER also exhibits a nodule regulation phenotype though it is not yet known how this relates to the other AON receptor kinases.

Apical Meristems

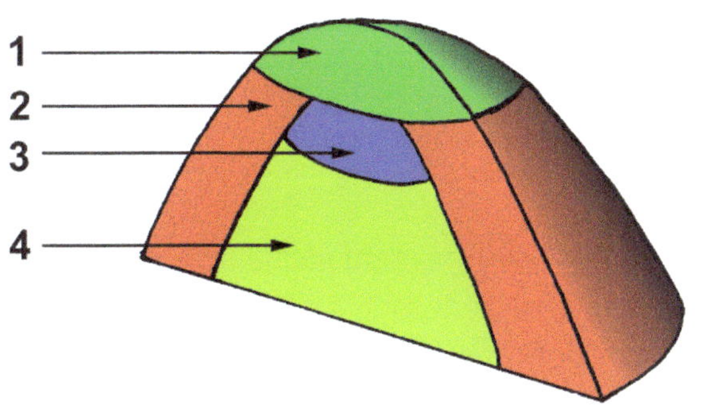

Organisation of an apical meristem (growing tip)
1 - Central zone
2 - Peripheral zone
3 - Medullary (i.e. central) meristem
4 - Medullary tissue

The number of layers varies according to plant type. In general the outermost layer is called the tunica while the innermost layers are the corpus. In monocots, the tunica determine the physical characteristics of the leaf edge and margin. In dicots, layer two of the corpus determine the characteristics of the edge of the leaf. The corpus and tunica play a critical part of the plant physical appearance as all plant cells are formed from the meristems. Apical meristems are found in two locations: the root and the stem. Some Arctic plants have an apical meristem in the lower/middle parts of the plant. It is thought that this kind of meristem evolved because it is advantageous in Arctic conditions.

Shoot Apical Meristems

Shoot apical meristems of *Crassula ovata* (left). Fourteen days later, leaves have developed (right).

The source of all above-ground organs. Cells at the shoot apical meristem summit serve as stem cells to the surrounding peripheral region, where they proliferate rapidly and are incorporated into differentiating leaf or flower primordia.

The shoot apical meristem is the site of most of the embryogenesis in flowering plants. Primordia of leaves, sepals, petals, stamens and ovaries are initiated here at the rate of one every time interval, called a plastochron. It is where the first indications that flower development has been evoked are manifested. One of these indications might be the loss of apical dominance and the release of otherwise dormant cells to develop as auxiliary shoot meristems, in some species in axils of primordia as close as two or three away from the apical dome. The shoot apical meristem consists of 4 distinct cell groups:

- Stem cells
- The immediate daughter cells of the stem cells
- A subjacent organising centre
- Founder cells for organ initiation in surrounding regions

The four distinct zones mentioned above are maintained by a complex signalling pathway. In *Arabidopsis thaliana*, 3 interacting *CLAVATA* genes are required to regulate the size of the stem cell reservoir in the shoot apical meristem by controlling the rate of cell division. CLV1 and CLV2 are predicted to form a receptor complex (of the LRR receptor-like kinase family) to which CLV3 is a ligand. CLV3 shares some homology with the ESR proteins of maize, with a short 14 amino acid region being conserved between the proteins. Proteins that contain these conserved regions have been grouped into the CLE family of proteins.

CLV1 has been shown to interact with several cytoplasmic proteins that are most likely involved in downstream signalling. For example, the CLV complex has been found to be associated with Rho/

Rac small GTPase-related proteins. These proteins may act as an intermediate between the CLV complex and a mitogen-activated protein kinase (MAPK), which is often involved in signalling cascades. KAPP is a kinase-associated protein phosphatase that has been shown to interact with CLV1. KAPP is thought to act as a negative regulator of CLV1 by dephosphorylating it.

Another important gene in plant meristem maintenance is *WUSCHEL* (shortened to *WUS*), which is a target of CLV signalling. *WUS* is expressed in the cells below the stem cells of the meristem and its presence prevents the differentiation of the stem cells. CLV1 acts to promote cellular differentiation by repressing *WUS* activity outside of the central zone containing the stem cells. *SHOOT MERISTEMLESS* (*STM*) also acts to prevent the differentiation of stem cells by repressing the expression of MYB genes that are involved in cellular differentiation.

Root Apical Meristem

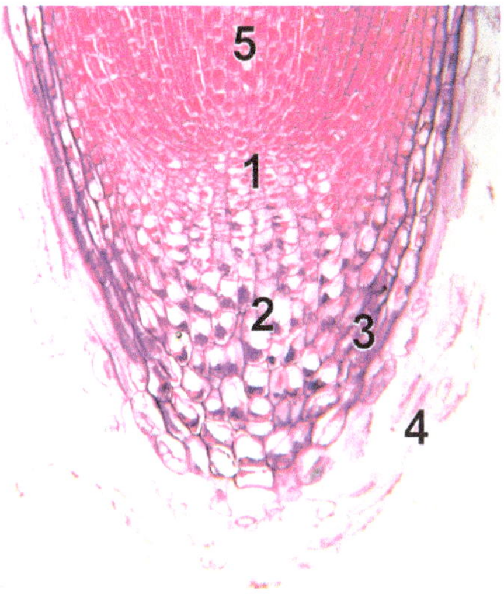

10x microscope image of root tip with meristem
1 - quiescent center
2 - calyptrogen (live rootcap cells)
3 - rootcap
4 - sloughed off dead rootcap cells
5 - procambium

Unlike the shoot apical meristem, the root apical meristem produces cells in two dimensions. It harbors two pools of stem cells around an organizing center called the quiescent center (QC) cells and together produce most of the cells in an adult root. At its apex, the root meristem is covered by the root cap, which protects and guides its growth trajectory. Cells are continuously sloughed off the outer surface of the root cap. The QC cells are characterized by their low mitotic activity. Evidence suggests that the QC maintains the surrounding stem cells by preventing their differentiation, via signal(s) that are yet to be discovered. This allows a constant supply of new cells in the meristem required for continuous root growth. Recent findings indicate that QC can also act as a reservoir of stem cells to replenish whatever is lost or damaged. Root apical meristem and tissue patterns become established in the embryo in the case of the primary root, and in the new lateral root primordium in the case of secondary roots.

Intercalary Meristem

In angiosperms, intercalary meristems occur only in monocot (in particular, grass) stems at the base of nodes and leaf blades. Horsetails also exhibit intercalary growth. Intercalary meristems are capable of cell division, and they allow for rapid growth and regrowth of many monocots. Intercalary meristems at the nodes of bamboo allow for rapid stem elongation, while those at the base of most grass leaf blades allow damaged leaves to rapidly regrow. This leaf regrowth in grasses evolved in response to damage by grazing herbivores, but is more familiar to us in response to lawnmowers.

Floral Meristem

When plants begin the developmental process known as flowering, the shoot apical meristem is transformed into an inflorescence meristem, which goes on to produce the floral meristem, which produces the familiar sepals, petals, stamens, and carpels of the flower.

In contrast to vegetative apical meristems and some exflorescence meristems, floral meristems are responsible for determinate growth, the limited growth of the flower to a particular size and form. The transition from shoot meristem to floral meristem requires floral meristem identity genes, that both specify the floral organs and cause the termination of the production of stem cells. *AGAMOUS* (*AG*) is a floral homeotic gene required for floral meristem termination and necessary for proper development of the stamens and carpels. *AG* is necessary to prevent the conversion of floral meristems to inflorescence shoot meristems, but is not involved in the transition from shoot to floral meristem. AG is turned on by the floral meristem identity gene *LEAFY* (*LFY*) and *WUS* and is restricted to the centre of the floral meristem or the inner two whorls. This way floral identity and region specificity is achieved. WUS activates AG by binding to a consensus sequence in the AG's second intron and LFY binds to adjacent recognition sites. Once AG is activated it represses expression of WUS leading to the termination of the meristem.

Through the years, scientists have manipulated floral meristems for economics reasons. An example is the mutant tobacco plant "Maryland Mammoth." In 1936, the department of agriculture of Switzerland performed several scientific tests with this plant. "Maryland Mammoth" is peculiar in this sense that it grows much faster than other tobacco plants.

Apical Dominance

Apical dominance is phenomenon where one meristem prevents or inhibits the growth of other meristems. As a result, the plant will have one clearly defined main trunk. For example, in trees, the tip of the main trunk bears the dominant meristem. Therefore, the tip of the trunk grows rapidly and is not shadowed by branches. If the dominant meristem is cut off, one or more branch tips will assume dominance. The branch will start growing faster and the new growth will be vertical. Over the years, the branch may begin to look more and more like an extension of the main trunk. Often several branches will exhibit this behaviour after the removal of apical meristem, leading to a bushy growth.

The mechanism of apical dominance is based on the plant hormone auxin. It is produced in the apical meristem and transported towards the roots in the cambium. If apical dominance is complete, it prevents any branches from forming as long as the apical meristem is active. If the dominance is incomplete, side branches will develop.

Recent investigations into apical dominance and the control of branching have revealed a new plant hormone family termed strigolactones. These compounds were previously known to be involved in seed germination and communication with mycorrhizal fungi and are now shown to be involved in inhibition of branching.

Diversity in Meristem Architectures

Is the mechanism of being *indeterminate* conserved in the SAMs of the plant world? The SAM contains a population of stem cells that also produce the lateral meristems while the stem elongates. It turns out that the mechanism of regulation of the stem cell number might indeed be evolutionarily conserved. The *CLAVATA* gene *CLV2* responsible for maintaining the stem cell population in *Arabidopsis thaliana* is very closely related to the Maize gene *FASCIATED EAR 2*(*FEA2*) also involved in the same function. Similarly, in Rice, the *FON1-FON2* system seems to bear a close relationship with the CLV signaling system in *Arabidopsis thaliana*. These studies suggest that the regulation of stem cell number, identity and differentiation might be an evolutionarily conserved mechanism in monocots, if not in angiosperms. Rice also contains another genetic system distinct from *FON1-FON2*, that is involved in regulating stem cell number. This example underlines the innovation that goes about in the living world all the time.

Role of the KNOX-family Genes

Note the long spur of the above flower. Spurs attract pollinators and confer pollinator specificity. *(Flower:Linaria dalmatica)*

Complex leaves of *C. hirsuta* are a result of KNOX gene expression

Genetic screens have identified genes belonging to the KNOX family in this function. These genes essentially maintain the stem cells in an undifferentiated state. The KNOX family has undergone quite a bit of evolutionary diversification, while keeping the overall mechanism more or less similar. Members of the KNOX family have been found in plants as diverse as Arabidopsis thaliana, rice, barley and tomato. KNOX-like genes are also present in some algae, mosses, ferns and gymnosperms. Misexpression of these genes leads to formation of interesting morphological features. For example, among members of *Antirrhinae*, only the species of genus Antirrhinum lack a structure called spur in the floral region. A spur is considered an evolutionary innovation because it defines pollinator specificity and attraction. Researchers carried out transposon mutagenesis in *Antirrhinum majus*, and saw that some insertions led to formation of spurs that were very similar to the other members of *Antirrhinae*, indicating that the loss of spur in wild *Antirrhinum majus* populations could probably be an evolutionary innovation.

The KNOX family has also been implicated in leaf shape evolution *(See below for a more detailed discussion)*. One study looked at the pattern of KNOX gene expression in *A. thaliana*, that has simple leaves and *Cardamine hirsuta*, a plant having complex leaves. In *A. thaliana*, the KNOX genes are completely turned off in leaves, but in *C.hirsuta*, the expression continued, generating complex leaves. Also, it has been proposed that the mechanism of KNOX gene action is conserved across all vascular plants, because there is a tight correlation between KNOX expression and a complex leaf morphology.

Primary Meristems

Apical meristems may differentiate into three kinds of primary meristem:

- Protoderm: lies around the outside of the stem and develops into the epidermis.
- Procambium: lies just inside of the protoderm and develops into primary xylem and primary phloem. It also produces the vascular cambium, and cork cambium, secondary meristems. The cork cambium further differentiates into the phelloderm (to the inside) and the phellem, or cork (to the outside). All three of these layers (cork cambium, phellem and phelloderm) constitute the periderm. In roots, the procambium can also give rise to the pericycle, which produces lateral roots in eudicots.
- Ground meristem: develops into the cortex and the pith. Composed of parenchyma, collenchyma and sclerenchyma cells.

These meristems are responsible for primary growth, or an increase in length or height, which were discovered by scientist Joseph D. Carr of North Carolina in 1943.

Secondary meristems

There are two types of secondary meristems, these are also called the *lateral meristems* because they surround the established stem of a plant and cause it to grow laterally (i.e., larger in diameter).

- Vascular cambium, which produces secondary xylem and secondary phloem. This is a process that may continue throughout the life of the plant. This is what gives rise to wood in plants. Such plants are called arborescent. This does not occur in plants that do not go through secondary growth (known as herbaceous plants).

- Cork cambium, which gives rise to the periderm, which replaces the epidermis.

Indeterminate Growth of Meristems

Though each plant grows according to a certain set of rules, each new root and shoot meristem can go on growing for as long as it is alive. In many plants, meristematic growth is potentially indeterminate, making the overall shape of the plant not determinate in advance. This is the primary growth. Primary growth leads to lengthening of the plant body and organ formation. All plant organs arise ultimately from cell divisions in the apical meristems, followed by cell expansion and differentiation. Primary growth gives rise to the apical part of many plants.

The growth of nitrogen fixing nodules on legume plants such as soybean and pea is either determinate or indeterminate. Thus, soybean (or bean and Lotus japonicus) produce determinate nodules (spherical), with a branched vascular system surrounding the central infected zone. Often, Rhizobium infected cells have only small vacuoles. In contrast, nodules on pea, clovers, and Medicago truncatula are indeterminate, to maintain (at least for some time) an active meristem that yields new cells for Rhizobium infection. Thus zones of maturity exist in the nodule. Infected cells usually possess a large vacuole. The plant vascular system is branched and peripheral.

Cloning

Under appropriate conditions, each shoot meristem can develop into a complete new plant or clone. Such new plants can be grown from shoot cuttings that contain an apical meristem. Root apical meristems are not readily cloned, however. This cloning is called asexual reproduction or vegetative reproduction and is widely practiced in horticulture to mass-produce plants of a desirable genotype. This process is also known as mericloning.

Propagating through cuttings is another form of vegetative propagation that initiates root or shoot production from secondary meristematic cambial cells. This explains why basal 'wounding' of shoot-borne cuttings often aids root formation.

Pollen

Pollen is a fine to coarse powdery substance comprising pollen grains which are male microgametophytes of seed plants, which produce male gametes (sperm cells). Pollen grains have a hard coat made of sporopollenin that protects the gametophytes during the process of their movement from the stamens to the pistil of flowering plants or from the male cone to the female cone of coniferous plants. If pollen lands on a compatible pistil or female cone, it germinates, producing a pollen tube that transfers the sperm to the ovule containing the female gametophyte. Individual pollen grains are small enough to require magnification to see detail. The study of pollen is called palynology and is highly useful in paleoecology, paleontology, archaeology, and forensics.

Tulip anther with many grains of pollen

Closeup image of a cactus flower and its stamens

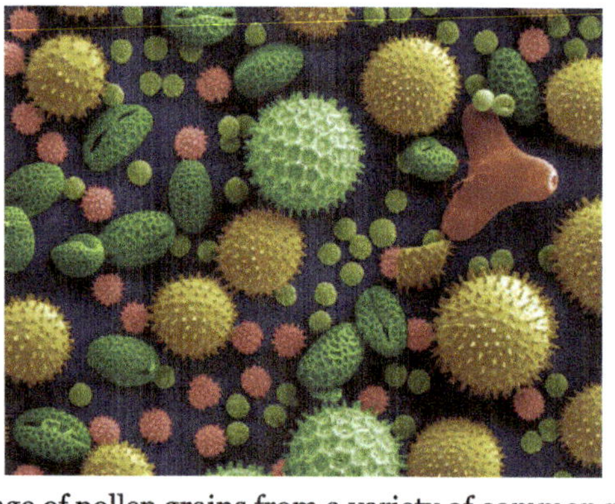

Scanning electron microscope image of pollen grains from a variety of common plants: sunflower (*Helianthus annuus*), morning glory (*Ipomoea purpurea*), prairie hollyhock (*Sidalcea malviflora*), oriental lily (*Lilium auratum*), evening primrose (*Oenothera fruticosa*), and castor bean (*Ricinus communis*).

Pollen in plants is used for transferring haploid male genetic material from the anther of a single flower to the stigma of another in cross-pollination. In a case of self-pollination, this process takes place from the anther of a flower to the stigma of the same flower.

The Structure and Formation of Pollen

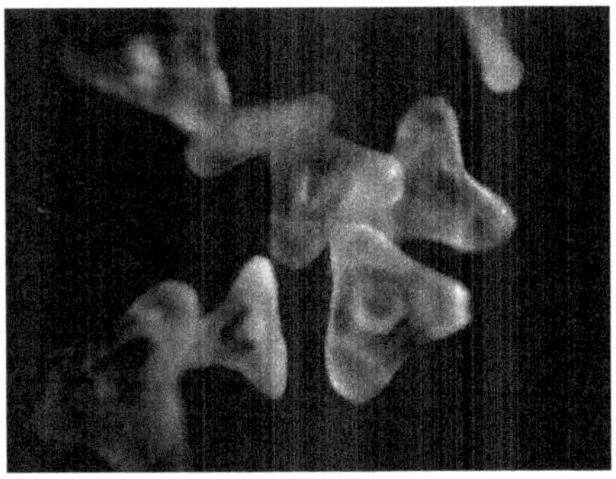

Triporate pollen of *Oenothera speciosa*

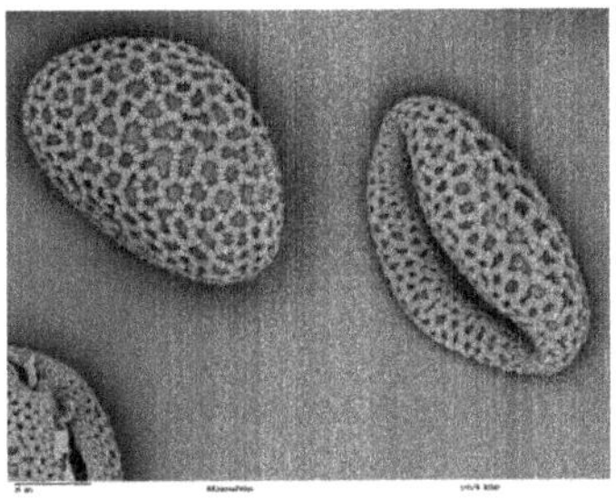

Pollen of *Lilium auratum* showing single sulcus (monosulcate)

Arabis pollen has three colpi and prominent surface structure.

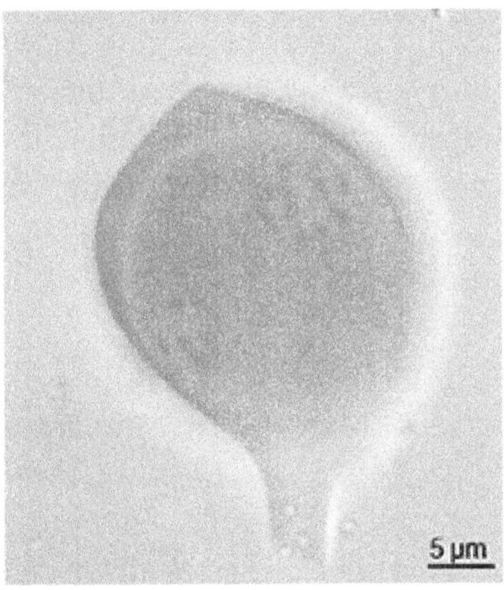

Pollens/Microspores of *Lycopersicon esculentum* at coenocytic tetrad stage of development observed through oil immersion microscope; the chromosomes of what will become four pollen grains can be seen.

Apple Pollen Under Microscopy

Pollen itself is not the male gamete. Each pollen grain contains vegetative (non-reproductive) cells (only a single cell in most flowering plants but several in other seed plants) and a generative (reproductive) cell. In flowering plants the vegetative tube cell produces the pollen tube, and the generative cell divides to form the two sperm cells.

Formation

Pollen is produced in the 'microsporangium' (contained in the anther of an angiosperm flower, male cone of a coniferous plant, or male cone of other seed plants). Pollen grains come in a wide variety of shapes (most often spherical), sizes, and surface markings characteristic of the species (see electron micrograph, right). Pollen grains of pines, firs, and spruces are winged. The smallest pollen grain, that of the forget-me-not (*Myosotis* spp.), is around 6 μm (0.006 mm) in diameter. Wind-borne pollen grains can be as large as about 90–100 μm.

In angiosperms, during flower development the anther is composed of a mass of cells that appear undifferentiated, except for a partially differentiated dermis. As the flower develops, four groups of sporogenous cells form within the anther. The fertile sporogenous cells are surrounded by layers of sterile cells that grow into the wall of the pollen sac. Some of the cells grow into nutritive cells that supply nutrition for the microspores that form by meiotic division from the sporogenous cells.

In a process called microsporogenesis, four haploid microspores are produced from each diploid sporogenous cell (microsporocyte, pollen mother cell or meiocyte), after meiotic division. After the formation of the four microspores, which are contained by callose walls, the development of the pollen grain walls begins. The callose wall is broken down by an enzyme called callase and the freed pollen grains grow in size and develop their characteristic shape and form a resistant outer wall called the exine and an inner wall called the intine. The exine is what is preserved in the fossil record. Two basic types of microsporogenesis are recognised, simultaneous and successive. In simultaneous microsporogenesis meiotic steps I and II are completed prior to cytokinesis, whereas in successive microsporogenesis cytokinesis follows. While there may be a continuum with intermediate forms, the type of microsporogenesis has systematic significance. The predominant form amongst the monocots is successive, but there are important exceptions.

In the microgametogenesis, the unicellular microspores undergoes mitosis and develops into mature microgametophytes containing the gametes. In some flowering plants, germination of the pollen grain often begins before it leaves the microsporangium, with the generative cell forming the two sperm cells.

Structure

Except in the case of some submerged aquatic plants, the mature pollen-grain has a double wall. The vegetative and generative cells are surrounded by a thin delicate wall of unaltered cellulose called the endospore or intine, and a tough resistant outer cuticularized wall composed largely of sporopollenin called the exospore or exine. The exine often bears spines or warts, or is variously sculptured, and the character of the markings is often of value for identifying genus, species, or even cultivar or individual. The spines may be less than a micron in length (spinulus, plural spinuli) referred to as spinulose (scabrate), or longer than a micron (echina, echinae) referred to as echinate. Various terms also describe the sculpturing such as reticulate, a net like appearance consisting of elements (murus, muri) separated from each other by a lumen (plural lumina).

The pollen wall protects the sperm while the pollen grain is moving from the anther to the stigma; it protects the vital genetic material from drying out and solar radiation. The pollen grain surface is covered with waxes and proteins, which are held in place by structures called sculpture elements on the surface of the grain. The outer pollen wall, which prevents the pollen grain from shrinking and crushing the genetic material during desiccation, is composed of two layers. These two layers are the tectum and the foot layer, which is just above the intine. The tectum and foot layer are separated by a region called the columella, which is composed of strengthening rods. The outer wall is constructed with a resistant biopolymer called sporopollenin.

Pollen apertures are regions of the pollen wall that may involve exine thinning or a significant reduction in exine thickness. They allow shrinking and swelling of the grain caused by changes in moisture content. Elongated apertures or furrows in the pollen grain are called colpi (singular: colpus) or sulci (singular: sulcus). Apertures that are more circular are called pores. Colpi, sulci and pores are major features in the identification of classes of pollen. Pollen may be referred to as inaperturate (apertures absent) or aperturate (apertures present). The aperture may have a lid (operculum), hence is described as operculate. However the term inaperturate covers a wide range of morphological types, such as functionally inaperturate (cryptoaperturate) and omniaperturate. Inaperaturate pollen grains often have thin walls, which facilitates pollen tube germination at any position.

The orientation of furrows (relative to the original tetrad of microspores) classifies the pollen as sulcate or colpate. Sulcate pollen has a furrow across the middle of what was the outer face when the pollen grain was in its tetrad. If the pollen has only a single sulcus, it is described as monosulcate. Colpate pollen has furrows other than across the middle of the outer faces. Eudicots have pollen with three colpi (tricolpate) or with shapes that are evolutionarily derived from tricolpate pollen. The evolutionary trend in plants has been from monosulcate to polycolpate or polyporate pollen.

Pollination

European honey bee carrying pollen in a pollen basket back to the hive

Marmalade hoverfly, pollen on its face and legs, sitting on a rockrose.

Africanized honey bees visiting yellow *Opuntia engelmannii* cactus

The transfer of pollen grains to the female reproductive structure (*pistil* in angiosperms) is called pollination. This transfer can be mediated by the wind, in which case the plant is described as anemophilous (literally wind-loving). Anemophilous plants typically produce great quantities of very lightweight pollen grains, sometimes with air-sacs. Non-flowering seed plants (e.g. pine trees) are characteristically anemophilous. Anemophilous flowering plants generally have inconspicuous flowers. Entomophilous (literally insect-loving) plants produce pollen that is relatively heavy, sticky and protein-rich, for dispersal by insect pollinators attracted to their flowers. Many insects and some mites are specialized to feed on pollen, and are called palynivores.

In non-flowering seed plants, pollen germinates in the pollen chamber, located beneath the micropyle, underneath the integuments of the ovule. A pollen tube is produced, which grows into the nucellus to provide nutrients for the developing sperm cells. Sperm cells of Pinophyta and Gnetophyta are without flagella, and are carried by the pollen tube, while those of Cycadophyta and Ginkgophyta have many flagella.

When placed on the stigma of a flowering plant, under favorable circumstances, a pollen grain puts forth a pollen tube, which grows down the tissue of the style to the ovary, and makes its way along the placenta, guided by projections or hairs, to the micropyle of an ovule. The nucleus of the tube cell has meanwhile passed into the tube, as does also the generative nucleus, which divides (if it hasn't already) to form two sperm cells. The sperm cells are carried to their destination in the tip of the pollen-tube.

Pollen in the Fossil Record

Pollen's sporopollenin outer sheath affords it some resistance to the rigours of the fossilisation process that destroy weaker objects; it is also produced in huge quantities. There is an extensive fossil record of pollen grains, often disassociated from their parent plant. The discipline of palynology is devoted to the study of pollen, which can be used both for biostratigraphy and to gain information about the abundance and variety of plants alive — which can itself yield important information about paleoclimates. Pollen is first found in the fossil record in the late Devonian period and increases in abundance until the present day.

Allergy to Pollen

Nasal allergy to pollen is called pollinosis, and allergy specifically to grass pollen is called hay fever. Generally, pollens that cause allergies are those of anemophilous plants (pollen is dispersed by air currents.) Such plants produce large quantities of lightweight pollen (because wind dispersal is random and the likelihood of one pollen grain landing on another flower is small), which can be carried for great distances and are easily inhaled, bringing it into contact with the sensitive nasal passages.

In the US, people often mistakenly blame the conspicuous goldenrod flower for allergies. Since this plant is entomophilous (its pollen is dispersed by animals), its heavy, sticky pollen does not become independently airborne. Most late summer and fall pollen allergies are probably caused by ragweed, a widespread anemophilous plant.

Arizona was once regarded as a haven for people with pollen allergies, although several ragweed species grow in the desert. However, as suburbs grew and people began establishing irrigated lawns and gardens, more irritating species of ragweed gained a foothold and Arizona lost its claim of freedom from hay fever.

Anemophilous spring blooming plants such as oak, birch, hickory, pecan, and early summer grasses may also induce pollen allergies. Most cultivated plants with showy flowers are entomophilous and do not cause pollen allergies.

The number of people in the United States affected by hay fever is between 20 and 40 million, and such allergy has proven to be the most frequent allergic response in the nation. There are certain evidential suggestions pointing out hay fever and similar allergies to be of hereditary origin. Individuals who suffer from eczema or are asthmatic tend to be more susceptible to developing long-term hay fever.

In Denmark, decades of rising temperatures cause pollen to appear earlier and in greater numbers, as well as introduction of new species such as ragweed.

The most efficient way to handle a pollen allergy is by preventing contact with the material. Individuals carrying the ailment may at first believe that they have a simple summer cold, but hay fever becomes more evident when the apparent cold does not disappear. The confirmation of hay fever can be obtained after examination by a general physician.

Treatment

Antihistamines are effective at treating mild cases of pollinosis, this type of non-prescribed drugs includes loratadine, cetirizine and chlorphenamine. They do not prevent the discharge of histamine, but it has been proven that they do prevent a part of the chain reaction activated by this biogenic amine, which considerably lowers hay fever symptoms.

Decongestants can be administered in different ways such as tablets and nasal sprays.

Allergy immunotherapy (AIT) treatment involves administering doses of allergens to accustom the body to pollen, thereby inducing specific long-term tolerance. Allergy immunotherapy can be administered orally (as sublingual tablets or sublingual drops), or by injections under the skin (subcutaneous). Discovered by Leonard Noon and John Freeman in 1911, allergy immunotherapy represents the only causative treatment for respiratory allergies.

Nutrition

Most major classes of predatory and parasitic arthropods contain species that eat pollen, despite the common perception that bees are the primary pollen-consuming arthropod group. Many other Hymenoptera other than bees consume pollen as adults, though only a small number feed on pollen as larvae (including some ant larvae). Spiders are normally considered carnivores but pollen is an important source of food for several species, particularly for spiderlings, which catch pollen on their webs. It is not clear how spiderlings manage to eat pollen however, since their mouths are not large enough to consume pollen grains. Some predatory mites also feed on pollen, with some species being able to subsist solely on pollen, such as *Euseius tularensis*, which feeds on the pollen of dozens of plant species. Members of some beetle families such as Mordellidae and Melyridae feed almost exclusively on pollen as adults, while various lineages within larger families such as Curculionidae, Chrysomelidae, Cerambycidae, and Scarabaeidae are pollen specialists even though most members of their families are not (e.g., only 36 of 40000 species of ground beetles, which are typically predatory, have been shown to eat pollen—but this is thought to be a severe underestimate as the feeding habits are only known for 1000 species). Similarly, Ladybird beetles mainly eat insects, but many species also eat pollen, as either part or all of their diet. Hemiptera are mostly herbivores or omnivores but pollen feeding is known (and has only been well studied in the Anthocoridae). Many adult flies, especially Syrphidae, feed on pollen, and three UK syrphid species feed strictly on pollen (syrphids, like all flies, cannot eat pollen directly due to the structure of their mouthparts, but can consume pollen contents that are dissolved in a fluid). Some species of fungus, including *Fomes fomentarius*, are able to break down grains of pollen as a secondary nutrition source that is particularly high in nitrogen.

Some species of *Heliconius* butterflies consume pollen as adults, which appears to be a valuable nutrient source, and these species are more distasteful to predators than the non-pollen consuming species.

Although bats, butterflies and hummingbirds are not pollen eaters *per se*, their consumption of nectar in flowers is an important aspect of the pollination process.

In humans

A variety of producers have started selling bee pollen for human consumption, often marketed as a food (rather than a dietary supplement). The largest constituent is carbohydrates, with protein content ranging from 7 to 35 percent depending on the plant species collected by bees.

Honey produced by bees from natural sources contains pollen derived p-coumaric acid, an antioxidant.

The U.S. Food and Drug Administration (FDA) has not found any harmful effects of bee pollen consumption, except from the usual allergies. However, FDA does not allow bee pollen marketers in the United States to make health claims about their produce, as no scientific basis for these has ever been proven. Furthermore, there are possible dangers not only from allergic reactions but also from contaminants such as pesticides and from fungi and bacteria growth related to poor storage procedures. A manufacturers's claim that pollen collecting helps the bee colonies is also controversial.

Pine pollen (송화가루, Songhwa Garu) is traditionally consumed in Korea as an ingredient in sweets and beverages.

Parasites

The growing industries in pollen harvesting for human and bee consumption rely on harvesting pollen baskets from honey bees as they return to their hives using a *pollen trap*. When this pollen has been tested for parasites, it has been found that a multitude of pollinator viruses and eukaryotic parasites are present in the pollen. It is currently unclear if the parasites are introduced by the bee that collected the pollen or if it is from contamination to the flower. Though this is not likely to pose a risk to humans, it is a major issue for the bumblebee rearing industry that relies on thousands of tonnes of honey bee collected pollen per year. Several sterilization methods have been employed, though no method has been 100% effective at steralizing, without reducing the nutritional value, of the pollen

Forensic Palynology

An SEM micrograph of Redbud pollen. Scanning electron microscopes are major instruments in palynology.

In forensic biology, pollen can tell a lot about where a person or object has been, because regions of the world, or even more particular locations such a certain set of bushes, will have a distinctive collection of pollen species. Pollen evidence can also reveal the season in which a particular object picked up the pollen. Pollen has been used to trace activity at mass graves in Bosnia, catch a burglar who brushed against a *Hypericum* bush during a crime, and has even been proposed as an additive for bullets to enable tracking them.

Wood

Wood is a porous and fibrous structural tissue found in the stems and roots of trees, and other woody plants. It has been used for thousands of years for both fuel and as a construction material. It is an organic material, a natural composite of cellulose fibers (which are strong in tension) embedded in a matrix of lignin which resists compression. Wood is sometimes defined as only the secondary xylem in the stems of trees, or it is defined more broadly to include the same type of tissue elsewhere such as in the roots of trees or shrubs. In a living tree it performs a support function, enabling woody plants to grow large or to stand up by themselves. It also conveys water and nutrients between the leaves, other growing tissues, and the roots. Wood may also refer to other plant materials with comparable properties, and to material engineered from wood, or wood chips or fiber.

In 2005, the Earth contained about 434 billion cubic meters of growing stock forest, 47% of which was commercial. As an abundant, carbon-neutral renewable resource, woody materials have been of intense interest as a source of renewable energy. In 1991 approximately 3.5 billion cubic meters of wood were harvested. Dominant uses were for furniture and building construction.

History

A 2011 discovery in the Canadian province of New Brunswick uncovered the earliest known plants to have grown wood, approximately 395 to 400 million years ago. Wood can be dated by carbon dating and in some species by dendrochronology to make inferences about when a wooden object was created.

People have used wood for millennia for many purposes, primarily as a fuel or as a construction material for making houses, tools, weapons, furniture, packaging, artworks, and paper. The year-to-year variation in tree-ring widths and isotopic abundances gives clues to the prevailing climate at that time.

Physical Properties

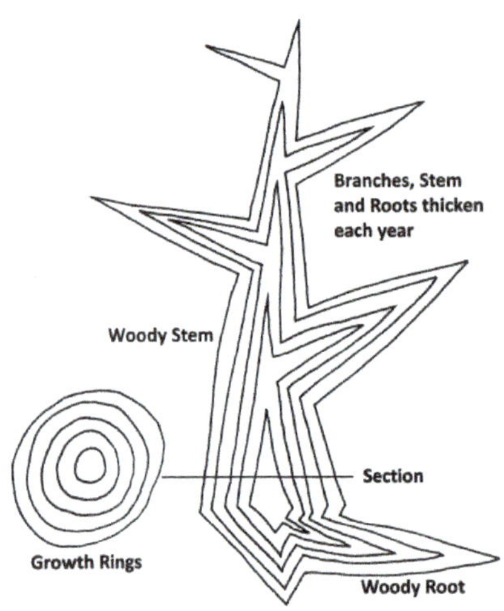

Diagram of secondary growth in a tree showing idealized vertical and horizontal sections. A new layer of wood is added in each growing season, thickening the stem, existing branches and roots, to form a growth ring.

Growth Rings

Wood, in the strict sense, is yielded by trees, which increase in diameter by the formation, between the existing wood and the inner bark, of new woody layers which envelop the entire stem, living branches, and roots. This process is known as secondary growth; it is the result of cell division in the vascular cambium, a lateral meristem, and subsequent expansion of the new cells. These cells then go on to form thickened secondary cell walls, composed mainly of cellulose, hemicellulose and lignin. Where the differences between the four seasons are distinct, growth can occur in a discrete annual or seasonal pattern, leading to growth rings; these can usually be most clearly seen on the end of a log, but are also visible on the other surfaces. If the distinctiveness between seasons is annual, these growth rings are referred to as annual rings. Where there is little seasonal difference growth rings are likely to be indistinct or absent.

If there are differences within a growth ring, then the part of a growth ring nearest the center of the tree, and formed early in the growing season when growth is rapid, is usually composed of wider elements. It is usually lighter in color than that near the outer portion of the ring, and is known as earlywood or springwood. The outer portion formed later in the season is then known as the latewood or summerwood. However, there are major differences, depending on the kind of wood (see below).

Knots

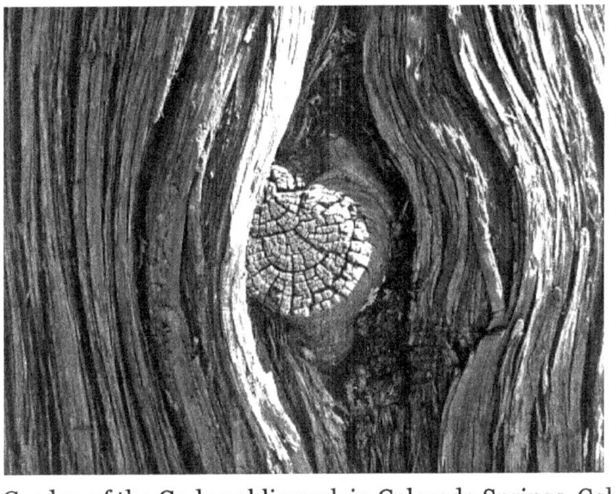

A knot on a tree at the Garden of the Gods public park in Colorado Springs, Colorado (October 2006)

A knot is a particular type of imperfection in a piece of wood; it will affect the technical properties of the wood, usually reducing the local strength and increasing the tendency for splitting along the wood grain, but may be exploited for visual effect. In a longitudinally sawn plank, a knot will appear as a roughly circular "solid" (usually darker) piece of wood around which the grain of the rest of the wood "flows" (parts and rejoins). Within a knot, the direction of the wood (grain direction) is up to 90 degrees different from the grain direction of the regular wood.

In the tree a knot is either the base of a side branch or a dormant bud. A knot (when the base of a side branch) is conical in shape (hence the roughly circular cross-section) with the inner tip at the point in stem diameter at which the plant's vascular cambium was located when the branch formed as a bud.

During the development of a tree, the lower limbs often die, but may remain attached for a time, sometimes years. Subsequent layers of growth of the attaching stem are no longer intimately joined with the dead limb, but are grown around it. Hence, dead branches produce knots which are not attached, and likely to drop out after the tree has been sawn into boards.

In grading lumber and structural timber, knots are classified according to their form, size, soundness, and the firmness with which they are held in place. This firmness is affected by, among other factors, the length of time for which the branch was dead while the attaching stem continued to grow.

Wood knot in vertical section

Knots materially affect cracking and warping, ease in working, and cleavability of timber. They are defects which weaken timber and lower its value for structural purposes where strength is an important consideration. The weakening effect is much more serious when timber is subjected to forces perpendicular to the grain and/or tension than where under load along the grain and/or compression. The extent to which knots affect the strength of a beam depends upon their position, size, number, and condition. A knot on the upper side is compressed, while one on the lower side is subjected to tension. If there is a season check in the knot, as is often the case, it will offer little resistance to this tensile stress. Small knots, however, may be located along the neutral plane of a beam and increase the strength by preventing longitudinal shearing. Knots in a board or plank are least injurious when they extend through it at right angles to its broadest surface. Knots which occur near the ends of a beam do not weaken it. Sound knots which occur in the central portion one-fourth the height of the beam from either edge are not serious defects.

—Samuel J. Record, The Mechanical Properties of Wood

Knots do not necessarily influence the stiffness of structural timber, this will depend on the size and location. Stiffness and elastic strength are more dependent upon the sound wood than upon localized defects. The breaking strength is very susceptible to defects. Sound knots do not weaken wood when subject to compression parallel to the grain.

In some decorative applications, wood with knots may be desirable to add visual interest. In applications where wood is painted, such as skirting boards, fascia boards, door frames and furniture, resins present in the timber may continue to 'bleed' through to the surface of a knot for months or even years after manufacture and show as a yellow or brownish stain. A knot primer paint or solution (knotting), correctly applied during preparation, may do much to reduce this problem but it is difficult to control completely, especially when using mass-produced kiln-dried timber stocks.

Heartwood and Sapwood

A section of a Yew branch showing 27 annual growth rings, pale sapwood, dark heartwood, and pith (center dark spot). The dark radial lines are small knots.

Heartwood (or duramen) is wood that as a result of a naturally occurring chemical transformation has become more resistant to decay. Heartwood formation is a genetically programmed process that occurs spontaneously. Some uncertainty exists as to whether the wood dies during heartwood formation, as it can still chemically react to decay organisms, but only once.

Heartwood is often visually distinct from the living sapwood, and can be distinguished in a cross-section where the boundary will tend to follow the growth rings. For example, it is sometimes much darker. However, other processes such as decay or insect invasion can also discolor wood, even in woody plants that do not form heartwood, which may lead to confusion.

Sapwood (or alburnum) is the younger, outermost wood; in the growing tree it is living wood, and its principal functions are to conduct water from the roots to the leaves and to store up and give back according to the season the reserves prepared in the leaves. However, by the time they become competent to conduct water, all xylem tracheids and vessels have lost their cytoplasm and the cells are therefore functionally dead. All wood in a tree is first formed as sapwood. The more leaves a tree bears and the more vigorous its growth, the larger the volume of sapwood required. Hence trees making rapid growth in the open have thicker sapwood for their size than trees of the same species growing in dense forests. Sometimes trees (of species that do form heartwood) grown in the open may become of considerable size, 30 cm (12 in) or more in diameter, before any heartwood begins to form, for example, in second-growth hickory, or open-grown pines.

The term *heartwood* derives solely from its position and not from any vital importance to the tree. This is evidenced by the fact that a tree can thrive with its heart completely decayed. Some species begin to form heartwood very early in life, so having only a thin layer of live sapwood, while in others the change comes slowly. Thin sapwood is characteristic of such species as chestnut, black locust, mulberry, osage-orange, and sassafras, while in maple, ash, hickory, hackberry, beech, and pine, thick sapwood is the rule. Others never form heartwood.

No definite relation exists between the annual rings of growth and the amount of sapwood. Within the same species the cross-sectional area of the sapwood is very roughly proportional to the size of the crown of the tree. If the rings are narrow, more of them are required than where they are wide. As the tree gets larger, the sapwood must necessarily become thinner or increase materially in volume. Sapwood is thicker in the upper portion of the trunk of a tree than near the base, because the age and the diameter of the upper sections are less.

When a tree is very young it is covered with limbs almost, if not entirely, to the ground, but as it grows older some or all of them will eventually die and are either broken off or fall off. Subsequent growth of wood may completely conceal the stubs which will however remain as knots. No matter how smooth and clear a log is on the outside, it is more or less knotty near the middle. Consequently, the sapwood of an old tree, and particularly of a forest-grown tree, will be freer from knots than the inner heartwood. Since in most uses of wood, knots are defects that weaken the timber and interfere with its ease of working and other properties, it follows that a given piece of sapwood, because of its position in the tree, may well be stronger than a piece of heartwood from the same tree.

It is remarkable that the inner heartwood of old trees remains as sound as it usually does, since in many cases it is hundreds, and in a few instances thousands, of years old. Every broken limb or root, or deep wound from fire, insects, or falling timber, may afford an entrance for decay, which, once started, may penetrate to all parts of the trunk. The larvae of many insects bore into the trees and their tunnels remain indefinitely as sources of weakness. Whatever advantages, however, that sapwood may have in this connection are due solely to its relative age and position.

If a tree grows all its life in the open and the conditions of soil and site remain unchanged, it will make its most rapid growth in youth, and gradually decline. The annual rings of growth are for many years quite wide, but later they become narrower and narrower. Since each succeeding ring is laid down on the outside of the wood previously formed, it follows that unless a tree materially increases its production of wood from year to year, the rings must necessarily become thinner as the trunk gets wider. As a tree reaches maturity its crown becomes more open and the annual wood production is lessened, thereby reducing still more the width of the growth rings. In the case of forest-grown trees so much depends upon the competition of the trees in their struggle for light and nourishment that periods of rapid and slow growth may alternate. Some trees, such as southern oaks, maintain the same width of ring for hundreds of years. Upon the whole, however, as a tree gets larger in diameter the width of the growth rings decreases.

Different pieces of wood cut from a large tree may differ decidedly, particularly if the tree is big and mature. In some trees, the wood laid on late in the life of a tree is softer, lighter, weaker, and more even-textured than that produced earlier, but in other trees, the reverse applies. This may or may not correspond to heartwood and sapwood. In a large log the sapwood, because of the time in the life of the tree when it was grown, may be inferior in hardness, strength, and toughness to equally sound heartwood from the same log. In a smaller tree, the reverse may be true.

Color

In species which show a distinct difference between heartwood and sapwood the natural color of heartwood is usually darker than that of the sapwood, and very frequently the contrast is conspicuous (see section of yew log above). This is produced by deposits in the heartwood of chemical substances, so that a dramatic color difference does not mean a dramatic difference in the mechanical properties of heartwood and sapwood, although there may be a dramatic chemical difference.

The wood of Coast Redwood is distinctively red.

Some experiments on very resinous Longleaf Pine specimens indicate an increase in strength, due to the resin which increases the strength when dry. Such resin-saturated heartwood is called "fat lighter". Structures built of fat lighter are almost impervious to rot and termites; however they are very flammable. Stumps of old longleaf pines are often dug, split into small pieces and sold as kindling for fires. Stumps thus dug may actually remain a century or more since being cut. Spruce impregnated with crude resin and dried is also greatly increased in strength thereby.

Since the latewood of a growth ring is usually darker in color than the earlywood, this fact may be used in judging the density, and therefore the hardness and strength of the material. This is particularly the case with coniferous woods. In ring-porous woods the vessels of the early wood often appear on a finished surface as darker than the denser latewood, though on cross sections of heartwood the reverse is commonly true. Otherwise the color of wood is no indication of strength.

Abnormal discoloration of wood often denotes a diseased condition, indicating unsoundness. The black check in western hemlock is the result of insect attacks. The reddish-brown streaks so common in hickory and certain other woods are mostly the result of injury by birds. The discoloration is merely an indication of an injury, and in all probability does not of itself affect the properties of the wood. Certain rot-producing fungi impart to wood characteristic colors which thus become symptomatic of weakness; however an attractive effect known as spalting produced by this process is often considered a desirable characteristic. Ordinary sap-staining is due to fungal growth, but does not necessarily produce a weakening effect.

Water Content

Water occurs in living wood in three locations, namely:

1. in the cell walls,
2. in the protoplasmic contents of the cells, and
3. as free water in the cell cavities and spaces, especially of the xylem.

In heartwood it occurs only in the first and last forms. Wood that is thoroughly air-dried retains 8–16% of the water in the cell walls, and none, or practically none, in the other forms. Even oven-dried wood retains a small percentage of moisture, but for all except chemical purposes, may be considered absolutely dry.

The general effect of the water content upon the wood substance is to render it softer and more pliable. A similar effect occurs in the softening action of water on rawhide, paper, or cloth. Within certain limits, the greater the water content, the greater its softening effect.

Drying produces a decided increase in the strength of wood, particularly in small specimens. An extreme example is the case of a completely dry spruce block 5 cm in section, which will sustain a permanent load four times as great as a green (undried) block of the same size will.

The greatest strength increase due to drying is in the ultimate crushing strength, and strength at elastic limit in endwise compression; these are followed by the modulus of rupture, and stress at elastic limit in cross-bending, while the modulus of elasticity is least affected.

Structure

Wood is a heterogeneous, hygroscopic, cellular and anisotropic material. It consists of cells, and the cell walls are composed of micro-fibrils of cellulose (40% – 50%) and hemicellulose (15% – 25%) impregnated with lignin (15% – 30%).

Sections of tree trunk

In coniferous or softwood species the wood cells are mostly of one kind, tracheids, and as a result the material is much more uniform in structure than that of most hardwoods. There are no vessels ("pores") in coniferous wood such as one sees so prominently in oak and ash, for example.

The structure of hardwoods is more complex. The water conducting capability is mostly taken care of by vessels: in some cases (oak, chestnut, ash) these are quite large and distinct, in others (buckeye, poplar, willow) too small to be seen without a hand lens. In discussing such woods it is customary to divide them into two large classes, *ring-porous* and *diffuse-porous*.

In ring-porous species, such as ash, black locust, catalpa, chestnut, elm, hickory, mulberry, and oak, the larger vessels or pores (as cross sections of vessels are called) are localized in the part of the growth ring formed in spring, thus forming a region of more or less open and porous tissue.

The rest of the ring, produced in summer, is made up of smaller vessels and a much greater proportion of wood fibers. These fibers are the elements which give strength and toughness to wood, while the vessels are a source of weakness.

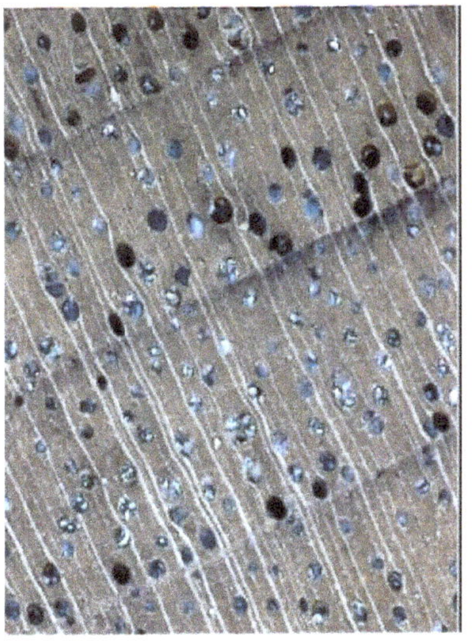

Magnified cross-section of Black Walnut, showing the vessels, rays (white lines) and annual rings: this is intermediate between diffuse-porous and ring-porous, with vessel size declining gradually

In diffuse-porous woods the pores are evenly sized so that the water conducting capability is scattered throughout the growth ring instead of being collected in a band or row. Examples of this kind of wood are alder, basswood, birch, buckeye, maple, willow,and the Populus species such as aspen, cottonwood and poplar. Some species, such as walnut and cherry, are on the border between the two classes, forming an intermediate group.

Earlywood and Latewood

In Softwood

Earlywood and latewood in a softwood; radial view, growth rings closely spaced in Rocky Mountain Douglas-fir

In temperate softwoods there often is a marked difference between latewood and earlywood. The latewood will be denser than that formed early in the season. When examined under a microscope the cells of dense latewood are seen to be very thick-walled and with very small cell cavities, while those formed first in the season have thin walls and large cell cavities. The strength is in the walls, not the cavities. Hence the greater the proportion of latewood the greater the density and strength. In choosing a piece of pine where strength or stiffness is the important consideration, the principal thing to observe is the comparative amounts of earlywood and latewood. The width of ring is not nearly so important as the proportion and nature of the latewood in the ring.

If a heavy piece of pine is compared with a lightweight piece it will be seen at once that the heavier one contains a larger proportion of latewood than the other, and is therefore showing more clearly demarcated growth rings. In white pines there is not much contrast between the different parts of the ring, and as a result the wood is very uniform in texture and is easy to work. In hard pines, on the other hand, the latewood is very dense and is deep-colored, presenting a very decided contrast to the soft, straw-colored earlywood.

It is not only the proportion of latewood, but also its quality, that counts. In specimens that show a very large proportion of latewood it may be noticeably more porous and weigh considerably less than the latewood in pieces that contain but little. One can judge comparative density, and therefore to some extent strength, by visual inspection.

No satisfactory explanation can as yet be given for the exact mechanisms determining the formation of earlywood and latewood. Several factors may be involved. In conifers, at least, rate of growth alone does not determine the proportion of the two portions of the ring, for in some cases the wood of slow growth is very hard and heavy, while in others the opposite is true. The quality of the site where the tree grows undoubtedly affects the character of the wood formed, though it is not possible to formulate a rule governing it. In general, however, it may be said that where strength or ease of working is essential, woods of moderate to slow growth should be chosen.

In Ring-porous Woods

Earlywood and latewood in a ring-porous wood (ash) in a Fraxinus excelsior; tangential view, wide growth rings

In ring-porous woods each season's growth is always well defined, because the large pores formed early in the season abut on the denser tissue of the year before.

In the case of the ring-porous hardwoods there seems to exist a pretty definite relation between the rate of growth of timber and its properties. This may be briefly summed up in the general statement that the more rapid the growth or the wider the rings of growth, the heavier, harder, stronger, and stiffer the wood. This, it must be remembered, applies only to ring-porous woods such as oak, ash, hickory, and others of the same group, and is, of course, subject to some exceptions and limitations.

In ring-porous woods of good growth it is usually the latewood in which the thick-walled, strength-giving fibers are most abundant. As the breadth of ring diminishes, this latewood is reduced so that very slow growth produces comparatively light, porous wood composed of thin-walled vessels and wood parenchyma. In good oak these large vessels of the earlywood occupy from 6 to 10 percent of the volume of the log, while in inferior material they may make up 25% or more. The latewood of good oak is dark colored and firm, and consists mostly of thick-walled fibers which form one-half or more of the wood. In inferior oak, this latewood is much reduced both in quantity and quality. Such variation is very largely the result of rate of growth.

Wide-ringed wood is often called "second-growth", because the growth of the young timber in open stands after the old trees have been removed is more rapid than in trees in a closed forest, and in the manufacture of articles where strength is an important consideration such "second-growth" hardwood material is preferred. This is particularly the case in the choice of hickory for handles and spokes. Here not only strength, but toughness and resilience are important. The results of a series of tests on hickory by the U.S. Forest Service show that:

> "The work or shock-resisting ability is greatest in wide-ringed wood that has from 5 to 14 rings per inch (rings 1.8-5 mm thick), is fairly constant from 14 to 38 rings per inch (rings 0.7–1.8 mm thick), and decreases rapidly from 38 to 47 rings per inch (rings 0.5–0.7 mm thick). The strength at maximum load is not so great with the most rapid-growing wood; it is maximum with from 14 to 20 rings per inch (rings 1.3–1.8 mm thick), and again becomes less as the wood becomes more closely ringed. The natural deduction is that wood of first-class mechanical value shows from 5 to 20 rings per inch (rings 1.3–5 mm thick) and that slower growth yields poorer stock. Thus the inspector or buyer of hickory should discriminate against timber that has more than 20 rings per inch (rings less than 1.3 mm thick). Exceptions exist, however, in the case of normal growth upon dry situations, in which the slow-growing material may be strong and tough."

The effect of rate of growth on the qualities of chestnut wood is summarized by the same authority as follows:

> "When the rings are wide, the transition from spring wood to summer wood is gradual, while in the narrow rings the spring wood passes into summer wood abruptly. The width of the spring wood changes but little with the width of the annual ring, so that the narrowing or broadening of the annual ring is always at the expense of the summer wood. The narrow vessels of the summer wood make it richer in wood substance than the spring wood composed of wide vessels. Therefore, rapid-growing specimens with wide rings have more

wood substance than slow-growing trees with narrow rings. Since the more the wood substance the greater the weight, and the greater the weight the stronger the wood, chestnuts with wide rings must have stronger wood than chestnuts with narrow rings. This agrees with the accepted view that sprouts (which always have wide rings) yield better and stronger wood than seedling chestnuts, which grow more slowly in diameter."

In Diffuse-porous Woods

In the diffuse-porous woods, the demarcation between rings is not always so clear and in some cases is almost (if not entirely) invisible to the unaided eye. Conversely, when there is a clear demarcation there may not be a noticeable difference in structure within the growth ring.

In diffuse-porous woods, as has been stated, the vessels or pores are even-sized, so that the water conducting capability is scattered throughout the ring instead of collected in the earlywood. The effect of rate of growth is, therefore, not the same as in the ring-porous woods, approaching more nearly the conditions in the conifers. In general it may be stated that such woods of medium growth afford stronger material than when very rapidly or very slowly grown. In many uses of wood, total strength is not the main consideration. If ease of working is prized, wood should be chosen with regard to its uniformity of texture and straightness of grain, which will in most cases occur when there is little contrast between the latewood of one season's growth and the earlywood of the next.

Monocot Wood

Trunks of the coconut palm, a monocot, in Java. From this perspective these look not much different from trunks of a dicot or conifer

Structural material that resembles ordinary, "dicot" or conifer wood in its gross handling characteristics is produced by a number of monocot plants, and these also are colloquially called wood. Of these, bamboo, botanically a member of the grass family, has considerable economic importance, larger culms being widely used as a building and construction material in their own right and, these days, in the manufacture of engineered flooring, panels and veneer. Another major plant group that produce material that often is called wood are the palms. Of much less importance are plants such as *Pandanus, Dracaena* and *Cordyline*. With all this material, the structure and composition of the structural material is quite different from ordinary wood.

Specific Gravity

The single most revealing property of wood as an indicator of wood quality is specific gravity (Timell 1986), as both pulp yield and lumber strength are determined by it. Specific gravity is the ratio of the mass of a substance to the mass of an equal volume of water; density is the ratio of a mass of a quantity of a substance to the volume of that quantity and is expressed in mass per unit substance, e.g., grams per milliliter (g/cm^3 or g/ml). The terms are essentially equivalent as long as the metric system is used. Upon drying, wood shrinks and its density increases. Minimum values are associated with green (water-saturated) wood and are referred to as *basic specific gravity* (Timell 1986).

Wood Density

Wood density is determined by multiple growth and physiological factors compounded into "one fairly easily measured wood characteristic" (Elliott 1970).

Age, diameter, height, radial growth, geographical location, site and growing conditions, silvicultural treatment, and seed source all to some degree influence wood density. Variation is to be expected. Within an individual tree, the variation in wood density is often as great as or even greater than that between different trees (Timell 1986). Variation of specific gravity within the bole of a tree can occur in either the horizontal or vertical direction.

Hard and Soft Woods

It is common to classify wood as either softwood or hardwood. The wood from conifers (e.g. pine) is called softwood, and the wood from dicotyledons (usually broad-leaved trees, e.g. oak) is called hardwood. These names are a bit misleading, as hardwoods are not necessarily hard, and softwoods are not necessarily soft. The well-known balsa (a hardwood) is actually softer than any commercial softwood. Conversely, some softwoods (e.g. yew) are harder than many hardwoods.

There is a strong relationship between the properties of wood and the properties of the particular tree that yielded it. The density of wood varies with species. The density of a wood correlates with its strength (mechanical properties). For example, mahogany is a medium-dense hardwood that is excellent for fine furniture crafting, whereas balsa is light, making it useful for model building. One of the densest woods is black ironwood.

Chemistry of Wood

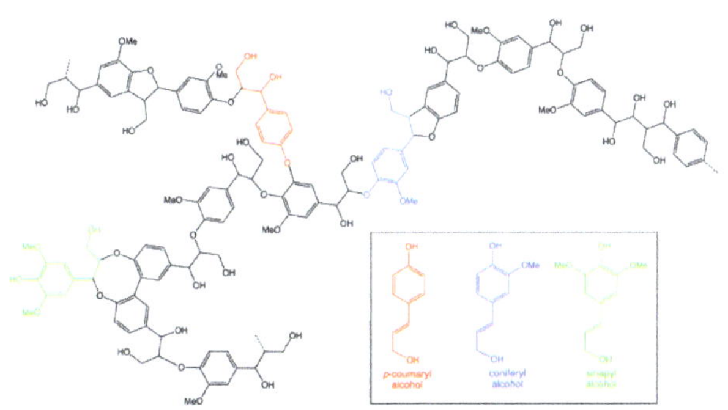

Chemical structure of lignin, which comprises about 25% of wood dry matter and is responsible for many of its properties.

The chemical composition of wood varies from species to species, but is approximately 50% carbon, 42% oxygen, 6% hydrogen, 1% nitrogen, and 1% other elements (mainly calcium, potassium, sodium, magnesium, iron, and manganese) by weight. Wood also contains sulfur, chlorine, silicon, phosphorus, and other elements in small quantity.

Aside from water, wood has three main components. Cellulose, a crystalline polymer derived from glucose, constitutes about 41–43%. Next in abundance is hemicellulose, which is around 20% in deciduous trees but near 30% in conifers. It is mainly five-carbon sugars that are linked in an irregular manner, in contrast to the cellulose. Lignin is the third component at around 27% in coniferous wood vs. 23% in deciduous trees. Lignin confers the hydrophobic properties reflecting the fact that it is based on aromatic rings. These three components are interwoven, and direct covalent linkages exist between the lignin and the hemicellulose. A major focus of the paper industry is the separation of the lignin from the cellulose, from which paper is made.

In chemical terms, the difference between hardwood and softwood is reflected in the composition of the constituent lignin. Hardwood lignin is primarily derived from sinapyl alcohol and coniferyl alcohol. Softwood lignin is mainly derived from coniferyl alcohol.

Extractives

Aside from the lignocellulose, wood consists of a variety of low molecular weight organic compounds, called *extractives*. The wood extractives are fatty acids, resin acids, waxes and terpenes. For example, rosin is exuded by conifers as protection from insects. The extraction of these organic materials from wood provides tall oil, turpentine, and rosin.

Uses

Fuel

Wood has a long history of being used as fuel, which continues to this day, mostly in rural areas of the world. Hardwood is preferred over softwood because it creates less smoke and burns longer. Adding a woodstove or fireplace to a home is often felt to add ambiance and warmth.

Construction

Wood has been an important construction material since humans began building shelters, houses and boats. Nearly all boats were made out of wood until the late 19th century, and wood remains in common use today in boat construction. Elm in particular was used for this purpose as it resisted decay as long as it was kept wet (it also served for water pipe before the advent of more modern plumbing).

Wood to be used for construction work is commonly known as *lumber* in North America. Elsewhere, *lumber* usually refers to felled trees, and the word for sawn planks ready for use is *timber*. In Medieval Europe oak was the wood of choice for all wood construction, including beams, walls, doors, and floors. Today a wider variety of woods is used: solid wood doors are often made from poplar, small-knotted pine, and Douglas fir.

The churches of Kizhi, Russia are among a handful of World Heritage Sites built entirely of wood, without metal joints. See Kizhi Pogost for more details.

The Saitta House, Dyker Heights, Brooklyn, New York built in 1899 is made of and decorated in wood.

New domestic housing in many parts of the world today is commonly made from timber-framed construction. Engineered wood products are becoming a bigger part of the construction industry. They may be used in both residential and commercial buildings as structural and aesthetic materials.

In buildings made of other materials, wood will still be found as a supporting material, especially in roof construction, in interior doors and their frames, and as exterior cladding.

Wood is also commonly used as shuttering material to form the mold into which concrete is poured during reinforced concrete construction.

Wood Flooring

Wood can be cut into straight planks and made into a wood flooring.

Engineered Wood

Engineered wood products, glued building products "engineered" for application-specific performance requirements, are often used in construction and industrial applications. Glued engineered wood products are manufactured by bonding together wood strands, veneers, lumber or other forms of wood fiber with glue to form a larger, more efficient composite structural unit.

These products include glued laminated timber (glulam), wood structural panels (including plywood, oriented strand board and composite panels), laminated veneer lumber (LVL) and other structural composite lumber (SCL) products, parallel strand lumber, and I-joists. Approximately 100 million cubic meters of wood was consumed for this purpose in 1991. The trends suggest that particle board and fiber board will overtake plywood.

Wood unsuitable for construction in its native form may be broken down mechanically (into fibers or chips) or chemically (into cellulose) and used as a raw material for other building materials, such as engineered wood, as well as chipboard, hardboard, and medium-density fiberboard (MDF). Such wood derivatives are widely used: wood fibers are an important component of most paper, and cellulose is used as a component of some synthetic materials. Wood derivatives can also be used for kinds of flooring, for example laminate flooring.

Furniture and Utensils

Wood has always been used extensively for furniture, such as chairs and beds. It is also used for tool handles and cutlery, such as chopsticks, toothpicks, and other utensils, like the wooden spoon.

Next Generation Wood Products

Further developments include new lignin glue applications, recyclable food packaging, rubber tire replacement applications, anti-bacterial medical agents, and high strength fabrics or composites. As scientists and engineers further learn and develop new techniques to extract various components from wood, or alternatively to modify wood, for example by adding components to wood, new more advanced products will appear on the marketplace. Moisture content electronic monitoring can also enhance next generation wood protection.

In the Arts

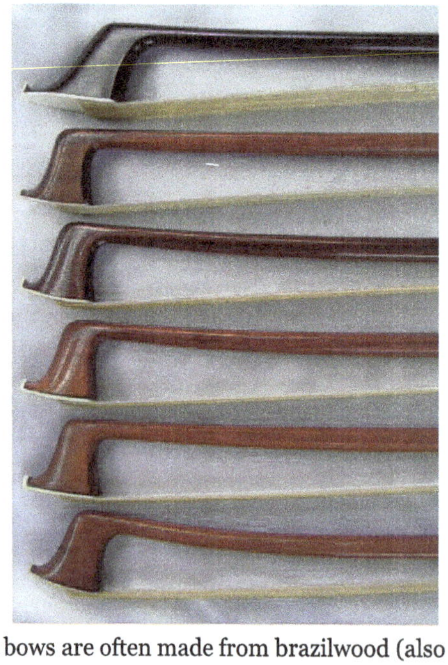

Stringed instrument bows are often made from brazilwood (also called pernambuco).

Wood has long been used as an artistic medium. It has been used to make sculptures and carvings for millennia. Examples include the totem poles carved by North American indigenous people from conifer trunks, often Western Red Cedar (*Thuja plicata*), and the Millennium clock tower, now housed in the National Museum of Scotland in Edinburgh. It is also used in woodcut printmaking, and for engraving.

Certain types of musical instruments, such as those of the violin family, the guitar, the clarinet and recorder, the xylophone, and the marimba, are traditionally made mostly or entirely of wood. The choice of wood may make a significant difference to the tone and resonant qualities of the instrument, and tonewoods have widely differing properties, ranging from the hard and dense african blackwood (used for the bodies of clarinets) to the light but resonant European spruce (*Picea abies*), which is traditionally used for the soundboards of violins. The most valuable tonewoods, such as the ripple sycamore (*Acer pseudoplatanus*), used for the backs of violins, combine acoustic properties with decorative color and grain which enhance the appearance of the finished instrument.

Despite their collective name, not all woodwind instruments are made entirely of wood. The reeds used to play them, however, are usually made from *Arundo donax*, a type of monocot cane plant.

Sports and Recreational Equipment

Many types of sports equipment are made of wood, or were constructed of wood in the past. For example, cricket bats are typically made of white willow. The baseball bats which are legal for use in Major League Baseball are frequently made of ash wood or hickory, and in recent years have been constructed from maple even though that wood is somewhat more fragile. NBA courts have been traditionally made out of parquetry.

Many other types of sports and recreation equipment, such as skis, ice hockey sticks, lacrosse sticks and archery bows, were commonly made of wood in the past, but have since been replaced with more modern materials such as aluminium, fiberglass, carbon fiber, titanium, and composite materials. One noteworthy example of this trend is the golf club commonly known as the *wood*, the head of which was traditionally made of persimmon wood in the early days of the game of golf, but is now generally made of synthetic materials.

Bacterial Degradation

Little is known about the bacteria that degrade cellulose. Symbiotic bacteria in Xylophaga may play a role in the degradation of sunken wood; while bacteria such as Alphaproteobacteria, Flavobacteria, Actinobacteria, Clostridia, and Bacteroidetes have been detected in wood submerged over a year.

References

- Raven, P. H.; Evert, R. F. and Eichhorn, S. E. (2005) Biology of Plants (7th edition) W. H. Freeman, New York, page 9, ISBN 0-7167-1007-2
- Evert, Ray Franklin and Esau, Katherine (2006) Esau's Plant anatomy: meristems, cells, and tissues of the plant body - their structure, function and development Wiley, Hoboken, New Jersey, page xv, ISBN 0-471-73843-3

- Raven, PH; Evert, RF; Eichhorm, SE (1999). Biology of Plants (6th ed.). New York: W.H. Freeman. ISBN 9780716762843.
- Stewart, W.N.; Rothwell, G.W. (1993). Palaeobotany and the evolution of plants. Cambridge, U.K.: Cambridge University Press. ISBN 0521382947
- Heywood, V.H.; Brummitt, R.K.; Culham, A.; Seberg, O. (2007). Flowering plant families of the world. New York: Firefly books. p. 287. ISBN 9781554072064.
- Galun, Esra (2007). Plant Patterning: Structural and Molecular Genetic Aspects. World Scientific Publishing Company. p. 333. ISBN 9789812704085
- Singh, Gurcharan (2004-01-01). "Palynology". Plant Systematics: An Integrated Approach. p. 142. ISBN 9781578083510. Retrieved 23 January 2014.
- "The Pollen Feeders". Relationships of Natural Enemies and Non-Prey Foods. 7. 2009. pp. 87–11. doi:10.1007/978-1-4020-9235-0_6. ISBN 978-1-4020-9234-3.
- Schwarze, Francis W. M. R.; Engels, Julia; Mattheck, Claus (2000). Fungal Strategies of Wood Decay in Trees. Springer. p. 61. ISBN 978-3-540-67205-0.
- Mimms, Agneta; Michael J. Kuckurek; Jef A. Pyiatte; Elizabeth E. Wright (1993). Kraft Pulping. A Compilation of Notes. TAPPI Press. pp. 6–7. ISBN 0-89852-322-2.
- Fiebach, Klemens; Grimm, Dieter (2000). "Resins, Natural". Ullmann's Encyclopedia of Industrial Chemistry. doi:10.1002/14356007.a23_073. ISBN 978-3-527-30673-2.
- Siewertsen, Bjarne. "Hård nyser for allergikere i varm fremtid" (English: Hard sneeze for allergic people in warm future) Danish Meteorological Institute, 18 April 2015. Retrieved: 19 April 2015.
- Simpson, Michael G. (2011). "Palynology". Plant Systematics. Academic Press. pp. 453–464. ISBN 978-0-08-051404-8. Retrieved 6 January 2014.

Permissions

All chapters in this book are published with permission under the Creative Commons Attribution Share Alike License or equivalent. Every chapter published in this book has been scrutinized by our experts. Their significance has been extensively debated. The topics covered herein carry significant information for a comprehensive understanding. They may even be implemented as practical applications or may be referred to as a beginning point for further studies.

We would like to thank the editorial team for lending their expertise to make the book truly unique. They have played a crucial role in the development of this book. Without their invaluable contributions this book wouldn't have been possible. They have made vital efforts to compile up to date information on the varied aspects of this subject to make this book a valuable addition to the collection of many professionals and students.

This book was conceptualized with the vision of imparting up-to-date and integrated information in this field. To ensure the same, a matchless editorial board was set up. Every individual on the board went through rigorous rounds of assessment to prove their worth. After which they invested a large part of their time researching and compiling the most relevant data for our readers.

The editorial board has been involved in producing this book since its inception. They have spent rigorous hours researching and exploring the diverse topics which have resulted in the successful publishing of this book. They have passed on their knowledge of decades through this book. To expedite this challenging task, the publisher supported the team at every step. A small team of assistant editors was also appointed to further simplify the editing procedure and attain best results for the readers.

Apart from the editorial board, the designing team has also invested a significant amount of their time in understanding the subject and creating the most relevant covers. They scrutinized every image to scout for the most suitable representation of the subject and create an appropriate cover for the book.

The publishing team has been an ardent support to the editorial, designing and production team. Their endless efforts to recruit the best for this project, has resulted in the accomplishment of this book. They are a veteran in the field of academics and their pool of knowledge is as vast as their experience in printing. Their expertise and guidance has proved useful at every step. Their uncompromising quality standards have made this book an exceptional effort. Their encouragement from time to time has been an inspiration for everyone.

The publisher and the editorial board hope that this book will prove to be a valuable piece of knowledge for students, practitioners and scholars across the globe.

Index